Point Set Theory

PURE AND APPLIED MATHEMATICS

A Program of Monographs, Textbooks, and Lecture Notes

EXECUTIVE EDITORS

MONOGRAPHS AND TEXTBOOKS IN
PURE AND APPLIED MATHEMATICS

67. *J. K. Beem and P. E. Ehrlich*, Global Lorentzian Geometry (1981)
68. *D. L. Armacost*, The Structure of Locally Compact Abelian Groups (1981)
69. *J. W. Brewer and M. K. Smith, eds.*, Emmy Noether: A Tribute to Her Life and Work (1981)
70. *K. H. Kim*, Boolean Matrix Theory and Applications (1982)
71. *T. W. Wieting*, The Mathematical Theory of Chromatic Plane Ornaments (1982)
72. *D. B. Gauld*, Differential Topology: An Introduction (1982)
73. *R. L. Faber*, Foundations of Euclidean and Non-Euclidean Geometry (1983)
74. *M. Carmeli*, Statistical Theory and Random Matrices (1983)
75. *J. H. Carruth, J. A. Hildebrant, and R. J. Koch*, The Theory of Topological Semigroups (1983)
76. *R. L. Faber*, Differential Geometry and Relativity Theory: An Introduction (1983)
77. *S. Barnett*, Polynomials and Linear Control Systems (1983)
78. *G. Karpilovsky*, Commutative Group Algebras (1983)
79. *F. Van Oystaeyen and A. Verschoren*, Relative Invariants of Rings: The Commutative Theory (1983)
80. *I. Vaisman*, A First Course in Differential Geometry (1984)
81. *G. W. Swan*, Applications of Optimal Control Theory in Biomedicine (1984)
82. *T. Petrie and J. D. Randall*, Transformation Groups on Manifolds (1984)
83. *K. Goebel and S. Reich*, Uniform Convexity, Hyperbolic Geometry, and Nonexpansive Mappings (1984)
84. *T. Albu and C. Năstăsescu*, Relative Finiteness in Module Theory (1984)
85. *K. Hrbacek and T. Jech*, Introduction to Set Theory, Second Edition, Revised and Expanded (1984)
86. *F. Van Oystaeyen and A. Verschoren*, Relative Invariants of Rings: The Noncommutative Theory (1984)
87. *B. R. McDonald*, Linear Algebra Over Commutative Rings (1984)
88. *M. Namba*, Geometry of Projective Algebraic Curves (1984)
89. *G. F. Webb*, Theory of Nonlinear Age-Dependent Population Dynamics (1985)
90. *M. R. Bremner, R. V. Moody, and J. Patera*, Tables of Dominant Weight Multiplicities for Representations of Simple Lie Algebras (1985)
91. *A. E. Fekete*, Real Linear Algebra (1985)
92. *S. B. Chae*, Holomorphy and Calculus in Normed Spaces (1985)
93. *A. J. Jerri*, Introduction to Integral Equations with Applications (1985)
94. *G. Karpilovsky*, Projective Representations of Finite Groups (1985)
95. *L. Narici and E. Beckenstein*, Topological Vector Spaces (1985)
96. *J. Weeks*, The Shape of Space: How to Visualize Surfaces and Three-Dimensional Manifolds (1985)
97. *P. R. Gribik and K. O. Kortanek*, Extremal Methods of Operations Research (1985)
98. *J.-A. Chao and W. A. Woyczynski, eds.*, Probability Theory and Harmonic Analysis (1986)
99. *G. D. Crown, M. H. Fenrick, and R. J. Valenza*, Abstract Algebra (1986)
100. *J. H. Carruth, J. A. Hildebrant, and R. J. Koch*, The Theory of Topological Semigroups, Volume 2 (1986)

Other Volumes in Preparation

Point Set Theory

John C. Morgan II
California State Polytechnic University
Pomona, California

CRC Press
Taylor & Francis Group
Boca Raton London New York

CRC Press is an imprint of the
Taylor & Francis Group, an **informa** business

Published 1990 by CRC Press
Taylor & Francis Group
6000 Broken Sound Parkway NW, Suite 300
Boca Raton, FL 33487-2742

© 1990 by Taylor & Francis Group
CRC Press is an imprint of Taylor & Francis Group, an Informa business

First issued in paperback 2019

No claim to original U.S. Government works

ISBN-13: 978-0-367-45094-6 (pbk)
ISBN-13: 978-0-8247-8178-1 (hbk)

Visit the Taylor & Francis Web site at
http://www.taylorandfrancis.com

and the CRC Press Web site at
http://www.crcpress.com

Dedicated with gratitude to David Blackwell

"Reasonable people sometimes do foolish things."

Preface

Various schemes for classifying point sets have evolved from investigations of Baire, Lebesgue, Hausdorff, Marczewski, and others. The purpose of this volume is to provide a systematic treatment of properties common to these classifications. By carrying general topological concepts to a higher level of abstraction, an extensive unification of analogies between Baire category and Lebesgue measure is effected. Transference of several of these analogies to Hausdorff measure, Hausdorff dimension, and other areas is also obtained within the general framework presented.

Although the reader is presumed to have prior knowledge only of undergraduate abstract set theory and real analysis courses, some (optional) discussion involving Lebesgue measure, Hausdorff measure, and other topics is included in order to motivate and illustrate the abstract concepts. *Point Set Theory* may be used as a basic text for a course in the modern theory of real functions and a foundation for further research in this area, as a comprehensive reference on classical aspects of point set topology and Lebesgue measure, and as a supplement to courses in General Topology and Measure Theory.

John C. Morgan II

Contents

Point Set Theory

1

Category Bases

I. INITIAL CONCEPTS

A. Preliminary terminology

The following notation will be utilized for the set of elements indicated.

$\mathbb{B}$: 0, 1
$\mathbb{N}$: nonzero natural numbers
$\mathbb{I}$: integers
$\mathbb{Q}$: rational numbers
$\mathbb{Q}^+$: positive rational numbers
$\mathbb{R}$: real numbers
$\mathbb{R}^+$: positive real numbers
$\mathbb{R}^*$: extended real numbers (i.e., $\mathbb{R} \cup \{\pm\infty\}$)

The term "ordinal number" will signify only nonzero ordinal numbers.
Of special importance are

ω: first ordinal number of power $\aleph_0$
Ω: first ordinal number of power $\aleph_1$
Λ: first ordinal number of power $2^{\aleph_0}$

An ordinal number α is called odd if it is representable in the form $a = \beta + \gamma$, where β is 0 or a limit ordinal and γ is an odd natural number; otherwise, α is called an even ordinal number.

For each $n \in \mathbb{N}$ we utilize the following symbols for the quantities given:

$\mathbb{B}^n$: n-tuples $\sigma = \langle \mu_1, \ldots, \mu_n \rangle$ with terms $\mu_k \in \mathbb{B}$ for all $k = 1, \ldots, n$

For $\sigma = \langle \mu_1, \ldots, \mu_m \rangle \in \mathbb{B}^m$ and $\tau = \langle v_1, \ldots, v_n \rangle \in \mathbb{B}^n$, we denote

$$\sigma\tau = \langle \mu_1, \ldots, \mu_m, v_1, \ldots, v_n \rangle$$

In particular, for $\sigma = \langle \mu_1, \ldots, \mu_m \rangle \in \mathbb{B}^m$ and $b \in \mathbb{B}$, we denote

$$\sigma b = \langle \mu_1, \ldots, \mu_m, b \rangle$$

For $\sigma = \langle \mu_1, \ldots, \mu_m \rangle \in \mathbb{B}^m$ and $m < n$,

$\mathbb{B}_\sigma^n$: n-tuples $\tau = \langle v_1, \ldots, v_n \rangle \in \mathbb{B}^n$ with $v_k = \mu_k$ for all $k = 1, \ldots, m$

In addition, we denote

$$\mathbb{B}^F = \bigcup_{n=1}^{\infty} \mathbb{B}^n$$

$\mathbb{B}^\infty$: sequences $\mu = \langle \mu_k \rangle_{k \in \mathbb{N}}$ with terms $\mu_k \in \mathbb{B}$ for all $k \in \mathbb{N}$

For an element $\mu = \langle \mu_k \rangle_{k \in \mathbb{N}}$ of $\mathbb{B}^\infty$ and $n \in \mathbb{N}$, we denote

$$\mu|n = \langle \mu_1, \ldots, \mu_n \rangle$$

For $\sigma \in \mathbb{B}^n$ we denote

$\mathbb{B}_\sigma^\infty$: sequences $v = \langle v_k \rangle_{k \in \mathbb{N}}$ in $\mathbb{B}^\infty$ with $v|n = \sigma$

Replacing $\mathbb{B}$ by $\mathbb{N}$ or $\mathbb{R}$ we obtain similar notation for $\mathbb{N}^n$, $\mathbb{N}_\sigma^n$, $\mathbb{N}^F$, $\mathbb{N}^\infty$, $\mathbb{N}_\sigma^\infty$, $\mathbb{R}^n$, $\mathbb{R}_\sigma^n$, $\mathbb{R}^F$, $\mathbb{R}^\infty$, $\mathbb{R}_\sigma^\infty$. For notational convenience, we shall also denote the set $\mathbb{N}^\infty$ by $\mathbb{Z}$.

We shall be concerned primarily below with certain subsets of n-dimensional Euclidean space $\mathbb{R}^n$, for $n \geq 1$. The distance between two points $x = \langle x_1, \ldots, x_n \rangle$, $y = \langle y_1, \ldots, y_n \rangle$ of $\mathbb{R}^n$, denoted by $d(x,y)$, is defined by

$$d(x,y) = \sqrt{\sum_{i=1}^{n} (x_i - y_i)^2}$$

We note that for points $x, y, z \in \mathbb{R}^n$ the distance function has the properties

(i) $d(x,y) \geq 0$.
(ii) $d(x,y) = 0$ if and only if $x = y$.
(iii) $d(x,y) = d(y,x)$.
(iv) $d(x,z) \leq d(x,y) + d(y,z)$.

The diameter of a set $S \subset \mathbb{R}^n$, symbolized by diam(S), is defined to be 0 if $S = \emptyset$ and otherwise by

$$\text{diam}(S) = \sup\{d(x,y): x, y \in S\}$$

A set whose diameter is a finite number is called bounded.

The distance between sets S and T, denoted by $d(S,T)$, is defined to be 0 if at least one of the sets is empty and otherwise by

$$d(S,T) = \inf\{d(x,y): x \in S \text{ and } y \in T\}$$

A set E of the form

$$E = \{\langle x_1, \ldots, x_n \rangle \in \mathbb{R}^n : a_i < x_i < b_i \text{ for } i = 1, \ldots, n\}$$

where $a_1, \ldots, a_n, b_1, \ldots, b_n \in \mathbb{R}$ and $a_i < b_i$ for $i = 1, \ldots, n$, is called an open rectangle. A set E of the form

$$E = \{\langle x_1, \ldots, x_n \rangle \in \mathbb{R}^n : a_i \leqslant x_i \leqslant b_i \text{ for } i = 1, \ldots, n\}$$

where $a_1, \ldots, a_n, b_1, \ldots, b_n \in \mathbb{R}$ and $a_i < b_i$ for $i = 1, \ldots, n$, is called a closed rectangle. The points $e = \langle e_1, \ldots, e_n \rangle$, with $e_i \in \{a_i, b_i\}$ for each index i, are called the vertices of the rectangle E. A point $e = \langle e_1, \ldots, e_n \rangle$ is called a rational point if all the terms e_i are rational numbers. The $2n$ sets

$$A_k = \{x \in E : x_k = a_k\} \qquad B_k = \{x \in E : x_k = b_k\}$$

for $k = 1, \ldots, n$, are called the faces of the closed rectangle E. Each rectangle E determines the $2n$ hyperplanes

$$L_k = \{x \in \mathbb{R}^n : x_k = a_k\} \qquad M_k = \{x \in \mathbb{R} : x_k = b_k\}$$

($k = 1, \ldots, n$), where L_k is parallel to M_k for each k and L_k is perpendicular to L_m for $k \neq m$. Conversely, any set of $2n$ different hyperplanes, consisting of n pairs (L_k, M_k) with L_k parallel to M_k and L_k perpendicular to L_m for $k \neq m$, determines an open rectangle and a closed rectangle.

In the case $n = 1$, open rectangles and closed rectangles are called open intervals and closed intervals, symbolized as usual by

$$(a,b) = \{x \in \mathbb{R} : a < x < b\}$$
$$[a,b] = \{x \in \mathbb{R} : a \leqslant x \leqslant b\}$$

A subset of $\mathbb{R}^n$ is called an open set if it is representable as the union of a family (possibly empty) of open rectangles. We note that each open set is representable as the union of a countable family of open rectangles (resp., closed rectangles) with rational vertices. This implies that the family of all open sets, denoted by $\mathscr{G}$, has the power $2^{\aleph_0}$ of the continuum.

The complement of an open set in $\mathbb{R}^n$ is called a closed set. The family of all closed sets, denoted by $\mathscr{F}$, also has the power of the continuum.

By a line segment (resp., open line segment, closed line segment) in $\mathbb{R}^n$ we mean a bounded interval (resp., open interval, closed interval) of a straight line in $\mathbb{R}^n$.

A family $\mathscr{H}$ of subsets of a set X is called a cover of a set $S \subset X$ if $S \subset \cup \mathscr{H}$. If $\mathscr{H}$ is a cover of a set S and $\mathscr{H}'$ is a subfamily of $\mathscr{H}$ that is also a cover of S, then $\mathscr{H}'$ is called a subcover of $\mathscr{H}$. A family of open subsets of $\mathbb{R}^n$ that is a cover of a set S is called an open cover of S.

A set $S \subset \mathbb{R}^n$ is called compact if every open cover of S contains a finite subcover of S. We recall that a set in $\mathbb{R}^n$ is compact if and only if it is closed and bounded. We note that the union (or intersection) of a finite number of compact sets is itself a compact set; any closed subset of a compact set is a compact set; and if S, T are any two nonempty, disjoint compact sets in $\mathbb{R}^n$, then $d(S,T) > 0$. We note further that if a function f is defined and continuous on a compact set $S \subset \mathbb{R}^n$, then $f(S)$ is a compact set.

We now define a special closed set contained in the unit interval $[0,1]$ known as the Cantor set. An open interval (c,d) contained in a closed interval $[a,b]$ is called centrally located if $c - a = b - d$. We remove from $[0,1]$ the centrally located open interval of length $\ell_1 = 1/3$; remove from the two closed intervals remaining, $[0,1/3]$ and $[2/3,1]$, their centrally located open intervals of length $\ell_2 = 1/9$; and continue this process indefinitely, removing at the kth stage the centrally located open intervals of length $\ell_k = 1/3^k$ from the 2^{k-1} closed intervals remaining. In actuality, we have removed the open intervals

$$\left(\frac{1}{3},\frac{2}{3}\right), \quad \left(\frac{1}{9},\frac{2}{9}\right), \quad \left(\frac{7}{9},\frac{8}{9}\right), \quad \left(\frac{1}{27},\frac{2}{27}\right), \quad \left(\frac{7}{27},\frac{8}{27}\right), \quad \left(\frac{19}{27},\frac{20}{27}\right), \quad \left(\frac{25}{27},\frac{26}{27}\right), \quad \cdots$$

the sum of whose length is

$$\frac{1}{3} + \frac{2}{9} + \frac{4}{27} + \cdots = \sum_{k=1}^{\infty} \frac{2^{k-1}}{3^k} = 1$$

The set of points in $[0,1]$ not removed is called the Cantor set and will be denoted by C. Arithmetically, the Cantor set consists of all real numbers x in $[0,1]$ having a representation of the form

$$x = \sum_{k=1}^{\infty} \frac{x_k}{3^k}$$

where $x_k \in \{0,2\}$ for each k. This set is an uncountable compact set that contains no interval.

B. Lebesgue measure

Generalizing the length of an interval, the area of an open rectangle

$$E = \{\langle x_1,\ldots,x_n\rangle \in \mathbb{R}^n : a_i < x_i < b_i \quad \text{for } i = 1,\ldots,n\}$$

or closed rectangle

$$E = \{\langle x_1,\ldots,x_n\rangle \in \mathbb{R}^n : a_i \leqslant x_i \leqslant b_i \quad \text{for } i = 1,\ldots,n\}$$

is defined to be the number

$$\tau(E) = \prod_{i=1}^{n} (b_i - a_i)$$

In addition, we define $\tau(\varnothing) = 0$.

Let $\mathscr{R}$ denote the family consisting of $\varnothing$ and all open rectangles in $\mathbb{R}^n$. With each set $S \subset \mathbb{R}^n$ we associate a nonnegative extended real number $\mu(S)$, called the Lebesgue measure[1] of S. We define $\mu(S)$ to be the infimum of the numbers $\Sigma_{j=1}^{\infty} \tau(E_j)$, where $\langle E_j\rangle_{j\in\mathbb{N}}$ is an arbitrary sequence of sets in $\mathscr{R}$ that covers S.

The following properties are valid for subsets of $\mathbb{R}^n$:

(a) Every countable set has Lebesgue measure zero.
(b) If I is a closed rectangle, then $\mu(I) = \tau(I)$.

Proof. For any number $\varepsilon > 0$, there is an open rectangle E such that $I \subset E$ and $\tau(I) < \tau(E) \leqslant \tau(I) + \varepsilon$. Hence, $\mu(I) \leqslant \tau(E) \leqslant \tau(I) + \varepsilon$, which implies that $\mu(I) \leqslant \tau(I) + \varepsilon$.

Conversely, suppose that ε is any given positive number and let $\langle E_j\rangle_{j\in\mathbb{N}}$ be a sequence in $\mathscr{R}$ that covers I and satisfies $\Sigma_{j=1}^{\infty} \tau(E_j) < \mu(I) + \varepsilon$. By the compactness property, there is an index $n_0 \in \mathbb{N}$ such that $I \subset \bigcup_{j=1}^{n_0} E_j$. Without loss of generality, we may assume that each of the sets E_j $(j = 1,\ldots,n)$ is nonempty. The hyperplanes determined the rectangles $E_1,\ldots,E_{n_0}$ and I determine open rectangles $D_{j,1},\ldots,D_{j,m_j}$ contained in E_j, for each $j = 1,\ldots,n$ and determine open rectangles $I_1,\ldots,I_p$ contained in I. As each rectangle I_i coincides with at least one of the rectangles $D_{j,k}$ for some j,k, we have

$$\tau(I) = \sum_{i=1}^{p} \tau(I_i) \leqslant \sum_{j=1}^{n_0} \sum_{k=1}^{m_j} \tau(D_{j,k}) = \sum_{j=1}^{n_0} \tau(E_j) < \mu(I) + \varepsilon$$

But $\tau(I) < \mu(I) + \varepsilon$ for every $\varepsilon > 0$ implies that $\tau(I) \leqslant \mu(I)$.

We conclude that $\mu(I) = \tau(I)$.

[1]Most textbooks on measure theory use the term "outer measure" here, instead of "measure." We are in accordance with the terminology of Rogers (1970).

NOTE The Cantor set is an uncountable compact subset of $\mathbb{R}$ that has Lebesgue measure zero. One can also construct uncountable compact subsets of the unit interval $[0,1]$ that contain no intervals and have positive Lebesgue measure. For example, if the construction process above for the Cantor set is modified to remove open intervals of length $\ell_k = 1/b^k$, where $b > 3$ is a given number, then the sum of the lengths of the intervals removed will be $1/(b-2)$ and the set remaining will be an uncountable compact set with Lebesgue measure

$$1 - \frac{1}{b-2} = \frac{b-3}{b-2}$$

(c) If $S \subset T$, then $\mu(S) \leqslant \mu(T)$.

(d) $\mu(\bigcup_{n=1}^{\infty} S_n) \leqslant \sum_{n=1}^{\infty} \mu(S_n)$ for every sequence of sets $\langle S_n \rangle_{n \in \mathbb{N}}$.

Proof. Suppose that ε is any given positive number. For each $n \in \mathbb{N}$, let $\langle E_{nj} \rangle_{j \in \mathbb{N}}$ be a sequence of sets in $\mathscr{R}$ that covers S_n with $\sum_{j=1}^{\infty} \tau(E_{nj}) \leqslant \mu(S_n) + \varepsilon/2^n$. The totality of sets E_{nj} forms a single sequence $\langle D_m \rangle_{m \in \mathbb{N}}$ of sets in $\mathscr{R}$ that covers $\bigcup_{n=1}^{\infty} S_n$ with

$$\mu\left(\bigcup_{n=1}^{\infty} S_n\right) \leqslant \sum_{m=1}^{\infty} \tau(D_m) \leqslant \sum_{n=1}^{\infty} \sum_{j=1}^{\infty} \tau(E_{nj}) \leqslant \sum_{n=1}^{\infty} \left[\mu(S_n) + \frac{\varepsilon}{2^n} \right] = \left[\sum_{n=1}^{\infty} \mu(S_n) \right] + \varepsilon$$

This implies that $\mu(\bigcup_{n=1}^{\infty} S_n) \leqslant \sum_{n=1}^{\infty} \mu(S_n)$.

For any positive number α, let

$$\mathscr{R}_\alpha = \{ E \in \mathscr{R} : \text{diam}(E) \leqslant \alpha \}$$

and for each set $S \subset \mathbb{R}^n$, define $\mu_\alpha(S)$ to be the infimum of the numbers $\sum_{j=1}^{\infty} \tau(E_j)$, where $\langle E_j \rangle_{j \in \mathbb{N}}$ is any sequence of sets in $\mathscr{R}_\alpha$ that covers S. Then we have

(e) $\mu_\alpha(S) = \mu(S)$ for every $S \subset \mathbb{R}^n$.

Proof. It is clear that $\mu(S) \leqslant \mu_\alpha(S)$, since the infimum of a nonempty subset of a set cannot be smaller than the infimum of the set itself.

Suppose that ε is any given positive number. Let $\langle E_j \rangle_{j \in \mathbb{N}}$ be a sequence of sets in $\mathscr{R}$ that covers S with $\sum_{k=1}^{\infty} \tau(E_j) \leqslant \mu(S) + \varepsilon/2$. For each $j \in \mathbb{N}$, we can find a sequence $\langle E_{jk} \rangle_{k \in \mathbb{N}}$ of sets in $\mathscr{R}_\alpha$ that covers E_j with $\sum_{k=1}^{\infty} \tau(E_{jk}) \leqslant \tau(E_j) + \varepsilon/2^{j+1}$. The totality of sets E_{jk} forms a single sequence $\langle D_m \rangle_{m \in \mathbb{N}}$ of sets in $\mathscr{R}_\alpha$ that covers S with

$$\mu_\alpha(S) \leqslant \sum_{m=1}^{\infty} \tau(D_m) = \sum_{j=1}^{\infty} \sum_{k=1}^{\infty} \tau(E_{jk}) \leqslant \sum_{j=1}^{\infty} \left[\tau(E_j) + \frac{\varepsilon}{2^{j+1}} \right] \leqslant \mu(S) + \varepsilon$$

The number ε being arbitrary, it follows that $\mu_\alpha(S) \leqslant \mu(S)$. We thus conclude that $\mu_\alpha(S) = \mu(S)$.

(f) If $n \in \mathbb{N}$ and $F_1, F_2, \ldots, F_n$ are disjoint compact sets, then

$$\mu\left(\bigcup_{i=1}^{n} F_i\right) = \sum_{i=1}^{n} \mu(F_i)$$

Proof. We shall only treat the case $n = 2$; the general case then follows by mathematical induction. In view of (d), we have $\mu(F_1 \cup F_2) \leqslant \mu(F_1) + \mu(F_2)$. Therefore, it will suffice to establish that $\mu(F_1) + \mu(F_2) \leqslant \mu(F_1 \cup F_2) + \varepsilon$ for every number $\varepsilon > 0$.

Let $\alpha = d(F_1, F_2)$. According to (e), there is a sequence $\langle E_j \rangle_{j \in \mathbb{N}}$ of sets in $\mathscr{R}_\alpha$ such that $F_1 \cup F_2 \subset \bigcup_{j=1}^{\infty} E_j$ and $\sum_{j=1}^{\infty} \tau(E_j) \leqslant \mu(F_1 \cup F_2) + \varepsilon$. Let $I = \{j \in \mathbb{N} : F_1 \cap E_j \neq 0\}$ and let $J = \mathbb{N} - I$. Then $F_1 \subset \bigcup_{j \in I} E_j$ and $F_2 \subset \bigcup_{j \in J} E_j$. Hence,

$$\mu(F_1) + \mu(F_2) \leqslant \sum_{j \in I} \tau(E_j) + \sum_{j \in J} \tau(E_j) = \sum_{j=1}^{\infty} \tau(E_j) \leqslant \mu(F_1 \cup F_2) + \varepsilon$$

(g) Every family of disjoint compact sets of positive Lebesgue measure is countable.

Proof. Let $S = \{F_\alpha : \alpha \in I\}$ be any family of disjoint compact sets of positive Lebesgue measure. For each $k \in \mathbb{N}$, let D_k be the closed rectangle having vertices $\langle e_1, \ldots, e_n \rangle$ with $e_i \in \{-k, k\}$ for each $i = 1, \ldots, n$. Then

$$I = \{\alpha : \mu(F_\alpha) > 0\} = \bigcup_{k=1}^{\infty} \bigcup_{m=1}^{\infty} \left\{\alpha : \mu(F_\alpha \cap D_k) > \frac{1}{m}\right\}$$

We show that the set

$$\left\{\alpha : \mu(F_\alpha \cap D_k) > \frac{1}{m}\right\}$$

has finitely many elements [less than $(2k)^n m$] for all $k, m \in \mathbb{N}$. Indeed, suppose that this set has at least $(2k)^n m$ elements. Then it follows from (c) and (f) that $\mu(D_k) > (2k)^n$. Now for any number $\varepsilon > 0$, we can cover the closed rectangle D_k with $2n + 1$ open rectangles $E_1, E_2 \ldots, E_{2n+1}$ with $\sum_{j=1}^{2n+1} \tau(E_j) \leqslant (2k)^n + \varepsilon$; e.g., let E_1 be the open rectangle having the same vertices as D_k and let $E_2, \ldots, E_{2n+1}$ be open rectangles covering the $2n$ faces of D_k, each of which has area $\tau(E_j) \leqslant \varepsilon/2n$. But this implies that $\mu(D_k) \leqslant (2k)^n$, yielding a contradiction. The given set must thus have fewer than $(2k)^n m$ elements. In view of the foregoing equality, the index set I must be a countable set. Therefore, the family S is countable.

We limit our discussion of Lebesgue measure to the foregoing properties, which are sufficient to verify that the hypotheses of theorems in Chapters 1 to 5 are satisfied for Lebesgue measure. However, some additional comments made below require additional properties of Lebesgue measure. For further information concerning measure and dimension concepts, see Carathéodory (1914), Hausdorff (1919), Saks (1937), Taylor (1965), Kingman and Taylor (1966), Kuratowski (1966), Royden (1968), Munroe (1971), Billingsley (1986), and especially Oxtoby (1980).

C. Axioms and examples

DEFINITION   A pair of $(X, \mathscr{C})$, where X is a set and $\mathscr{C}$ is a family of subsets of X, is called a category base if the nonempty sets in $\mathscr{C}$, called regions, satisfy the following axioms:

1. Every point of X belongs to some region; i.e., $X = \cup \mathscr{C}$.
2. Let A be a region and let $\mathscr{D}$ be any nonempty family of disjoint regions that has power less than the power of $\mathscr{C}$.
   (a) If $A \cap (\cup \mathscr{D})$ contains a region, then there is a region $D \in \mathscr{D}$ such that $A \cap D$ contains a region.
   (b) If $A \cap (\cup \mathscr{D})$ contains no region, then there is a region $B \subset A$ that is disjoint from every region in $\mathscr{D}$.

Unless otherwise specified, the definitions, lemmas, theorems, and so on, below are formulated in terms of a fixed category base $(X, \mathscr{C})$, subsets of X, and points in X. If more than one category base $(X, \mathscr{C})$ is under consideration, then, when necessary for clarity, the regions in $\mathscr{C}$ will be called $\mathscr{C}$-regions.

If A and B are regions with $B \subset A$, then B is called a subregion of A.

EXAMPLE X is a set and $\mathscr{C}$ is a family of subsets of X closed under finite intersections with $X = \cup \mathscr{C}$. Of particular interest are the situations where $\mathscr{C}$ is the family

   (i) consisting only of X. (A category base for which X itself is the only region is called a trivial base.)
   (ii) of all subsets of X (called the discrete base for X).
   (iii) of all subsets of X whose complement is a finite set (called the cofinite base for X).
   (iv) of all subsets of X whose complement is a countable set (called the cocountable base for X).

EXAMPLE $X = \bigcup_{\alpha \in I} X_\alpha$ is a decomposition of a nonempty set X into nonempty disjoint sets X_α and $\mathscr{C} = \{X_\alpha : \alpha \in I\}$.

The most important examples of category bases are the following two, which motivate the general theory.

EXAMPLE $X = \mathbb{R}^n$ and $\mathscr{C}$ is the family of all closed rectangles.

EXAMPLE $X = \mathbb{R}^n$ and $\mathscr{C}$ is the family of all compact sets of positive Lebesgue measure.

Proof. Clearly, $X = \cup\mathscr{C}$, since every closed rectangle is in $\mathscr{C}$.

Suppose that $A \in \mathscr{C}$ and $\mathscr{D}$ is a nonempty family of disjoint sets in $\mathscr{C}$ that has power less than the power of $\mathscr{C}$. According to property (g) above, we can write $\mathscr{D} = \{D_n : n \in I\}$, where $I \subset \mathbb{N}$.

If $A \cap (\bigcup_{n \in I} D_n)$ contains a set in $\mathscr{C}$, then, by properties (c) and (d) above, there is an index $n_0 \in I$ for which $\mu(A \cap D_{n_0}) > 0$ and consequently, $A \cap D_{n_0}$ contains a region in $\mathscr{C}$.

Suppose that $A \cap (\bigcup_{n \in I} D_n)$ contains no set in $\mathscr{C}$. Then it follows from property (c) that $\mu(A \cap D_n) = 0$ for every $n \in I$. Let $\alpha = \mu(A)$. By virtue of property (d), we can cover the set $A \cap (\bigcup_{n \in I} D_n)$ by a sequence $\langle E_j \rangle_{j \in \mathbb{N}}$ of sets in $\mathscr{R}$ such that $\Sigma_{j=1}^{\infty} \tau(E_j) < \alpha$. Applying property (d) again, as well as property (c), we obtain

$$\alpha = \mu(A) \leqslant \mu\left[A \cap \left(\bigcup_{j=1}^{\infty} E_j\right)\right] + \mu\left(A - \bigcup_{j=1}^{\infty} E_j\right) < \alpha + \mu\left(A - \bigcup_{j=1}^{\infty} E_j\right)$$

The set $B = A - \bigcup_{j=1}^{\infty} E_j$, being a closed subset of the compact set A, is thus a set in $\mathscr{C}$ that is contained in A and is disjoint from every set D_n.

DEFINITION A family of sets is called a disjoint family if the intersection of any two different sets belonging to the family is empty.

DEFINITION A family $\mathscr{S}$ of subsets of a set X is said to satisfy the countable chain condition (abbreviated CCC) if every disjoint subfamily of $\mathscr{S}$ is countable.

In the last two examples above, as well as for the trivial, cofinite, and cocountable bases, the family $\mathscr{C}$ satisfies CCC.

For all the examples above, the cardinality restriction on the family of regions $\mathscr{D}$ in Axiom 2 is unnecessary. This restriction is required, however, in order to include certain examples where $\mathscr{C}$ does not satisfy CCC. Some of these examples follow.[1]

NOTATION We denote by CH the Continuum Hypothesis: $2^{\aleph_0} = \aleph_1$.

EXAMPLE (Assume CH.) Let $X = \mathbb{R}^n$ and let $\mathscr{C}$ consist of all sets representable in the form $A - M$, where A is a closed line segment and M is a countable set.

NOTATION We denote by $\mathscr{H}_c$ the family of all monotone increasing, continuous functions h mapping the set of nonnegative real numbers into

---

[1]Other examples are given in Morgan (1984, 1986). See also Chapter 5, Section I.E.

itself such that $h(0) = 0$ and $h(t) > 0$ whenever $t > 0$. We denote by μ^h the Hausdorff measure on $X = \mathbb{R}^n$ determined by a function $h \in \mathscr{H}_c$.[1]

For a real number p with $0 < p \leqslant n$, the Hausdorff measure μ^h determined by the function $h(t) = t^p$ is called the Hausdorff p-dimensional measure.

EXAMPLE   (Assume CH.[2]) Let $X = \mathbb{R}^n$, let $h \in \mathscr{H}_c$, and let $\mathscr{C}$ be the family of all closed sets that have positive measure for the Hausdorff measure μ^h.

EXAMPLE   (Assume CH.) Let $X = \mathbb{R}^n$, let p be a real number with $0 \leqslant p < n$, and let $\mathscr{C}$ be the family of all closed sets whose Hausdorff dimension is larger than p.

EXAMPLE   Let $X = \mathbb{R}^n$ and let $\mathscr{C}$ be the family of all closed sets that have positive Hausdorff measure for some $h \in \mathscr{H}_c$. [To show that $(X, \mathscr{C})$ is a category base, one uses the facts that every uncountable closed set contains a set in $\mathscr{C}$,[3] that a countable union of sets of power less than the power of the continuum has power less than that of the continuum, and that every uncountable closed set contains continuum many disjoint uncountable closed sets.[4]]

D. Regional properties

DEFINITION   Let π denote a given property of sets. A set S is said to have the property π in a region A if the set $S \cap A$ has the property π. A set has the property π everywhere in a region A if it has the property π in every subregion of A. A set has the property π everywhere if it has the property π in every region.

For instance, let $\mathscr{C}$ be the category base of all closed rectangles (or all open rectangles) in $\mathbb{R}^n$ and let π denote the property that a set be nonempty. Then a set has the property π everywhere if and only if every closed rectangle (or every open rectangle) contains at least one point of the set; such a set is said to be everywhere dense in $\mathbb{R}^n$.

---

[1]The basic properties of Hausdorff measure and Hausdorff dimension are found in Carathéodory (1914), Hausdorff (1919), Munroe (1971), Rogers (1970), and Billingsley (1986).
[2]Or, that Martin's Axiom holds [cf. Ostaszewski (1974)].
[3]See Rogers (1970, p. 67).
[4]See Chapter 5, Section I.C.

E. Local properties

By a neighborhood of a point we mean any region containing that point. Every region is thus a neighborhood of each of its own points. The following notion of localization generalizes the classical notion of a set in $\mathbb{R}^n$ having a certain property in all sufficiently small neighborhoods of a given point.[1]

DEFINITION Let π be a property of sets satisfying the condition
 (*) If a set S has the property π in a region A, then S has the property π in every subregion of A.
A set is said to have the property π locally at a given point if every neighborhood of the point contains a neighborhood of the point in which the set has the property π.

In some instances this definition has a simpler equivalent form.

THEOREM 1 If $(X,\mathscr{C})$ is a category base closed under finite intersections and π satisfies the condition (*), then a set has the property π locally at a given point if and only if there exists a neighborhood of the point in which the set has the property π.

For example, suppose that $\mathscr{C}$ is the category base consisting of the empty set and all open rectangles in $\mathbb{R}^n$. If π denotes the property that a set be finite, then a set S has the property π locally at a given point x if and only if there is an open rectangle containing x which contains at most a finite number of points of S. A point x at which this property π does not hold locally is called a limit point for the set S. That is, a point $x \in \mathbb{R}^n$ is a limit point for a set S if every open rectangle containing x contains infinitely many points of S.

F. Basically hereditary families

DEFINITION A family $\mathscr{S}$ of subsets of a set X is called hereditary if every subset of a member of $\mathscr{S}$ is also a member of $\mathscr{S}$.

For example, the families of all finite sets, all countable sets, and all sets not containing the number 1 are hereditary families of subsets of $\mathbb{R}$.

DEFINITION A family $\mathscr{N}$ of regions is called basically hereditary if every subregion of a region belonging to $\mathscr{N}$ contains a region that belongs to $\mathscr{N}$.

We leave to the reader the proof of the following fact.

THEOREM 2 If $(X,\mathscr{C})$ is a category base, $\mathscr{N}$ is a basically hereditary family of regions in $\mathscr{C}$, and $Y = \cup\mathscr{N}$, then $(Y,\mathscr{N})$ is a category base.

[1] Cf. Cantor (1966, pp. 140, 149, 264–270).

II. SINGULAR, MEAGER, AND ABUNDANT SETS

A. Definitions and examples

With respect to a given category base $(X,\mathscr{C})$, the subsets of X are classified according to the following scheme:

DEFINITION A set is a *singular set* if every region contains a subregion that is disjoint from the set. A set that is representable as a countable union of singular sets is called a *meager set*. A set that is not meager is called an *abundant set*. A set whose complement is meager is called a *comeager set*.

When necessary for the sake of clarity, singular, meager, and abundant sets for a category base $(X,\mathscr{C})$ will be called $\mathscr{C}$-singular, $\mathscr{C}$-meager, and $\mathscr{C}$-abundant sets, respectively. The families of singular sets and meager sets will be denoted by $\mathfrak{S}(\mathscr{C})$ and $\mathfrak{M}(\mathscr{C})$, respectively.

We note that if $\mathscr{C}$ is finite, then every meager set is singular.

EXAMPLE If $\mathscr{C}$ is a trivial base, then the empty set is the only meager set and every nonempty set is abundant.

EXAMPLE If $\mathscr{C}$ is a discrete base, then the empty set is the only meager set and every nonempty set is abundant.

EXAMPLE If $\mathscr{C}$ is the cofinite base for an uncountable set X, then the singular, meager, and abundant sets coincide with the finite, countable, and uncountable sets, respectively.

EXAMPLE If $\mathscr{C}$ is the cocountable base for an uncountable set X, then the singular sets and meager sets both coincide with the countable sets, while the abundant sets coincide with the uncountable sets.

EXAMPLE If $X = \bigcup_{\alpha \in I} X_\alpha$ is a decomposition of a set X into nonempty disjoint sets X_α and $\mathscr{C} = \{X_\alpha : \alpha \in I\}$, then every nonempty set is abundant.

EXAMPLE If $X = \mathbb{R}^n$ and $\mathscr{C}$ is the family of all closed rectangles, then the singular sets, meager sets, and abundant sets are called nowhere dense sets, sets of the first category, and sets of the second category, respectively,[1] while the comeager sets are called residual sets.[2] Every finite set is nowhere dense, every countable set is of the first category, and every closed rectangle is a set of the second category. The Cantor set is an uncountable nowhere dense set.

---

[1]Cf. Baire (1898, 1899a, 1899b).
[2]Denjoy (1915).

EXAMPLE If $X = \mathbb{R}^n$ and $\mathscr{C}$ is the family of all compact sets with positive Lebesgue measure, then it can be shown[1] that the meager sets are the same as the singular sets and coincide with the sets having Lebesgue measure zero. The abundant sets for this category base are called sets of positive Lebesgue outer measure.

EXAMPLE (Assume CH.) If $X = \mathbb{R}^n$, h is a given function in $\mathscr{H}_c$ and $\mathscr{C}$ is the family of all closed sets that have positive measure for the Hausdorff measure μ^h, then the singular and meager sets both coincide with the sets that have no subsets of finite, positive outer measure, or equivalently, the sets every subset of which is a measurable set with measure 0 or ∞.[2] If μ^h is σ-finite (i.e., X is representable as a countable union of measurable sets of finite measure), then the singular and meager sets coincide with the sets of measure zero. If μ^h is non-σ-finite, then, assuming CH, there exist uncountable meager sets that do not have measure zero.[3]

EXAMPLE (Assume CH.) If $X = \mathbb{R}^n$, $0 \leqslant p < n$, and $\mathscr{C}$ is the family of all closed sets whose Hausdorff dimension is larger than p, then the singular and meager sets both coincide with the sets that have no subset of finite, positive Hausdorff q-dimensional outer measure for any number $q > p$. The set of Liouville numbers in $\mathbb{R}$ is a meager set for every p.[4]

EXAMPLE If $X = \mathbb{R}^n$ and $\mathscr{C}$ is the family of all closed sets that have positive Hausdorff measure for some $h \in \mathscr{H}_c$, then the singular and meager sets are the same as the sets that have no subset of finite, positive Hausdorff outer measure for any $h \in \mathscr{H}_c$. Included among the meager sets are the sets that have μ^h-measure zero for every $h \in \mathscr{H}_c$. Such sets were investigated by Besicovitch, who constructed an uncountable set of this type.[5]

DEFINITION A family $\mathscr{I}$ of subsets of a set X is called an ideal if it is hereditary and is closed under finite unions. An ideal that is closed under arbitrary countable unions is called a σ-ideal.

It is a simple matter to prove the following fact:

THEOREM 1 The family of all singular sets forms an ideal and the family of all meager sets forms a σ-ideal.

---

[1]See the references given in Section I.B.
[2]Cf. Morgan (1985).
[3]See Corollary 17 in Chapter 2, Section III.
[4]See Oxtoby (1980, Chap. 2).
[5]Besicovitch (1934); see also Rogers (1970), Chap. 2, Sec. 3.6).

B. Intersections of regions

The intersection of two regions is not necessarily a region. However, we do
have a positive result concerning the intersection.

THEOREM 2   The intersection of any two regions either contains a region
or is a singular set.

Proof. Assume that C, D are regions and that $C \cap D$ contains no region. This
implies, of course, that $C \neq D$ and hence that the family $\mathscr{C}$ contains at least
two distinct regions. Let E be any region.

If $E \cap C$ contains no region, then we apply Axiom 2b with $A = E$ and
$\mathscr{D} = \{C\}$ to obtain the existence of a region $B \subset E—C$ that is disjoint from
$C \cap D$.

If $E \cap C$ contains a region F, then $F \cap D \subset C \cap D$ and consequently
$F \cap D$ contains no region. Applying Axiom 2b with $A = F$ and $\mathscr{D} = \{D\}$, we
obtain the existence of a region $B \subset F—D$ that is disjoint from $C \cap D$.

We thus see that $C \cap D$ is a singular set.

C. Principal lemmas

We give here two basic lemmas that are utilized to establish several theorems
below.

LEMMA 3   If $(X, \mathscr{C})$ is a category base, $\mathscr{N}$ is a subfamily of $\mathscr{C}$ with the
property that each region in $\mathscr{C}$ contains a region in $\mathscr{N}$, and $Y = \cup \mathscr{N}$, then
$(Y, \mathscr{N})$ is a category base and the $\mathscr{N}$-singular sets coincide with the $\mathscr{C}$-singular
subsets of Y. In addition, if U is a subset of Y and $Y—U$ is $\mathscr{N}$-singular, then
$X—U$ is $\mathscr{C}$-singular.

Proof. By Theorem 2 in Section I, $(Y, \mathscr{N})$ is a category base. From the
assumption that each region in $\mathscr{C}$ contains a region in $\mathscr{N}$, it follows that the
$\mathscr{N}$-singular sets and $\mathscr{C}$-singular subsets of Y coincide. Finally,

$$X—U = (X—Y) \cup (Y—U)$$

is the union of two $\mathscr{C}$-singular sets whenever $Y—U$ is $\mathscr{N}$-singular.

LEMMA 4   If $(Y, \mathscr{N})$ is a category base, then there exists a disjoint sub-
family $\mathscr{M}$ of $\mathscr{N}$ such that $Y— \cup \mathscr{M}$ is a singular set. Moreover, $\mathscr{M}$ may be so
selected that for every region N, there exists a region $M \in \mathscr{M}$ such that $N \cap M$
contains a region.

Proof.[1] We determine the family $\mathcal{M}$ by transfinite induction. Let

$$N_1, N_2, \ldots, N_\alpha, \ldots \qquad (\alpha < \Theta)$$

be an enumeration of all regions, where Θ is the smallest ordinal number whose power is the same as that of $\mathcal{N}-\{\varnothing\}$.

Set $M_1 = N_1$. Assume that $1 < \alpha < \Theta$ and the sets M_β have already been defined for all ordinal numbers $\beta < \alpha$. If the set $E_\alpha = N_\alpha \cap (\bigcup_{\beta<\alpha} M_\beta)$ contains no region, then we apply Axiom 2b and define M_α to be a subregion of N_α that is disjoint from $\bigcup_{\beta<\alpha} M_\beta$. If the set E_α does contain a region, then we set $M_\alpha = M_1$.

Suppose that N is any given region. We show that there is a region $M \in \mathcal{M} = \{M_\beta : \beta < \Theta\}$ such that $N \cap M$ contains a region. With respect to the enumeration above, we have $N = N_\alpha$ for some $\alpha < \Theta$. If E_α contains no region, then we can take $M = M_\alpha$. If E_α does contain a region, then, according to Axiom 2a, there is an index $\beta < \alpha$ such that $N_\alpha \cap M_\beta$ contains a region and we take $M = M_\beta$.

It readily follows from the property just established that $Y - \cup \mathcal{M}$ is a singular set.

D.   Characterizations

In the notation $(X, \mathscr{S})$ the symbol $\mathscr{S}$ will always denote a family of subsets of the set X whose members are called $\mathscr{S}$-sets. Arising from $\mathscr{S}$ are the families of sets $\mathscr{S}_\sigma$, $\mathscr{S}_\delta$, $\mathscr{S}_{\sigma\delta}$, $\mathscr{S}_{\delta\sigma}$, defined by

$\mathscr{S}_\sigma$: all sets representable as the union of a countable subfamily of $\mathscr{S}$
$\mathscr{S}_\delta$: all sets representable as the intersection of a countable subfamily of $\mathscr{S}$
$\mathscr{S}_{\sigma\delta}$: all sets representable as the intersection of a countable subfamily of $\mathscr{S}_\sigma$
$\mathscr{S}_{\delta\sigma}$: all sets representable as the union of a countable subfamily of $\mathscr{S}_\delta$

NOTATION   For a category base $(X, \mathscr{C})$, the symbol $\mathscr{X}$ will denote the family of all sets whose complement is a member of $\mathscr{C}$.

THEOREM 5   If $\mathscr{C}$ satisfies CCC, then each singular set is contained in a singular $\mathscr{X}_\delta$-set and each meager set is contained in a meager $\mathscr{X}_{\delta\sigma}$-set.

Proof. Suppose that S is a singular set and let $\mathcal{N}$ be the family of all regions disjoint from S. Apply Lemmas 3 and 4 to obtain a countable family $\mathcal{M} = \{M_n : n \in I\}$ of disjoint regions in $\mathcal{N}$ such that $X - \bigcup_{n \in I} M_n$ is a singular set. Then we have

$$S \subset \bigcap_{n \in I} (X - M_n)$$

---

[1] A simplification of the original proof due to a referee of *Fundamenta Mathematicae* (1987).

and the first assertion is established. The second assertion regarding meager sets is an immediate consequence of the first.

If $\mathscr{C}$ is the category base of all open sets in $\mathbb{R}^n$, then the singular sets, meager sets, and abundant sets coincide with the nowhere dense sets, sets of the first category, and sets of the second category, respectively. In this instance, Theorem 5 yields:

(σ) Every nowhere dense set is contained in a closed, nowhere dense set.

(τ) Every set of the first category is contained in an $\mathscr{F}_\sigma$-set of the first category.

If $\mathscr{C}$ is the category base of all compact sets in $\mathbb{R}^n$ of positive Lebesgue measure, then Theorem 5 affirms the fact

(μ) Every set of Lebesgue measure zero is contained in a $\mathscr{G}_\delta$-set of Lebesgue measure zero.

We shall utilize Theorem 5 to characterize the singular and meager sets for an additional example.

DEFINITION $(X, \mathscr{A})$ is called a field if the family $\mathscr{A}$ is closed under finite unions and complementation. $(X, \mathscr{A})$ is called a σ-field if $\mathscr{A}$ is closed under countable unions and complementation.

If $\mathscr{S}$ is any family of subsets of X, then the family of all subsets of X is a σ-field containing the family $\mathscr{S}$. The intersection of all σ-fields $\mathscr{A}$ of subsets of X that contain $\mathscr{S}$ is itself a σ-field, called the σ-field generated by $\mathscr{S}$. This σ-field contains all the families of sets $\mathscr{S}, \mathscr{S}_\sigma, \mathscr{S}_\delta, \mathscr{S}_{\sigma\delta}, \mathscr{S}_{\delta\sigma}$.

DEFINITION $(X, \mathscr{S})$ is called an ideal (resp., σ-ideal) in a field $(X, \mathscr{A})$ if the following conditions are satisfied:

(a) $\mathscr{S} \subset \mathscr{A}$.

(b) $\mathscr{S}$ is closed under finite (resp., countable) unions.

(c) If $S \in \mathscr{S}$, $T \subset S$, and $T \in \mathscr{A}$, then $T \in \mathscr{S}$.

$(X, \mathscr{S})$ is called proper if $X \notin \mathscr{S}$.

An ideal (resp., σ-ideal) as defined earlier is such with respect to the field of all subsets of X.

THEOREM 6   If $(X, \mathscr{S})$ is a proper σ-ideal in a σ-field $(X, \mathscr{A})$ such that $\mathscr{C} = \mathscr{A} - \mathscr{S}$ satisfies CCC, then $(X, \mathscr{C})$ is a category base with

$$\mathfrak{S}(\mathscr{C}) = \{S : S \subset I \text{ for some } I \in \mathscr{S}\}$$

If $\mathscr{S}$ is not only a σ-ideal in $\mathscr{A}$, but a σ-ideal in the family of all subsets of X, then

$$\mathfrak{S}(\mathscr{C}) = \mathscr{S}$$

In any case, we have

$$\mathfrak{S}(\mathscr{C}) = \mathfrak{M}(\mathscr{C})$$

Proof. It is a simple matter to verify that $(X,\mathscr{C})$ is a category base. We shall first show that $\mathscr{I}$ coincides with the family of all singular sets in $\mathscr{X}$.

Suppose that $S \in \mathscr{I}$. Then $X-S \in \mathscr{A}$. Since $X \notin \mathscr{I}$, we must have $X-S \notin \mathscr{I}$. Hence, $X-S \in \mathscr{C}$ and $S \in \mathscr{X}$. For any region A, the set $A-S$ is a region that is disjoint from S. Thus, every set in $\mathscr{I}$ belongs to $\mathscr{X}$ and is singular.

Conversely, suppose that S is a singular set in $\mathscr{X}$. If $S = \varnothing$, then obviously $S \in \mathscr{I}$. Assume that $S \neq \varnothing$. Because no nonempty set in $\mathscr{C}$ can be singular, we cannot have $S \in \mathscr{C}$. But we do have $S \in \mathscr{A}$. Consequently, $S \in \mathscr{I}$.

As $\mathscr{C}$ is closed under countable unions, the family $\mathscr{X}$ is closed under countable intersections, so $\mathscr{X}_\delta = \mathscr{X}$. The singular $\mathscr{X}_\delta$-sets thus coincide with the sets in $\mathscr{I}$. From Theorem 5 we then obtain

$$\mathfrak{S}(\mathscr{C}) = \{S : S \subset I \text{ for some } I \in \mathscr{I}\}$$

The further assertions of the theorem are now readily verified.

E.   Distributions of sets in regions

DEFINITION   A set S is singular (resp., meager, abundant) in a region A if $S \cap A$ is a singular (resp., meager, abundant) set. A set is abundant everywhere in a region A if it is abundant in every subregion of A. A set is abundant everywhere if it is abundant in every region.

Clearly, a set that is singular (resp., meager) in a region A is also singular (resp., meager) in every subregion of A.

We note that for the category base of all compact sets of positive Lebesgue measure in $\mathbb{R}^n$, a set S is abundant everywhere if and only if for every open rectangle E we have

$$\mu(S \cap E) = \tau(E)$$

For $n = 1$, this condition states that

$$\mu[S \cap (a,b)] = b - a$$

for every open interval (a,b).

THEOREM 7   If a set S is abundant everywhere in a region A and A is abundant everywhere in a region B, then S is abundant everywhere in B.

Proof. Let C be any subregion of B. Since A is abundant everywhere in B, the set $A \cap C$ is abundant. By Theorem 2, there is a region $D \subset A \cap C$. From $D \subset A$ and the assumption that S is abundant everywhere in A it follows that

$S \cap D$ is an abundant set. From the inclusion $S \cap D \subset S \cap C$ we see that $S \cap C$ is also an abundant set. Therefore, S is abundant in every subregion of B.

In the next section we establish the most important result concerning the distribution of abundant sets.

F. The Fundamental Theorem

FUNDAMENTAL THEOREM Every abundant set is abundant everywhere in some region.[1]

Proof. Assume that S is a set which is not abundant everywhere in any region. Let

$$\mathcal{N} = \{A \in \mathscr{C} : S \cap A \text{ is meager}\}$$

Apply Lemmas 3 and 4 to obtain a subfamily $\mathcal{M}$ of $\mathcal{N}$ consisting of disjoint sets such that $X - \cup \mathcal{M}$ is $\mathscr{C}$-singular. Let $M_1, M_2, \ldots, M_\alpha, \ldots$ be a well-ordering of $\mathcal{M}$ and set $U = \bigcup_\alpha M_\alpha$. We shall show that $S \cap U$ is $\mathscr{C}$-meager.

For each α, set

$$S \cap M_\alpha = \bigcup_{n=1}^{\infty} S_{\alpha,n}$$

where the sets $S_{\alpha,n}$ are $\mathscr{C}$-singular. Then

$$S \cap U = \bigcup_{n=1}^{\infty} \bigcup_\alpha S_{\alpha,n}$$

and it suffices to show that for each n, the set

$$S_n = \bigcup_\alpha S_{\alpha,n}$$

is $\mathcal{N}$-singular, since this implies that S_n is $\mathscr{C}$-singular.

Suppose that A is any region in $\mathcal{N}$ and let α be the smallest index for which $A \cap M_\alpha$ contains a region $C \in \mathcal{N}$. We have $C \subset M_\alpha$ and, for all indices $\beta \neq \alpha$, $S_{\beta,n} \subset M_\beta$ and $M_\alpha \cap M_\beta = \varnothing$, which implies that $C \cap S_{\beta,n} = \varnothing$. Choose an $\mathcal{N}$-region $B \subset C - S_{\alpha,n}$. Then $B \subset A$ and $B \cap S_n = \varnothing$. Hence S_n is $\mathcal{N}$-singular.

From the inclusion

$$S = (S \cap U) \cup (S - U) \subset (S \cap U) \cup (X - U)$$

it follows that S is a $\mathscr{C}$-meager set.

[1]Cf. Baire (1899b), Lebesgue (1905, pp. 185–186), Banach (1930), Oxtoby (1957), Morgan (1974, 1977a), and Sander and Slipek (1979). See also Chapter 4, Section IV.C.

The following consequence of the Fundamental Theorem is used frequently below.

THEOREM 8   If a set is abundant in a given region, then it is abundant everywhere in some subregion of that region.

Proof. Let S be a set that is abundant in a region A. According to the Fundamental Theorem, the set $S \cap A$ is abundant everywhere in a region B. In particular, this implies that $A \cap B$ is an abundant set. By Theorem 2, $A \cap B$ contains a region C. Now, if D is any subregion of C, then D is a subregion of B, and since $S \cap A$ is abundant everywhere in B, the set $(S \cap A) \cap D$, as well as the set $S \cap D$, is abundant. Thus, S is abundant everywhere in the subregion C of A.

G. Local properties

DEFINITION   A set is locally singular at a given point if every neighborhood of the point contains a neighborhood of the point in which the set is singular.

THEOREM 9   A set is singular if and only if it is locally singular at every point.

Proof. If S is singular, then S is obviously locally singular at every point.

Conversely, suppose that S is locally singular at every point of X. If A is any region, then there is a region $B \subset A$ such that $S \cap B$ is singular, and consequently, there is a region $C \subset B$ that is disjoint from S. Let $\mathcal{N}$ denote the family of regions C that are disjoint from S. Applying Lemmas 3 and 4, we obtain a subfamily $\mathcal{M}$ of $\mathcal{N}$ such that $X — \cup \mathcal{M}$ is a singular set. From the inclusion $S \subset X — \cup \mathcal{M}$ we conclude that S is singular.

THEOREM 10   The set of all points of a set S at which S is locally singular is a singular set.

Proof. Let S_0 denote the set of all points in S at which S is locally singular and assume, to the contrary, that S_0 is not singular. There is then a region A such that for every region $C \subset A$ we have $C \cap S_0 \neq \emptyset$. Choose a point $x \in A \cap S$. Since S is locally singular at x, there is a region $B \subset A$ for which $B \cap S$ is a singular set. Accordingly, there is a region $C \subset B$ such that $C \cap S = \emptyset$. But this implies the contradiction $C \cap S_0 = \emptyset$. Therefore, S_0 must be a singular set.

DEFINITION   A set is locally meager at a given point if every neighborhood of the point contains a neighborhood of the point in which the set is meager. Otherwise, the set is said to be locally abundant at the point.

The next two theorems are generalizations of theorems of Banach.[1]

THEOREM 11 A set is meager if and only if it is locally meager at every point.

Proof. If S is a meager set, then S is obviously locally meager at every point.

Suppose, on the other hand, that S is an abundant set. By the Fundamental Theorem, S is abundant everywhere in some region A. Choosing any point $x \in A$, the set S is not locally meager at x.

THEOREM 12 The set of all points of a set S at which S is locally meager is a meager set.

Proof. Let S_1 denote the set of all points in S at which S is locally meager and assume, to the contrary, that S_1 is an abundant set. Let A be a region in which S_1 is abundant everywhere. Choose a point x belonging to the abundant set $A \cap S_1$ and let B be a subregion of A containing x such that $B \cap S$ is a meager set. From the inclusion $B \cap S_1 \subset B \cap S$ it follows that $B \cap S_1$ is also a meager set, contradicting the fact that S_1 is abundant everywhere in A. We conclude that S_1 is a meager set.

If a set S is abundant everywhere in a region A, then S is locally abundant at every point of A. However, the converse is not valid. For example, if $\mathscr{C}$ is the family of all compact sets of positive Lebesgue measure in $\mathbb{R}^n$, then any set in $\mathscr{C}$ is locally abundant at every point of $\mathbb{R}^n$, but need not be abundant in every region.

III. BAIRE SETS

A. Definition and examples

DEFINITION A set S is a Baire set if every region contains a subregion in which either S or $X{-}S$ is a meager set.

Restated, a set is a Baire set if there is no region in which both the set and its complement are abundant everywhere. We will denote the family of all Baire sets for $(X, \mathscr{C})$ by $\mathfrak{B}(\mathscr{C})$.

EXAMPLE If $\mathscr{C}$ is a trivial base, then $\varnothing$ and X are the only Baire sets.

EXAMPLE If $\mathscr{C}$ is a discrete base, then every set is a Baire set.

EXAMPLE If $\mathscr{C}$ is either the cofinite base or the cocountable base, then the countable sets and their complements comprise the Baire sets.

---

[1]Banach (1930) and Morgan (1977a).

EXAMPLE If $X = \bigcup_{\alpha \in I} X_\alpha$ is a decomposition of a set X into nonempty disjoint sets X_α and $\mathscr{C} = \{X_\alpha : \alpha \in I\}$, then the Baire sets coincide with the sets that are representable as the union of some subfamily of $\mathscr{C}$.

EXAMPLE If $X = \mathbb{R}^n$ and $\mathscr{C}$ is the family of all closed rectangles, then the Baire sets are called the sets that have the Baire property.[1]

EXAMPLE If $X = \mathbb{R}^n$ and $\mathscr{C}$ is the family of all compact sets of positive Lebesgue measure, then the Baire sets are known as the Lebesgue measurable sets.

Given any statement τ involving only abstract set-theoretical concepts and the terms "first category," "second category," and "set with the Baire property," one obtains a statement

$$\mu = \Phi(\tau)$$

concerning Lebesgue measure by applying the following substitution matrix to the statement τ:

$$\Phi = \begin{pmatrix} \text{first category,} & \text{measure zero} \\ \text{second category,} & \text{positive outer measure} \\ \text{set with the Baire property,} & \text{measurable set} \end{pmatrix}$$

For instance, the statement

(τ) A set S is a first category set if and only if every subset of S is a set with the Baire property.

is transformed into the statement

(μ) A set S is a measure zero set if and only if every subset of S is a measurable set.

Such a statement τ is called a Baire category statement and the corresponding statement μ is called its Lebesgue measure analogue. Inversely, any statement μ involving only abstract set-theoretical concepts and the Lebesgue measure-theoretic terms "measure zero," "positive outer measure," and "measurable set" can be transformed into a Baire category analogue τ.

We note that the particular analogous statements τ and μ given above are both true statements.[2] In fact, although there are valid Baire category statements whose Lebesgue measure analogue is invalid, and vice versa, there

[1] The term "Baire set" was apparently first used in Pettis (1951) to designate sets that have the Baire property (or satisfy the condition of Baire). The term "Baire set" was initially used in a different sense in Burstin (1914, p. 1537, fn. 2); see also Lebesgue (1905, pp. 184–190).
[2] See Theorem 13 in Chapter 5, Section II.

are more than 100 valid Baire category statements whose Lebesgue measure
analogue is also valid.[1] A large number of these analogies are unified in this
book within the context of the notion of a category base.[2] Of fundamental
importance in this unification is the observation that the Baire category
concepts are definable in terms of closed rectangles and the Lebesgue
measure concepts are similarly definable in terms of compact sets of positive
measure. We note that the characterization of the Lebesgue measurable sets,
as the sets having the property that every compact set of positive measure
contains a compact set of positive measure in which either the given set or its
complement has measure zero, has apparently never appeared in any
textbook on measure theory.[3]

EXAMPLE   (Assume CH.) If $X = \mathbb{R}^n$, h is a given function in $\mathscr{H}_c$ and $\mathscr{C}$ is
the family of all closed sets that have positive measure for the Hausdorff
measure μ^h, then the Baire sets coincide with the μ^h-measurable sets, although
the meager sets may not coincide with the sets of μ^h-measure zero.[4]

We note that the statement (μ) given above is not valid for non-σ-finite
Hausdorff measures. For such measures it can be shown, assuming CH, that
there are sets of infinite measure every subset of which is measurable.[5]
Nevertheless, the following general statement is valid for Baire category,
Lebesgue measure, Hausdorff measure, and other classification schemes.

(π) A set S is a meager set if and only if every subset of S is a Baire set.[6]

In formulating Hausdorff measure analogues of Baire category statements, the sets that contain no subset of finite, positive Hausdorff outer
measure correspond to the sets of the first category.

B.   Equivalent bases

The primary function of point set theory is to distinguish different types of
infinite sets. In this respect, both the cofinite and cocountable bases for an
uncountable set should be considered to be equivalent, since both of these

[1]See Oxtoby (1980), Morgan (1984, 1986), and Covington (1986) for a detailed
discussion of many of these analogies.
[2]See Szpilrajn (1930), Hausdorff (1936, pp. 250–251; 1969), Kondô (1936), Ho and
Naimpally (1979), and Covington (1986) regarding other unification efforts.
[3]This characterization is essentially given in Burstin (1914, Sec. 4, Erster Satz, p. 1539);
see also Covington (1986, pp. 59–61).
[4]See Morgan (1985).
[5]See Corollary 17 in Chapter 2, Section III.
[6]See Theorem 13 in Chapter 5, Section II.

category bases result in the dichotomy of countable versus uncountable within the family of infinite sets. However, we shall also require for equivalence that the families of Baire sets coincide.

DEFINITION Two category bases $(X,\mathscr{C})$ and $(X,\mathscr{D})$ are called equivalent if $\mathfrak{M}(\mathscr{C}) = \mathfrak{M}(\mathscr{D})$ and $\mathfrak{B}(\mathscr{C}) = \mathfrak{B}(\mathscr{D})$.

The following fact is often useful to establish the equivalence of two category bases.

THEOREM 1 If $(X,\mathscr{C})$ and $(X,\mathscr{D})$ are category bases such that each $\mathscr{C}$-region contains a $\mathscr{D}$-region and, conversely, each $\mathscr{D}$-region contains a $\mathscr{C}$-region, then $(X,\mathscr{C})$ and $(X,\mathscr{D})$ are equivalent.

Proof. It is a simple matter to verify that $\mathfrak{S}(\mathscr{C}) = \mathfrak{S}(\mathscr{D})$. Hence, $\mathfrak{M}(\mathscr{C}) = \mathscr{M}(\mathscr{D})$.

Suppose that $S \in \mathfrak{B}(\mathscr{C})$ and let D be any $\mathscr{D}$-region. We have to show that there is a $\mathscr{D}$-region $E \subset D$ such that either $E \cap S$ or $E \cap (X-S)$ is $\mathscr{D}$-meager. Assume to the contrary that for every $\mathscr{D}$-region $E \subset D$, both $E \cap S$ and $E \cap (X-S)$ are $\mathscr{D}$-abundant.

Let A be a $\mathscr{C}$-region contained in D. If B is any $\mathscr{C}$-region contained in A then, since B contains a $\mathscr{D}$-subregion of D, both $B \cap S$ and $B \cap (X-S)$ are $\mathscr{D}$-abundant. This means both $B \cap S$ and $B \cap (X-S)$ are $\mathscr{C}$-abundant everywhere in A, contradicting the assumption $S \in \mathfrak{B}(\mathscr{C})$. We must therefore have $S \in \mathfrak{B}(\mathscr{D})$.

By a similar argument, if $S \in \mathfrak{B}(\mathscr{D})$, then $S \in \mathfrak{B}(\mathscr{C})$. We thus conclude that $\mathfrak{B}(\mathscr{C}) = \mathfrak{B}(\mathscr{D})$.

We note that if $(X,\mathscr{C})$ is any category base, then $(X,\mathscr{C} \cup \{\varnothing\})$ is an equivalent category base. In particular, if X is a nonempty set and $\mathscr{C} = \{\{x\}:x \in X\}$, then $(X,\mathscr{C})$ is equivalent to a category base that is closed under finite intersection. Of course, it is also true that if $(X,\mathscr{C})$ is any category base, then $(X,\mathscr{C}-\{\varnothing\})$ is an equivalent category base. We note further that if $(X,\mathscr{C})$ is any category base, then $(X,\mathscr{C} \cup \{X\})$ is an equivalent category base.

EXAMPLE The cofinite base for a nonempty set X is equivalent to the cocountable base for X, although the singular sets may not agree.

EXAMPLE If X is a nonempty set and $\mathscr{C} = \{\{x\}:x \in X\}$, then $(X,\mathscr{C})$ is equivalent to the discrete base for X.

EXAMPLE The cofinite base for a denumerable set X is not equivalent to the discrete base for X, although these bases have the same Baire sets.

EXAMPLE For $X = \mathbb{R}^n$ the following category bases are equivalent:
(1) all closed rectangles.

(2) all closed rectangles with rational vertices.
(3) all open rectangles.
(4) all open sets.
(5) all sets of the form $A-M$, where A is a closed rectangle and M is a countable set.

For the first four of these bases the singular sets are the same as the nowhere dense sets, but this is not true of the last base. All five bases satisfy CCC.

EXAMPLE If $X = \mathbb{R}^n$, $\mathscr{C}$ is the family of all closed rectangles, and $\mathscr{D}$ is the family of all residual sets, then $(X,\mathscr{C})$ and $(X,\mathscr{D})$ are category bases that are not equivalent, although $\mathfrak{M}(\mathscr{C}) = \mathfrak{M}(\mathscr{D})$.

EXAMPLE The category base of all compact sets of positive Lebesgue measure in $\mathbb{R}^n$ is equivalent to that of all closed sets of positive Lebesgue measure.

EXAMPLE The category base of all closed sets of positive Hausdorff n-dimensional measure in $\mathbb{R}^n$ is equivalent to the category base of all compact sets of positive Lebesgue measure in $\mathbb{R}^n$.

C. Elementary properties

THEOREM 2 A necessary and sufficient condition that a set S be a Baire set is that whenever S is abundant everywhere in a region A, the set $A-S$ is meager.

Proof. If S is not a Baire set, then there exists a region A in which both S and $X-S$ are abundant everywhere. Consequently, S is abundant everywhere in A and $A-S$ is an abundant set, so the condition does not hold.

Conversely, suppose that the condition fails to hold. Let S be abundant everywhere in a region A and suppose that $A-S$ is an abundant set. The set $A-S$ is abundant everywhere in a region B. According to Theorem 2 in Section II, $A \cap B$ contains a region C. The sets S and $X-S$ being both abundant everywhere in C, we conclude that S is not a Baire set.

THEOREM 3 If a Baire set S is abundant in every region in which a given set T is abundant, then $T-S$ is a meager set.

Proof. Assume to the contrary that $T-S$ is an abundant set. Then $T-S$ is abundant everywhere in a region A. As $X-S$ is also abundant everywhere in A, it follows from Theorem 2 that $A \cap S$ is a meager set. Thus, there exists a region A in which T is abundant but S is not.

THEOREM 4 If $\langle S_n \rangle_{n \in \mathbb{N}}$ is a sequence of Baire sets each of which is abundant everywhere in a given region A, then $\bigcap_{n=1}^{\infty} S_n$ is abundant everywhere in A.

Proof. Let B be any subregion of A. From Theorem 2 it follows that the set

$$\bigcup_{n=1}^{\infty} (B - S_n) = B - \bigcap_{n=1}^{\infty} S_n$$

is a meager set. The region B being an abundant set, the set $B \cap (\bigcap_{n=1}^{\infty} S_n)$ must be an abundant set.

D. Essentially disjoint sets

The notion of essentially disjoint sets is a generalization of that of disjoint sets.

DEFINITION Two sets are called essentially disjoint if their intersection is a meager set. A family of sets is called essentially disjoint if each pair of different sets in the family is essentially disjoint.[1]

THEOREM 5 If $\mathscr{C}$ satisfies CCC, then every family of essentially disjoint, abundant Baire sets is countable.[2]

Proof. Assume that the conclusion is false. Then there exists a transfinite sequence $\langle S_\alpha \rangle_{\alpha < \Omega}$ of different essentially disjoint, abundant Baire sets. We define, by transfinite induction, a transfinite sequence $\langle A_\alpha \rangle_{\alpha < \Omega}$ of disjoint regions such that S_α is abundant everywhere in A_α, for each $\alpha < \Omega$.

By the Fundamental Theorem, we know that each set S_α is abundant everywhere in some region. Moreover, no two sets S_α can be abundant everywhere in the same region. For if $\alpha \neq \beta$ and both S_α and S_β are abundant everywhere in the same region A, then from the equality

$$B \cap S_\beta = [(B \cap S_\beta) \cap S_\alpha] \cup [(B \cap S_\beta) \cap (X - S_\alpha)]$$

it follows that $X - S_\alpha$ is abundant in every subregion B of A, contradicting the fact that S_α is a Baire set.

Let A_1 be a region in which S_1 is abundant everywhere. Assume that $1 < \alpha < \Omega$ and the sets A_β have already been defined for all ordinal numbers

---

[1] The term "almost disjoint" is generally used in the literature instead of "essentially disjoint." We have adopted the latter term because it seems semantically awkward to say: "disjoint sets are almost disjoint sets."

[2] That every family of disjoint sets in $\mathbb{R}^n$ which are of the second category and have the Baire property must be countable was apparently first noticed by Sierpiński [cf. Sélivanowski (1933, p. 26), and Sierpiński (1933 f, p. 31)].

$\beta < \alpha$. Let B_α be a region in which S_α is abundant everywhere. Then $B_\alpha \cap (\bigcup_{\beta < \alpha} A_\beta)$ contains no region. For suppose that it did contain a region C. From the inclusion $C \subset B_\alpha$, we know that $C \cap S_\alpha$ is abundant and hence $B_\alpha \cap (\bigcup_{\beta < \alpha} A_\beta)$ is also abundant. Applying Theorem 2 of Section II, there is an ordinal number $\beta < \alpha$ such that $B_\alpha \cap A_\beta$ contains a region D. But then both S_β and S_α are abundant everywhere in D, an impossibility! Thus, $B_\alpha \cap (\bigcup_{\beta < \alpha} A_\beta)$ must contain no region. We know, from the preceding paragraph, that the family $\mathscr{C}$ of all regions is uncountable. Using Axiom 2b, we define A_α to be a region contained in $B_\alpha - \bigcup_{\beta < \alpha} A_\beta$. We thus determine by transfinite induction an uncountable family of disjoint regions A_α, contradicting the hypothesis that $\mathscr{C}$ satisfies CCC.

The converse of this theorem is not valid.[1]

EXAMPLE Let $X = \{x : 1 \leqslant x < \Omega\}$ be the set of all countable ordinal numbers and let $\mathscr{C}$ be the family of all sets of the form

$$\{x \in X : \alpha + n \leqslant x < \beta\}$$

where α is 1 or a limit ordinal, n is a nonnegative integer, and β is the smallest limit ordinal larger than α.

For this category base, every region is a meager set. Hence, every subset of X is a Baire set and is meager. The abundant Baire sets vacuously satisfy CCC, but $\mathscr{C}$ does not satisfy CCC.

E.   Structure of Baire sets

THEOREM 6   The Baire sets form a σ-field that contains all regions and all meager sets.[2]

Proof. It follows from Theorem 2 in Section II that $\mathfrak{B}(\mathscr{C})$ contains all regions. It is also clear that $\mathfrak{B}(\mathscr{C})$ contains all meager sets and is closed under complementation.

Suppose that $S = \bigcup_{n=1}^\infty S_n$, where each set S_n is a Baire set, and suppose that S is abundant in a region A. Then for some index n, the set $A \cap S_n$ is abundant. According to Theorem 8 in Section II, $A \cap S_n$ is abundant everywhere in some subregion B of A. Since S_n is a Baire set and is abundant everywhere in B, the set $B \cap (X - S_n)$ is a meager set. But $X - S \subset X - S_n$. Hence, there is a subregion B of A such that $B \cap (X - S)$ is a meager set. This means that S is a Baire set.

We conclude that $\mathfrak{B}(\mathscr{C})$ is a σ-field.

---

[1]See, however, Theorem 3 in Chapter 2, Section I.

[2]Lebesgue (1905, pp. 186–187), Banach (1930), Szpilrajn (1935), and Morgan (1977a).

In the case that $\mathscr{C}$ satisfies CCC, the Baire sets have special representations of a simple form.

THEOREM 7 If $\mathscr{C}$ satisfies CCC, then the following statements are equivalent for a set S:

(i) S is a Baire set.
(ii) $S = (G—P) \cup R$, where G is a $\mathscr{C}_\sigma$-set and P,R are meager sets.
(iii) $S = (F—Q) \cup T$, where F is a $\mathscr{K}_\delta$-set and Q,T are meager sets.
(iv) S is the union of a $\mathscr{C}_{\sigma\delta}$-set and a meager set.
(v) S is the difference of a $\mathscr{K}_{\delta\sigma}$-set and a meager set.

Proof. (i) $\Rightarrow$ (ii). By a $\mathscr{C}_\sigma$-set we mean the union of a countable family of regions in $\mathscr{C}$. As the union of the empty family of regions, the empty set is thus a $\mathscr{C}_\sigma$-set. Therefore, condition (ii) is satisfied whenever S is a meager set.

Suppose that S is an abundant Baire set. Let $\mathscr{N}$ be the family of all regions in which S is abundant everywhere and let $Y = \cup \mathscr{N}$. We see from Theorem 2 in Section I that $(Y,\mathscr{N})$ is a category base.

By Lemma 4 in Section II there is a family

$$\mathscr{M} = \{M_\alpha : \alpha \in I\}$$

consisting of disjoint $\mathscr{N}$-regions such that $Y—\bigcup_{\alpha \in I} M_\alpha$ is an $\mathscr{N}$-singular set and having the property that for every set $N \in \mathscr{N}$ there is a set $M \in \mathscr{M}$ such that $N \cap M$ contains a set in $\mathscr{N}$. Using the fact that if $\{M_\alpha : \alpha \in I\}$ is any family of disjoint sets and $\{K_\alpha : \alpha \in I\}$ is a family of sets with $K_\alpha \subset M_\alpha$ for all $\alpha \in I$, then

$$\bigcup_{\alpha \in I}(M_\alpha—K_\alpha) = \left(\bigcup_{\alpha \in I} M_\alpha\right)—\left(\bigcup_{\alpha \in I} K_\alpha\right)$$

we obtain

$$S = \left[\left(\bigcup_{\alpha \in I} M_\alpha\right)—\bigcup_{\alpha \in I}(M_\alpha—S)\right] \cup \left(S—\bigcup_{\alpha \in I} M_\alpha\right)$$

We first show that the set

$$R = S—\bigcup_{\alpha \in I} M_\alpha$$

is $\mathscr{C}$-meager.

Assume to the contrary that R is $\mathscr{C}$-abundant. Let R be $\mathscr{C}$-abundant everywhere in a $\mathscr{C}$-region A. Then S is also $\mathscr{C}$-abundant everywhere in A, so $A \in \mathscr{N}$. Hence, there is an index $\alpha \in I$ such that $A \cap M_\alpha$ contains an $\mathscr{N}$-region B. Clearly, M_α is $\mathscr{C}$-abundant everywhere in B. From the inclusion $R \subset X—M_\alpha$ and the fact that R is $\mathscr{C}$-abundant everywhere in A, it follows that $X—M_\alpha$ is $\mathscr{C}$-abundant everywhere in B. This, however, contradicts the fact that the region M_α is a Baire set with respect to $\mathscr{C}$. Thus, R must be a $\mathscr{C}$-meager set.

Using now the hypothesis that $\mathscr{C}$ satisfies CCC, we know that the family $\mathscr{M}$ is countable. Consequently, the set

$$G = \bigcup_{\alpha \in I} M_\alpha$$

is a $\mathscr{C}_\sigma$-set. Utilizing Theorem 2, we see that the set

$$P = \bigcup_{\alpha \in I} (M_\alpha - S)$$

is $\mathscr{C}$-meager. Thus, we have the desired representation

$$S = (G - P) \cup R$$

(ii) $\Rightarrow$ (iv). By Theorem 5 of Section II there is a meager $\mathscr{K}_{\delta\sigma}$-set Q containing P and we have the appropriate representation

$$S = (G - Q) \cup [R \cup (S \cap (Q - P))]$$

(iv) $\Rightarrow$ (i). This is a simple consequence of Theorem 6.

Finally, the fact that (iii) and (v) are equivalent to (i) follows from (ii) and (iv) by considering complements.

If $\mathscr{C}$ is the category base of all open sets in $\mathbb{R}^n$, then $\mathscr{C}_\sigma = \mathscr{G}$. From the statements (i), (ii), (iv), and (v), we then obtain the equivalent statements:

(τ1) S has the Baire property.
(τ2) $S = (G - P) \cup R$, where G is an open set and P, R are sets of the first category.
(τ3) S is the union of a $\mathscr{G}_\delta$-set and a set of the first category.
(τ4) S is the difference of an $\mathscr{F}_\sigma$-set and a set of the first category.

If $\mathscr{C}$ is the category base of all compact sets of positive Lebesgue measure in $\mathbb{R}^n$, then $\mathscr{C}_\sigma = \mathscr{F}_\sigma$ and $\mathscr{K}_\delta = \mathscr{G}_\delta$. The statements (i) and (iii) yield the equivalent statements

(μ1) S is a Lebesgue measurable set.
(μ2) $S = (F - Q) \cup T$, where F is a $\mathscr{G}_\delta$-set and P, R are sets of Lebesgue measure zero.

In the proof of the implication (ii) $\Rightarrow$ (iv), the meager $\mathscr{K}_{\delta\sigma}$-set Q can be taken to be a $\mathscr{G}_\delta$-set of measure zero (cf. Section II.D). Accordingly, (μ1) and (μ2) are also equivalent to the statements

(μ3) S is the union of an $\mathscr{F}_\sigma$-set and a set of Lebesgue measure zero.
(μ4) S is the difference of a $\mathscr{G}_\delta$-set and a set of Lebesgue measure zero.

The following consequence of Theorem 7 is frequently utilized.

THEOREM 8 If $\mathscr{C}$ satisfies CCC, then $\mathfrak{B}(\mathscr{C})$ is the smallest σ-field containing all regions and all singular sets.

F. Non-Baire sets

It is a simple matter to give examples of category bases for which there are sets that are not Baire sets. For instance, if we take the trivial base with $X = \{1,2\}$ and $\mathscr{C} = \{X\}$, then neither of the sets $\{1\}$, $\{2\}$ is a Baire set. We see also in this example that the union of two sets which are not Baire sets may itself be a Baire set. There is, however, a special circumstance in which the union of sets which are not Baire sets will also fail to be a Baire set.

THEOREM 9 If $\{S_i : i \in I\}$ is a countable family of disjoint sets, at least one of which is not a Baire set, and $\{E_i : i \in I\}$ is a family of disjoint Baire sets such that $S_i \subset E_i$ for every $i \in I$, then $\bigcup_{i \in I} S_i$ is not a Baire set.[1]

Proof. Suppose that $i_0 \in I$ and the set S_{i_0} is not a Baire set. Then there exists a region A in every subregion of which both S_{i_0} and $X - S_{i_0}$ are abundant sets. The set $X - \bigcup_{i \neq i_0} E_i$, which contains S_{i_0}, is a Baire set that is abundant in every subregion of A. Hence, there is a region $B \subset A$ such that $B \cap (\bigcup_{i \neq i_0} E_i)$ is a meager set.

Let C be any subregion of B. The sets $C \cap S_{i_0}$ and $C \cap (X - S_{i_0})$ are both abundant sets. The set $C \cap (\bigcup_{i \in I} S_i)$, which contains $C \cap S_{i_0}$, is also an abundant set. Moreover, from the equality

$$C \cap (X - S_{i_0}) = \left[C \cap (X - S_{i_0}) \cap \left(\bigcup_{i \neq i_0} E_i \right) \right] \cup \left[C \cap (X - S_{i_0}) \cap \left(X - \bigcup_{i \neq i_0} E_i \right) \right]$$

we see that the set

$$C \cap (X - S_{i_0}) \cap \left(X - \bigcup_{i \neq i_0} E_i \right) = C \cap \left[X - \left(S_{i_0} \cup \bigcup_{i \neq i_0} E_i \right) \right]$$

is abundant. Hence, so also is the set $C \cap (X - \bigcup_{i \in I} S_i)$, which contains this set. We thus see that for every region $C \subset B$, both $C \cap (\bigcup_{i \in I} S_i)$ and $C \cap (X - \bigcup_{i \in I} S_i)$ are abundant sets. Therefore, $\bigcup_{i \in I} S_i$ is not a Baire set.

COROLLARY 10   If S is not a Baire set and T is a meager set, then $S \cup T$ is not a Baire set.

Proof. Take $I = \{1,2\}$, $S_1 = S$, $S_2 = T - S$, $E_1 = X - (T - S)$, and $E_2 = T - S$ in the theorem.

---

[1] Cf. Wilkosz (1920, Sec. 1, Teorema III), Pu (1972), and Albanese (1974, Lemma I).

G. Restrictions

NOTATION   If $S \subset X$ and $\mathscr{R}$ is a family of subsets of X, then

$$\mathscr{R} \cap S = \{R \cap S : R \in \mathscr{R}\}$$

DEFINITION   A category base $(S, \mathscr{E})$ is called a restriction of $(X, \mathscr{C})$ to S if $\mathfrak{M}(\mathscr{E}) = \mathfrak{M}(\mathscr{C}) \cap S$ and $\mathfrak{B}(\mathscr{E}) = \mathfrak{B}(\mathscr{C}) \cap S$.

Obviously, any two restrictions of $(X, \mathscr{C})$ to S are equivalent bases. We give here two specific restrictions that will be used later.

For any region $A \in \mathscr{C}$ define

$$\mathscr{C}_A = \{B \in \mathscr{C} : B \subset A\}$$

THEOREM 11 $(A, \mathscr{C}_A)$ is a category base with

$$\mathfrak{S}(\mathscr{C}_A) = \{S \in \mathfrak{S}(\mathscr{C}) : S \subset A\} = \mathfrak{S}(\mathscr{C}) \cap A$$

$$\mathfrak{M}(\mathscr{C}_A) = \{S \in \mathfrak{M}(\mathscr{C}) : S \subset A\} = \mathfrak{M}(\mathscr{C}) \cap A$$

$$\mathfrak{B}(\mathscr{C}_A) = \{S \in \mathfrak{B}(\mathscr{C}) : S \subset A\} = \mathfrak{B}(\mathscr{C}) \cap A$$

Proof. It is readily seen that $(A, \mathscr{C}_A)$ is a category base.

As every $\mathscr{C}$-singular subset of A is obviously $\mathscr{C}_A$-singular, we need only verify the reverse implication. Suppose that S is a $\mathscr{C}_A$-singular subset of A and let B be any $\mathscr{C}$-region.

If $A \cap B$ is $\mathscr{C}$-singular, then there is a $\mathscr{C}$-region $C \subset B$ such that $C \cap (A \cap B) = C \cap A = \varnothing$. From $S \subset A$ we obtain $C \cap S = \varnothing$.

Suppose, on the other hand, that $A \cap B$ is not $\mathscr{C}$-singular. Then there is a $\mathscr{C}$-region $D \subset A \cap B$. Since D is a $\mathscr{C}_A$-region and S is $\mathscr{C}_A$-singular, there is a $\mathscr{C}_A$-region $C \subset D$ such that $C \cap S = \varnothing$.

In any case, S is a $\mathscr{C}$-singular set and we have established

$$\mathfrak{S}(\mathscr{C}_A) = \{S \in \mathfrak{S}(\mathscr{C}) : S \subset A\}$$

From this equality we easily derive the equality

$$\mathfrak{M}(\mathscr{C}_A) = \{S \in \mathfrak{M}(\mathscr{C}) : S \subset A\}$$

Suppose now that $S \subset A$ and $S \in \mathfrak{B}(\mathscr{C})$. Let B be any $\mathscr{C}_A$-region. Then B is a $\mathscr{C}$-region. From the definition of Baire set, there is a $\mathscr{C}$-region $C \subset B$ such that either S or $X—S$ is a $\mathscr{C}$-meager set in C. The set C is a $\mathscr{C}_A$-region and, by virtue of the preceding equality, either S or $A—S$ is a $\mathscr{C}_A$-meager set in C. Hence, $S \in \mathfrak{B}(\mathscr{C}_A)$.

Conversely, suppose that $S \in \mathfrak{B}(\mathscr{C}_A)$ and let E be any $\mathscr{C}$-region.

If $A \cap E$ contains no $\mathscr{C}$-region, then there is a $\mathscr{C}$-region $B \subset E$ such that $B \cap A = \varnothing$. Since S is a subset of A, the set S is $\mathscr{C}$-meager in the $\mathscr{C}$-region B.

Assume, on the other hand, that $A \cap E$ contains a $\mathscr{C}$-region B. Then B is a $\mathscr{C}_A$-region. Because $S \in \mathfrak{B}(\mathscr{C}_A)$, there is a $\mathscr{C}_A$-region $C \subset B$ such that either S or A—S is a $\mathscr{C}_A$-meager set in C. Therefore, C is a $\mathscr{C}$-region contained in E with the property that either S or X—S is a $\mathscr{C}$-meager set in C.

In any case, we have $S \in \mathfrak{B}(\mathscr{C})$ and we have established

$$\mathfrak{B}(\mathscr{C}_A) = \{S \in \mathfrak{B}(\mathscr{C}) : S \subset A\}$$

It is a simple matter to verify the remaining asserted equalities.

THEOREM 12   If $\mathscr{C}$ satisfies CCC, S is an abundant set, and

$$\mathscr{C}^* = [\mathfrak{B}(\mathscr{C}) \cap S] - [\mathfrak{M}(\mathscr{C}) \cap S]$$

then $(S, \mathscr{C}^*)$ is a category base satisfying CCC with

$$\mathfrak{S}(\mathscr{C}^*) = \mathfrak{M}(\mathscr{C}^*) = \mathfrak{M}(\mathscr{C}) \cap S$$

$$\mathfrak{B}(\mathscr{C}^*) = \mathfrak{B}(\mathscr{C}) \cap S$$

Proof. The family $\mathscr{A} = \mathfrak{B}(\mathscr{C}) \cap S$ is a σ-field containing S and $\mathscr{I} = \mathfrak{M}(\mathscr{C}) \cap S$ is a σ-ideal in $\mathscr{A}$. To show that $(S, \mathscr{C}^*)$ is a category base we have only, in view of Theorem 6 of Section II, to show that $\mathscr{C}^* = \mathscr{A} - \mathscr{I}$ satisfies CCC.

Assume to the contrary that $\mathscr{C}^*$ does not satisfy CCC. Then there is a transfinite sequence $\langle E_\alpha \rangle_{\alpha < \Omega}$ of disjoint sets belonging to $\mathscr{A} - \mathscr{I}$. For each $\alpha < \Omega$ we have $E_\alpha = F_\alpha \cap S$, where $F_\alpha \in \mathfrak{B}(\mathscr{C}) - \mathfrak{M}(\mathscr{C})$. Define a transfinite sequence $\langle G_\alpha \rangle_{\alpha < \Omega}$ of sets by placing $G_1 = F_1$ and $G_\alpha = F_\alpha - \bigcup_{\beta < \alpha} F_\beta$ for $\alpha > 1$. The sets G_α are disjoint Baire sets and, in fact, they are abundant sets.

For since $E_\alpha \notin \mathfrak{M}(\mathscr{C}) \cap S$, each set E_α is abundant. For each $\alpha < \Omega$ we have $E_\alpha \subset F_\alpha$, and the sets E_α being disjoint, we have $E_\alpha \cap F_\beta = \varnothing$ for $\beta \neq \alpha$. This implies that $E_\alpha \subset G_\alpha$ for each α. Hence, the sets G_α are abundant.

Now, the existence of the transfinite sequence $\langle G_\alpha \rangle_{\alpha < \Omega}$ of disjoint abundant Baire sets contradicts Theorem 5. We thus conclude that $\mathscr{C}^*$ does satisfy CCC.

Utilizing Theorem 6 in Section II and Theorem 8 above, we obtain the asserted equalities.

H.   Localization

DEFINITION   A set is a Baire set locally at a given point if every neighborhood of the point contains a neighborhood of the point in which the set is a Baire set.

THEOREM 13   A set is a Baire set if and only if it is a Baire set locally at every point.

Proof. If S is a Baire set, then it follows from the fact that every region is a Baire set that S is a Baire set locally at every point.

Suppose, on the other hand, that S is a set that is not a Baire set. Then there is a region A in which both S and $X-S$ are abundant everywhere. Choose a point x in A. If B is any subregion of A that contains x, then for every region $C \subset B$, the sets $S \cap C = (S \cap B) \cap C$ and $(X-S) \cap C = [X-(S \cap B)] \cap C$ are both abundant, so the set $S \cap B$ is not a Baire set. Therefore, S is not a Baire set locally at the point x.

I. Approximate relations

We assume that $(X, \mathscr{I})$ is an ideal in a field $(X, \mathscr{A})$ and that all sets considered are elements of $\mathscr{A}$.

DEFINITION We say that a set S is contained in a set T modulo $\mathscr{I}$, expressed symbolically by $S \prec T$, if $S - T \in \mathscr{I}$.

The following properties are easily verified:

(α1) If $S \subset T$, then $S \prec T$.
(α2) If $S \in \mathscr{I}$, then $S \prec T$ for each set T.
(α3) If $S \prec T$ and $T \in \mathscr{I}$, then $S \in \mathscr{I}$.
(α4) $S \prec S$.
(α5) If $S \prec T$ and $T \prec U$, then $S \prec U$.
(α6) If $S \prec T$ and $U \prec V$, then $S \cap U \prec T \cap V$.
(α7) If $S \prec T$ and $U \prec V$, then $S \cup U \prec T \cup V$.
(α8) If $S \prec T$ and $U \prec V$, then $U - T \prec V - S$.

If $\mathscr{I}$ is a σ-ideal, then we also have

(α9) If $S_n \prec T_n$ for all $n \in \mathbf{N}$, then $\bigcap_{n=1}^{\infty} S_n \prec \bigcap_{n=1}^{\infty} T_n$.
(α10) If $S_n \prec T_n$ for all $n \in \mathbf{N}$, then $\bigcup_{n=1}^{\infty} S_n \prec \bigcup_{n=1}^{\infty} T_n$.

DEFINITION S and T are called equivalent modulo $\mathscr{I}$ if $S \prec T$ and $T \prec S$. This relationship is expressed symbolically by $S \approx T$.

We denote by $S \triangle T$ the symmetric difference of S and T; i.e.,

$$S \triangle T = (S - T) \cup (T - S)$$

Note that S and T are equivalent modulo $\mathscr{I}$ if and only if $S \triangle T \in \mathscr{I}$.

The following properties hold:

(β1) $S \approx S$.
(β2) If $S \approx T$ then $T \approx S$.
(β3) If $S \approx T$ and $T \approx U$, then $S \approx U$.
(β4) If $S \approx T$ and $S \in \mathscr{I}$, then $T \in \mathscr{I}$.

(β5) If $S \approx T$ and $U \approx V$, then $S \prec U$ if and only if $T \prec V$.
(β6) If $S \approx T$ and $U \approx V$, then $S \cap U \approx T \cap V$.
(β7) If $S \approx T$ and $U \approx V$, then $S \cup U \approx T \cup V$.
(β8) If $S \approx T$ and $U \approx V$, then S—$U \approx T$—V.
(β9) If $\bigcup_{\alpha \in I} (S_\alpha \Delta T_\alpha) \in \mathcal{I}$, then $\bigcup_{\alpha \in I} S_\alpha \approx \bigcup_{\alpha \in I} T_\alpha$.
(β10) $S \approx T$ if and only if $S = (T$—$P) \cup R$, where $P, R \in \mathcal{I}$.
(β11) $S \approx T$ if and only if $S = (T \cup P)$—R, where $P, R \in \mathcal{I}$.

The representation of S in (β10) is obtained from the equality

$$S = [T$—$(T$—$S)] \cup (S$—$T)$$

The other properties are readily verified.

If $\mathcal{I}$ is a σ-ideal, then we also have

(β12) If $S_n \approx T_n$ for all $n \in \mathbb{N}$, then $\bigcap_{n=1}^{\infty} S_n \approx \bigcap_{n=1}^{\infty} T_n$.
(β13) If $S_n \approx T_n$ for all $n \in \mathbb{N}$, then $\bigcup_{n=1}^{\infty} S_n \approx \bigcup_{n=1}^{\infty} T_n$.

Assume now that $(X, \mathscr{C})$ is a category base, $(X, \mathscr{A})$ is a σ-field, $\mathcal{I} = \mathfrak{M}(\mathscr{C})$, and $\mathfrak{B}(\mathscr{C}) \subset \mathscr{A}$. Then we say that S is essentially contained in T, or T essentially contains S, if $S \prec T$. We say that S is essentially equal to T if $S \approx T$, i.e., if the symmetric difference $S \Delta T$ is a meager set.

We note that

(γ1) If S is a Baire set and $S \approx T$, then T is a Baire set.

If $\mathscr{C}$ satisfies CCC, then we also have

(γ2) S is a Baire set if and only if S is essentially equal to a $\mathscr{C}_\sigma$-set.
(γ3) S is a Baire set if and only if S is essentially equal to a $\mathscr{K}_\delta$-set.

J.   Hulls of sets

We shall establish here a unified version of the following classical theorems concerning sets in $\mathbb{R}^n$:

(τ) Every set S is contained in an $\mathscr{F}_\sigma$-set E that is locally of the same category at each point as S.[1]
(μ) Every set S is contained in a $\mathscr{G}_\delta$-set E that has the same Lebesgue measure as S.

For this unification we utilize the abstract notion of a hull, formulated by Marczewski.[2]

---

[1]Cf. Nikodym (1925, Lemme II) [a correction to the proof of this lemma appears in Nikodym (1926, p. 294, fn. 1)].
[2]See Saks and Sierpiński (1928) and Szpilrajn (1930, 1933).

DEFINITION   A set E is a hull for a set S if the following conditions are satisfied:
(a) E is a Baire set.
(b) $S \subset E$.
(c) If F is a Baire set and $S \subset F$, then $E—F$ is a meager set.

From the definition one readily obtains the following:

(i) If E is a hull for a set S and F is a set containing S, then F is a hull for S if and only if $E \approx F$.
(ii) If E_n is a hull for S_n for all $n \in \mathbb{N}$, then $\bigcup_{n=1}^{\infty} E_n$ is a hull for $\bigcup_{n=1}^{\infty} S_n$.

We now obtain an extension of Theorem 5 of Section II that yields the desired unification.

THEOREM 14   If $\mathscr{C}$ satisfies CCC, then every set has a hull which is a $\mathscr{K}_{\delta\sigma}$-set.

Proof. Let S be any set. If every Baire set contained in $X—S$ is meager, then the set $E = X$ is the desired hull.

Assume, on the other hand, that there exists at least one abundant Baire set contained in $X—S$. Well-order all such sets and proceed by induction to select a family $\mathscr{M} = \{M_k : k \in I\}$ of disjoint abundant Baire sets contained in $X—S$ such that $(X—S)—\cup\mathscr{M}$ contains no abundant Baire set. By virtue of Theorem 5, the index set I is countable. Hence, $T = X—\cup\mathscr{M}$ is a Baire set containing S.

From the equivalence of conditions (i) and (v) of Theorem 7, we have $T = E—U$, where E is a $\mathscr{K}_{\delta\sigma}$-set containing S and U is a meager set.[1] It remains only to verify that condition (c) in the definition of a hull is satisfied.

Let F be any Baire set containing S. Then

$$(E—F)—U = T—F \subset T—S = (X—S)—\cup\mathscr{M}$$

which, according to the definition of the family $\mathscr{M}$, implies that the set $(E—F)—U$ is meager. From the equality

$$E—F = [(E—F) \cap U] \cup [(E—F)—U]$$

we conclude that $E—F$ is a meager set.

DEFINITION   A set E is essentially a hull, or is an essential hull, for a set S if the following conditions are satisfied:
(a) E is a Baire set.
(b) $S \prec E$.
(c) If F is a Baire set and $S \prec F$, then $E \prec F$.

[1]In the Lebesgue measure case, see Section III.E, statement (μ4).

We note the following properties:

(iii) If E is a hull for S, then E is essentially a hull for S.

(iv) If E is essentially a hull for a set S, then there is a meager set P such that $E \cup P$ is a hull for S; e.g., take $P = S—E$.

(v) If E is essentially a hull for S and F is a set that essentially contains S, then F is essentially a hull for S if and only if $E \approx F$.

(vi) If E_n is essentially a hull for S_n for each $n \in \mathbb{N}$, then $\bigcup_{n=1}^{\infty} E_n$ is essentially a hull for $\bigcup_{n=1}^{\infty} S_n$.

DEFINITION A set E is essentially a hull, or is an essential hull, for a family $\mathcal{S}$ of sets if the following conditions are satisfied:

(a) E is a Baire set.

(b) $S \prec E$ for every $S \in \mathcal{S}$.

(c) If F is a Baire set and $S \prec F$ for all $S \in \mathcal{S}$, then $E \prec F$.

THEOREM 15 Every family of Baire sets has an essential hull.[1]

Proof. Let $\mathcal{S}$ be a family of Baire sets. Define

$$\mathcal{N}_1 = \{A \in \mathcal{C} : \text{every set in } \mathcal{S} \text{ is meager in } A\}$$

$$\mathcal{N}_2 = \{A \in \mathcal{C} : \text{at least one set in } \mathcal{S} \text{ is abundant everywhere in } A\}$$

Note that the families $\mathcal{N}_1$ and $\mathcal{N}_2$ are disjoint. Define $\mathcal{N} = \mathcal{N}_1 \cup \mathcal{N}_2$ and set $Y = \cup \mathcal{N}$.

Each region in $\mathcal{C}$ contains a region in $\mathcal{N}$. Accordingly, we can apply Lemmas 3 and 4 in Section II to obtain a family $\mathcal{M}$ of disjoint regions in $\mathcal{N}$ such that $X—\cup\mathcal{M}$ is singular and having the property that for every region $A \in \mathcal{N}$ there is a region $B \in \mathcal{M}$ such that $A \cap B$ contains a region in $\mathcal{N}$. Because $\mathcal{M}$ is a disjoint family, we have

$$[\cup(\mathcal{M} \cap \mathcal{N}_1)] \cap [\cup(\mathcal{M} \cap \mathcal{N}_2)] = \emptyset$$

We show that the set $E = \cup(\mathcal{M} \cap \mathcal{N}_2)$ is an essential hull for $\mathcal{S}$.

(a) E is a Baire set.

Indeed, if A is any region in $\mathcal{C}$, then A contains a region in $\mathcal{N}$, so there exist regions $B \in \mathcal{M}$, $C \in \mathcal{N}$ such that $C \subset A \cap B$. If $B \in \mathcal{M} \cap \mathcal{N}_1$, then $C \subset \cup(\mathcal{M} \cap \mathcal{N}_1)$ and consequently $C \cap E = \emptyset$. On the other hand, if $B \in \mathcal{M} \cap \mathcal{N}_2$, then $C \subset \cup(\mathcal{M} \cap \mathcal{N}_2) = E$. Accordingly, E is a Baire set.

(b) For every set $S \in \mathcal{S}$, the set $S—E$ is meager.

Assuming that this is not so, there exists a set $S \in \mathcal{S}$ for which $S—E$ is abundant everywhere in a region $A \in \mathcal{N}_2$. Choose $B \in \mathcal{M}$ and $C \in \mathcal{N}_2$ so that $C \subset A \cap B$. The set S, being abundant in C, is abundant in B. This means that $B \in \mathcal{M} \cap \mathcal{N}_2$ and hence $B \subset E$. But then both E and $X—E$, which contains

[1]K. Schilling (personal communication, 1984).

S—E, are abundant everywhere in C, contradicting the fact that E is a Baire set.

(c) If F is a Baire set and S—F is meager for every $S \in \mathscr{S}$, then E—F is meager.

We prove the contrapositive. Suppose that E—F is abundant. Then E—F is abundant everywhere in some region $A \in \mathscr{N}$. Choose $B \in \mathscr{M}$ and $C \in \mathscr{N}$ with $C \subset A \cap B$. We first show that $B \in \mathscr{N}_2$.

Assume to the contrary that $B \in \mathscr{N}_1$. From the inclusion $B \subset \cup(\mathscr{M} \cap \mathscr{N}_1)$ we obtain $B \cap E = \varnothing$, which implies that $C \cap E = \varnothing$. This, however, contradicts the fact that E—F is abundant in C. We thus see that $B \in \mathscr{N}_2$.

From $B \in \mathscr{M} \cap \mathscr{N}_2$ we obtain the existence of a set $S \in \mathscr{S}$ that is abundant everywhere in B, hence also abundant everywhere in C. Since S is a Baire set, the set C—S is meager. The inclusion

$$C—F \subset (C—S) \cup (S—F)$$

and the fact that C—F is abundant yield the conclusion that S—F is abundant.

K. Quotient algebras

Assume that $(X, \mathscr{S})$ is any ideal in a field $(X, \mathscr{A})$.

The relation $\approx$ of equivalence modulo $\mathscr{S}$, being an equivalence relation, induces a partition of the sets in $\mathscr{A}$ into disjoint equivalence classes. We denote by $[S]$ or $[S]_{\mathscr{S}}$ the equivalence class containing a given set $S \in \mathscr{A}$. The following conditions are equivalent for sets $S, T \in \mathscr{A}$:

$$S \approx T \quad S \in [T] \quad [S] = [T]$$

The set of all these equivalence classes is denoted by $\mathscr{A}/\mathscr{S}$ and is called the quotient algebra of $\mathscr{A}$ modulo $\mathscr{S}$. An element $[S] \in \mathscr{A}/\mathscr{S}$ is called nonzero if $[S] \neq [\varnothing]$, i.e., if $S \notin \mathscr{S}$.

We define a relation $<$ between elements $[S], [T] \in \mathscr{A}/\mathscr{S}$ by

$$[S] < [T] \quad \text{if and only if} \quad S \prec T$$

Operations $\wedge$, $\vee$, $\sim$ (called meet, join, and difference, respectively) are defined for elements $[S], [T] \in \mathscr{A}/\mathscr{S}$ by

$$[S] \wedge [T] = [S \cap T]$$
$$[S] \vee [T] = [S \cup T]$$
$$[S] \sim [T] = [S—T]$$

By virtue of the properties $(\beta 5)$–$(\beta 8)$ given in Section III.I, the relation $<$ and the operations $\wedge$, $\vee$, $\sim$ are well-defined and do not depend upon the specific representative sets S, T chosen for the given equivalence classes. It can be shown that the operations $\wedge$, $\vee$, $\sim$ have the same general algebraic properties as the set-theoretical operations $\cap$, $\cup$, —.

DEFINITION A quotient algebra $\mathscr{A}/\mathscr{I}$ is called a complete quotient algebra if every nonempty set of elements in $\mathscr{A}/\mathscr{I}$ has a least upper bound in $\mathscr{A}/\mathscr{I}$ with respect to the relation $<$.

For complete quotient algebras the operations of meet and join can be defined, more generally, for any number of equivalence classes.[1]

Now, a quotient algebra $\mathscr{A}/\mathscr{I}$ is complete if and only if, for every nonempty family $\mathscr{S} \subset \mathscr{A}$, there exists a set E satisfying the following conditions:

(a) $E \in \mathscr{A}$.
(b) $S \prec E$ for every $S \in \mathscr{S}$.
(c) If $F \in \mathscr{A}$ and $S \prec F$ for every $S \in \mathscr{S}$, then $E \prec F$.

Hence, reformulating Theorem 15 in terms of quotient algebras, we obtain

THEOREM 16 The quotient algebra of all Baire sets modulo the ideal of meager sets is complete.[2]

DEFINITION Two quotient algebras $\mathscr{A}/\mathscr{I}$ and $\mathscr{B}/\mathscr{J}$ are called isomorphic if there is a one-to-one function ψ mapping $\mathscr{A}/\mathscr{I}$ onto $\mathscr{B}/\mathscr{J}$ under which the operations $\wedge$, $\vee$, $\approx$ are preserved; i.e., for all elements $[S], [T] \in \mathscr{A}/\mathscr{I}$ we have

$$\psi([S] \wedge [T]) = \psi([S]) \wedge \psi([T])$$

$$\psi([S] \vee [T]) = \psi([S]) \vee \psi([T])$$

$$\psi([S] \sim [T]) = \psi([S]) \sim \psi([T])$$

where the operations $\wedge$, $\vee$, $\sim$ are defined for elements of $\mathscr{A}/\mathscr{I}$ on the left side of the equality signs and for elements of $\mathscr{B}/\mathscr{J}$ on the right side.

We note the following fact:[3]

THEOREM 17 The quotient algebra of all sets in $\mathbb{R}^n$ having the Baire property modulo the ideal of sets of the first category is not isomorphic to the

[1] Cf. Sikorski (1964, Secs. 10, 18, 20, 21, 25); see also von Neumann (1960, App. 1).
[2] This theorem, due to K. Schilling (personal communication, 1984), generalizes a result of Birkhoff and Ulam (ca. 1935); cf. Sikorski (1949).
[3] See Sikorski (1964, p. 77) and Chapter 3, Section I.D.

quotient algebra of all Lebesgue measurable sets in $\mathbb{R}^n$ modulo the ideal of sets of Lebesgue measure zero.

L. Transfinite sequences of sets

Let $\langle S_\alpha \rangle_{\alpha < \Theta}$ be a transfinite sequence of sets.

DEFINITION $\langle S_\alpha \rangle_{\alpha < \Theta}$ is convergent to a set S, called a limit of the sequence, if there is an index $\gamma < \Omega$ for which the set $\bigcup_{\gamma < \alpha < \Theta} (S_\alpha \triangle S)$ is meager.[1] If $\langle S_\alpha \rangle_{\alpha < \Theta}$ converges to the empty set, then $\langle S_\alpha \rangle_{\alpha < \Theta}$ is called a null sequence.

The following properties are simple consequences of the definition.

(i) $\langle S_\alpha \rangle_{\alpha < \Theta}$ converges to S if and only if $\langle S_\alpha \triangle S \rangle_{\alpha < \Theta}$ is a null sequence.

(ii) If $\langle S_\alpha \rangle_{\alpha < \Theta}$ converges to S, then $\langle S_\alpha \rangle_{\alpha < \Theta}$ converges to T if and only if $S \approx T$.

(iii) Every transfinite subsequence of a convergent transfinite sequence is convergent to the same limit.

(iv) If $\langle S_\alpha \rangle_{\alpha < \Theta}$ is a null sequence and E is a subset of $\bigcup_{\alpha < \Theta} S_\alpha$ whose intersection with each set S_α is meager, then E is a meager set.

DEFINITION $\langle S_\alpha \rangle_{\alpha < \Theta}$ is called a constant sequence if there is a set S such that $S_\alpha = S$ for all $\alpha < \Theta$.

(v) Every constant transfinite sequence is convergent.

DEFINITION $\langle S_\alpha \rangle_{\alpha < \Theta}$ is called stationary (resp., essentially stationary) if there is an index $\gamma < \Omega$ such that $S_\alpha = S_\gamma$ (resp., $S_\alpha \approx S_\gamma$) for all indices $\alpha \geqslant \gamma$.

(vi) Every stationary transfinite sequence is convergent.

DEFINITION $\langle S_\alpha \rangle_{\alpha < \Theta}$ is called ascending (resp., descending) if $S_\alpha \subset S_\beta$ (resp., $S_\beta \subset S_\alpha$) whenever $\alpha < \beta < \Theta$. $\langle S_\alpha \rangle_{\alpha < \Theta}$ is called essentially ascending (resp., essentially descending) if $S_\alpha \prec S_\beta$ (resp., $S_\beta \prec S_\alpha$) whenever $\alpha < \beta < \Theta$.

THEOREM 18 If $\mathscr{C}$ satisfies CCC, then every essentially ascending transfinite sequence of Baire sets, as well as every essentially descending transfinite sequence of Baire sets, is essentially stationary.

Proof. Suppose that $\langle S_\alpha \rangle_{\alpha < \Theta}$ is an essentially ascending transfinite sequence of Baire sets. Place $T_1 = S_1$. For $1 < \alpha < \Theta$, define

$$T_\alpha = S_\alpha - S_{\alpha-1}$$

[1]Cf. Hausdorff (1936) and Sierpiński and Szpilrajn (1936a).

if α is a successor ordinal, and

$$T_\alpha = S_\alpha - \bigcup_{\beta < \alpha} S_\beta$$

if α is a limit ordinal. The family $\{T_\alpha : \alpha < \Theta\}$ is then a family of essentially disjoint Baire sets. According to Theorem 5, there exists a smallest index $\gamma < \Omega$ such that T_α is a meager set for all $\alpha \geqslant \gamma$. We show by transfinite induction that $S_\alpha \approx S_\gamma$ for all $\alpha \geqslant \gamma$.

Obviously, $S_\gamma \approx S_\gamma$. Assume that $\alpha > \gamma$ and we have already established $S_\beta \approx S_\gamma$ for all indices β with $\gamma \leqslant \beta < \alpha$. If α is a successor ordinal, then $S_\alpha \approx S_{\alpha-1}$ and since $S_{\alpha-1} \approx S_\gamma$, we have $S_\alpha \approx S_\gamma$. Suppose, then, that α is a limit ordinal. Because T_α is a meager set, we have $S_\alpha \approx \bigcup_{\beta < \alpha} S_\beta$. From the assumption that $S_\beta \approx S_\gamma$ for $\gamma \leqslant \beta < \alpha$ it follows that $\bigcup_{\gamma \leqslant \beta < \alpha} S_\beta \approx S_\gamma$. We also have $\bigcup_{\beta \leqslant \gamma} S_\beta \approx S_\gamma$. Hence,

$$S_\alpha \approx \bigcup_{\beta < \alpha} S_\beta = \left(\bigcup_{\beta < \gamma} S_\beta\right) \cup \left(\bigcup_{\gamma \leqslant \beta < \alpha} S_\beta\right) \approx \left(\bigcup_{\beta < \gamma} S_\beta\right) \cup S_\gamma = \bigcup_{\beta \leqslant \gamma} S_\beta \approx S_\gamma$$

Thus, $S_\alpha \approx S_\gamma$, as we wished to show.

If $\langle S_\alpha \rangle_{\alpha < \Theta}$ is an essentially descending transfinite sequence of Baire sets, then the complements of these sets form an essentially ascending transfinite sequence of Baire sets. By the result just established, there is an index $\gamma < \Omega$ such that $X - S_\alpha \approx X - S_\gamma$ for all $\alpha \geqslant \gamma$. This implies that $S_\alpha \approx S_\gamma$ for all $\alpha \geqslant \gamma$.

We next introduce the notion of an asymptotic hull for a transfinite sequence $\langle S_\alpha \rangle_{\alpha < \Theta}$. We define

$$\mathscr{R}_\beta = \{S_\alpha : \alpha \geqslant \beta\}$$

for all $\beta < \Theta$; i.e., $\mathscr{R}_\beta$ is the βth remainder of the sequence.

DEFINITION A set E is called an asymptotic hull for $\langle S_\alpha \rangle_{\alpha < \Theta}$ if the following conditions are satisfied:
(a) E is a Baire set.
(b) There exists an ordinal number $\gamma < \Omega$ such that for all ordinal numbers $\beta \geqslant \gamma$, there exists a countable family $\mathscr{S} \subset \mathscr{R}_\beta$ such that E is essentially a hull for the set $\cup(\mathscr{S} \cup \mathscr{T})$, for every countable family $\mathscr{T} \subset \mathscr{R}_\beta$.

THEOREM 19 If $\mathscr{C}$ satisfies CCC, then every transfinite sequence of sets has an asymptotic hull.[1]

Proof. For each $\beta < \Theta$, let $\mathscr{H}_\beta$ denote the family of all Baire sets F for which there exists a countable family $\mathscr{S} \subset \mathscr{R}_\beta$ such that F is essentially a hull of

[1]Cf. Prikry (1976) and Grzegorek (1979).

$\cup(\mathcal{S} \cup \mathcal{T})$ for every countable family $\mathcal{T} \subset \mathcal{R}_\beta$. We first show:

(i) Each of the families $\mathcal{H}_\beta$ is nonempty.

Let

$$\mathcal{T}_1, \mathcal{T}_2, \ldots, \mathcal{T}_\lambda, \ldots \qquad (\lambda < \Theta)$$

be a transfinite enumeration of all countable subfamilies of $\mathcal{R}_\beta$. By means of Theorem 14, we determine a hull F_λ for the set $\cup(\bigcup_{\xi \leqslant \lambda} \mathcal{T}_\xi)$, for each $\lambda < \Theta$. Then $\langle F_\lambda \rangle_{\lambda < \Theta}$ is an ascending transfinite sequence of Baire sets. According to Theorem 18, there exists a smallest ordinal number $\mu < \Omega$ such that $F_\lambda \approx F_\mu$ for all ordinal numbers λ with $\mu \leqslant \lambda < \Theta$. The family $\mathcal{S} = \bigcup_{\xi \leqslant \mu} \mathcal{T}_\xi$ is a countable subfamily of $\mathcal{R}_\beta$. In order to show that $F_\mu \in \mathcal{H}_\beta$, we have to show that F_μ is essentially a hull for the set $\cup[\bigcup_{\xi \leqslant \mu} (\mathcal{T}_\xi \cup \mathcal{T}_\nu)]$ for every ordinal number $\nu < \Theta$. Choosing $\lambda > \max\{\mu, \nu\}$, we have

$$\bigcup\left[\bigcup_{\xi \leqslant \mu} (\mathcal{T}_\xi \cup \mathcal{T}_\nu)\right] \subset \bigcup\left(\bigcup_{\xi \leqslant \lambda} \mathcal{T}_\xi\right) \subset F_\lambda \approx F_\mu$$

which implies that

$$\bigcup\left[\bigcup_{\xi \leqslant \mu} (\mathcal{T}_\xi \cup \mathcal{T}_\nu)\right] \prec F_\mu$$

Now, if G is any Baire set that essentially contains the set $\cup[\bigcup_{\xi \leqslant \mu} (\mathcal{T}_\xi \cup \mathcal{T}_\nu)]$, then G essentially contains the set $\cup(\bigcup_{\xi \leqslant \mu} \mathcal{T}_\xi)$. The set F_μ being a hull for the latter set, we have $F_\mu \prec G$. Therefore, $F_\mu \in \mathcal{H}_\beta$.

Having shown each family $\mathcal{H}_\beta$ is nonempty, we next show:

(ii) If $\beta \leqslant \gamma < \Theta$, $F \in \mathcal{H}_\beta$, and $G \in \mathcal{H}_\gamma$, then $G \prec F$.

Let $\mathcal{S}_\beta$ and $\mathcal{S}_\gamma$ be countable subfamilies of $\mathcal{R}_\beta$ and $\mathcal{R}_\gamma$, respectively, whose existence is assumed in order that $F \in \mathcal{H}_\beta$ and $G \in \mathcal{H}_\gamma$. Because $\beta \leqslant \gamma$, the set F is essentially a hull for $\bigcup(\mathcal{S}_\beta \cup \mathcal{S}_\gamma)$. Hence, the set $\bigcup(\mathcal{S}_\beta \cup \mathcal{S}_\gamma) - F$ is meager, as is also its subset $\bigcup\mathcal{S}_\gamma - F$. Since G is essentially a hull for $\bigcup\mathcal{S}_\gamma$ and $\bigcup\mathcal{S}_\gamma \prec F$, we must have $G \prec F$.

As a simple consequence of (ii) and the fact that if $F \approx G$, then F and G are essentially hulls for the same sets, we obtain:

(iii) If F and G are Baire sets, then F and G belong to the same family $\mathcal{H}_\beta$ if and only if $F \approx G$.

We now show

(iv) There exists a smallest ordinal number $\gamma < \Omega$ such that $\mathcal{H}_\beta = \mathcal{H}_\gamma$ for all ordinal numbers $\beta \geqslant \gamma$.

For each $\beta < \Theta$, choose a set $E_\beta \in \mathcal{H}_\beta$. According to (ii), $\langle E_\beta \rangle_{\beta < \Theta}$ is an essentially descending transfinite sequence of Baire sets. From Theorem 18, it follows that there exists a smallest ordinal number $\gamma < \Omega$ such that $E_\beta \approx E_\gamma$ for all ordinal numbers $\beta \geqslant \gamma$. Using (iii), we see that $\mathcal{H}_\beta = \mathcal{H}_\gamma$ for all $\beta \geqslant \gamma$.

The set $E = E_\gamma$ is an asymptotic hull for $\langle S_\alpha \rangle_{\alpha < \Theta}$.

M.   Operation $\mathscr{A}$. Constituents

Suppose that for each element $\sigma \in \mathbb{N}^F$ there is associated a set S_σ. The family $\{S_\sigma : \sigma \in \mathbb{N}^F\}$ of associated sets is called a determinant system. The set

$$S = \bigcup_{\mu \in \mathbb{Z}} \bigcap_{n=1}^\infty S_{\mu|n}$$

is called the nucleus of the determinant system $\{S_\sigma : \sigma \in \mathbb{N}^F\}$, or the result of performing operation $\mathscr{A}$ on the determinant system.

The operation $\mathscr{A}$ is a generalization of the union and intersection operations for a sequence of sets $\langle E_n \rangle_{n \in \mathbb{N}}$. If we set $S_{\mu|n} = E_{\mu_1}$, then $\bigcap_{n=1}^\infty S_{\mu|n} = E_{\mu_1}$ for each sequence $\mu = \langle \mu_1, \mu_2, \ldots \rangle \in \mathbb{Z}$ and consequently,

$$\bigcup_{\mu \in \mathbb{Z}} \bigcap_{n=1}^\infty S_{\mu|n} = \bigcup_{n=1}^\infty E_n$$

On the other hand, if we set $S_{\mu|n} = E_n$, then we have $\bigcap_{n=1}^\infty S_{\mu|n} = \bigcap_{n=1}^\infty E_n$ for each sequence $\mu = \langle \mu_1, \mu_2, \ldots \rangle \in \mathbb{Z}$ and hence

$$\bigcup_{\mu \in \mathbb{Z}} \bigcap_{n=1}^\infty S_{\mu|n} = \bigcap_{n=1}^\infty E_n$$

As Nikodym has established,[1] if the corresponding sets in two determinant systems are equivalent modulo a σ-ideal $\mathscr{I}$, then their nuclei are also equivalent modulo $\mathscr{I}$. That is, we have

THEOREM 20   With respect to a given σ-ideal $(X, \mathscr{I})$, if $\{S_\sigma : \sigma \in \mathbb{N}^F\}$ and $\{T_\sigma : \sigma \in \mathbb{N}^F\}$ are determinant systems of subsets of X with $S_\sigma \approx T_\sigma$ for all $\sigma \in \mathbb{N}^F$, then

$$\bigcup_{\mu \in \mathbb{Z}} \bigcap_{n=1}^\infty S_{\mu|n} \approx \bigcup_{\mu \in \mathbb{Z}} \bigcap_{n=1}^\infty T_{\mu|n}$$

Proof. Let

$$U = \bigcup_{\sigma \in \mathbb{N}^F} (S_\sigma \Delta T_\sigma) = \bigcup_{n=1}^\infty \bigcup_{\sigma \in \mathbb{N}^n} (S_\sigma \Delta T_\sigma)$$

By hypothesis, $S_\sigma \Delta T_\sigma \in \mathscr{I}$ for every $\sigma \in \mathbb{N}^F$. The union being a countable union, we have $U \in \mathscr{I}$.

For any sequence $\mu = \langle \mu_1, \mu_2, \ldots \rangle \in \mathbb{Z}$, we have

$$\left(\bigcap_{n=1}^\infty S_{\mu|n} \right) - \left(\bigcap_{n=1}^\infty T_{\mu|n} \right) \subset \bigcup_{n=1}^\infty (S_{\mu|n} - T_{\mu|n})$$

[1] Nikodym (1926).

and

$$\left(\bigcap_{n=1}^{\infty} T_{\mu|n}\right) - \left(\bigcap_{n=1}^{\infty} S_{\mu|n}\right) \subset \bigcup_{n=1}^{\infty} (T_{\mu|n} - S_{\mu|n})$$

Hence the set

$$\bigcup_{\mu \in Z} \left[\left(\bigcap_{n=1}^{\infty} S_{\mu|n}\right) \triangle \left(\bigcap_{n=1}^{\infty} T_{\mu|n}\right) \right]$$

which is a subset of U, is an element of $\mathcal{I}$. Applying property (β9) of Section III.I, we obtain

$$\bigcup_{\mu \in Z} \bigcap_{n=1}^{\infty} S_{\mu|n} \approx \bigcup_{\mu \in Z} \bigcap_{n=1}^{\infty} T_{\mu|n}$$

Generalizing the fact that the family of Baire sets is closed under countable unions and countable intersections, we shall now prove

THEOREM 21   The family of Baire sets is closed under operation $\mathcal{A}$.[1]

Proof. Let

$$S = \bigcup_{\mu \in Z} \bigcap_{n=1}^{\infty} S_{\mu|n}$$

be the nucleus of a determinant system $\{S_\sigma : \sigma \in \mathbb{N}^F\}$ consisting of Baire sets. The family of Baire sets being closed under finite intersections, we may assume, without loss of generality, that for each sequence $\mu \in Z$ and each $n \in \mathbb{N}$ we have

$$S_{\mu|n+1} \subset S_{\mu|n}$$

(Otherwise, setting

$$S'_{\mu|n} = \bigcap_{i=1}^{n} S_{\mu|i}$$

for all $\mu \in Z$ and all $n \in \mathbb{N}$, we obtain a determinant system $\{S'_\sigma : \sigma \in \mathbb{N}^F\}$ of Baire sets that satisfies this inclusion and whose nucleus is also S.)

In order to show that S is a Baire set it suffices to show: If A is any region in which S is abundant everywhere, then $A - S$ is a meager set. Assume therefore that A is such a region.

[1]Cf. Luzin and Sierpiński (1917a), Luzin and Sierpiński (1918, 1923), Nikodym (1925, 1926), Szpilrajn (1930; 1933; 1935, Sec. 2.5; 1937, Sec. 2.2), Saks (1937, Chap. II, Sec. 5), Rogers (1970, Sec. 1.7), and Morgan and Schilling (1987). For more general set-theoretical operations that preserve Baire sets, see Ljapunov (1949a; 1949b; 1953; 1973, Sec. 5), Schilling and Vaught (1983), and Schilling (1984); see also Kuratowski (1935, p. 537).

Suppose that $n \in \mathbb{N}$ and $\sigma \in \mathbb{N}^F$. Define

$$T_\sigma = \bigcup_{v \in Z} \bigcap_{k=1}^{\infty} S_{\sigma(v|k)}$$

where $S_{\sigma(v|k)} = S_{\mu_1 \ldots \mu_n v_1 \ldots v_k}$ for $\sigma = \langle \mu_1, \ldots, \mu_n \rangle$. We determine a particular family (possibly empty) $\mathscr{M}_\sigma^*$ of disjoint subregions of A such that T_σ is abundant everywhere in each region in $\mathscr{M}_\sigma^*$.[1] Let $\mathscr{N}_\sigma$ consist of all those regions in which T_σ is either meager or abundant everywhere. Then $(\bigcup \mathscr{N}_\sigma, \mathscr{N}_\sigma)$ is a category base. Applying Lemma 4 of Section II, we define $\mathscr{M}$ to be a disjoint subfamily of $\mathscr{N}_\sigma$ having the property that for every region $N \in \mathscr{N}_\sigma$ there is a region $M \in \mathscr{M}_\sigma$ such that $N \cap M$ contains a region in $\mathscr{N}_\sigma$. Set

$$\mathscr{M}_\sigma^* = \{M \in \mathscr{M}_\sigma : T_\sigma \text{ is abundant everywhere in } M\}$$

Now defining

$$R_\sigma = S_\sigma \cap (\bigcup \mathscr{M}_\sigma^*)$$

we have

$$R_\sigma \subset S_\sigma$$

Set

$$Q = A - \bigcup_{m=1}^{\infty} R_m$$

and for each $n \in \mathbb{N}$ and $\sigma \in \mathbb{N}^n$, set

$$Q_\sigma = R_\sigma - \bigcup_{m=1}^{\infty} R_{\sigma m}$$

We then have

$$A - S = A - \bigcup_{\mu \in Z} \bigcap_{n=1}^{\infty} S_{\mu|n} \subset A - \bigcup_{\mu \in Z} \bigcap_{n=1}^{\infty} R_{\mu|n}$$

$$\subset \left(A - \bigcup_{m=1}^{\infty} R_m\right) \cup \left[\bigcup_{\mu \in Z} \bigcap_{n=1}^{\infty} \left(R_{\mu|n} - \bigcup_{m=1}^{\infty} R_{(\mu|n)m}\right)\right]$$

$$= Q \cup \left(\bigcup_{n=1}^{\infty} \bigcup_\sigma Q_\sigma\right)$$

where σ varies over all sequences $\sigma = \langle \mu_1, \ldots, \mu_n \rangle \in \mathbb{N}^n$ for each $n \in \mathbb{N}$. Now, the totality of sets Q_σ is countable. Hence, in order to show that $A - S$ is a meager set, we have only to show that Q and all the sets Q_σ are meager sets.

[1] The families $\mathscr{N}_\sigma$, $\mathscr{M}_\sigma$, $\mathscr{M}_\sigma^*$ in this proof are merely families of regions indexed by σ, contrary to the notation of Section II.D.

Suppose that Q is not meager. Being a subset of A, the set Q is abundant everywhere in a region $B \subset A$. From the inclusion

$$S \subset \bigcup_{p=1}^{\infty} T_p$$

and the fact that S is abundant in B, it follows that there is an index p_1 such that T_{p_1} is abundant in B. There is then a subregion N of B in which T_{p_1} is abundant everywhere. According to the definition of the family $\mathcal{M}_{p_1}$, there exists a region $M \in \mathcal{M}_{p_1}$ such that $N \cap M$ contains a region C in which T_{p_1} is abundant everywhere. As $T_{p_1} \subset S_{p_1}$, the set S_{p_1} is also abundant everywhere in C. Now we have

$$S_{p_1} \cap C \subset R_{p_1} \subset \bigcup_{m=1}^{\infty} R_m$$

which implies that

$$Q \subset X - \bigcup_{m=1}^{\infty} R_m \subset X - (S_{p_1} \cap C)$$

Hence, Q being abundant everywhere in C, the set $X - (S_{p_1} \cap C)$ is abundant everywhere in C. Because S_{p_1} is also abundant everywhere in C and both S_{p_1} and $X - (S_{p_1} \cap C)$ are Baire sets, the set

$$S_{p_1} \cap [X - (S_{p_1} \cap C)] = S_{p_1} - C$$

is abundant in C. But this is impossible! We conclude that Q must be a meager set.

Suppose that $n \in N$ and $\sigma \in N^n$. To show that Q_σ is a meager set, we assume to the contrary that Q_σ is abundant. Then Q_σ is abundant everywhere in some region D.

The set Q_σ is abundant everywhere in some region $B \in \mathcal{N}_\sigma$. For if T_σ is meager in D, then $D \in \mathcal{N}_\sigma$, so we may take $B = D$. Whereas if T_σ is abundant in D, then T_σ is abundant everywhere in some region $B \subset D$, so $B \in \mathcal{N}_\sigma$ and Q_σ is abundant everywhere in B.

The set T_σ must also be abundant everywhere in B. For suppose that T_σ is meager in some region $B' \subset B$. Then there exists a region $M \in \mathcal{M}_\sigma$ and a region $B'' \in \mathcal{N}_\sigma$ such that $B'' \subset B' \cap M$. The set T_σ cannot be abundant everywhere in M and consequently, $M \notin \mathcal{M}_\sigma^*$. The regions in $\mathcal{M}_\sigma$ being disjoint, we have $M \cap (\bigcup \mathcal{M}_\sigma^*) = \varnothing$. Since $Q_\sigma \subset \bigcup \mathcal{M}_\sigma^*$, we have $B'' \cap Q_\sigma = \varnothing$. This implies that Q_σ is not abundant everywhere in B, which is a contradiction.

Having thus established that T_σ is abundant everywhere in B, we can replace Q, S, T_p, p_1, R_m in the argument above with Q_σ, T_σ, $T_{\sigma p}$, σp_1, $R_{\sigma m}$, respectively, to obtain the conclusion that Q_σ must be a meager set.

Suppose now that $\{S_\sigma : \sigma \in \mathbb{N}^F\}$ is a determinant system of arbitrary sets with nucleus S; i.e.,

$$S = \bigcup_{\mu \in Z} \bigcap_{n=1}^{\infty} S_{\mu|n}$$

For each element $\sigma \in \mathbb{N}^F$ define

(i) $$S_\sigma^1 = S_\sigma$$

and for ordinal numbers $\alpha < \Omega$, define

(ii) $$S_\sigma^{\alpha+1} = S_\sigma^\alpha \cap \left(\bigcup_{m=1}^{\infty} S_{\sigma m}^\alpha \right)$$

and

(iii) $$S_\sigma^\alpha = \bigcap_{\xi < \alpha} S_\sigma^\xi$$

whenever α is a limit ordinal. It can be shown by transfinite induction that

(iv) $$S_\sigma^\beta \subset S_\sigma^\alpha$$

if $\alpha \leqslant \beta < \Omega$. In addition, if $\mu = \langle \mu_1, \mu_2, \ldots \rangle \in Z$, then for every $k \in \mathbb{N}$ and every $\alpha < \Omega$ we have

(v) $$\bigcap_{n=1}^{\infty} S_{\mu|n} \subset S_{\mu|k}^\alpha$$

For each $\alpha < \Omega$, set

(vi) $$U_\alpha = \bigcup_{m=1}^{\infty} S_m^\alpha$$

(vii) $$V_\alpha = \bigcup_{\sigma \in \mathbb{N}'} (S_\sigma^\alpha - S_\sigma^{\alpha+1})$$

(viii) $$T_\alpha = U_\alpha - V_\alpha$$

Then we have the following result:[1]

THEOREM 22

$$S = \bigcap_{\alpha < \Omega} U_\alpha \quad \text{and} \quad S = \bigcup_{\alpha < \Omega} T_\alpha$$

Proof. If $x \in S$, then there exists a sequence $\mu = \langle \mu_1, \mu_2, \ldots \rangle \in Z$ such that $x \in \bigcap_{n=1}^{\infty} S_{\mu|n}$. For any ordinal number $\alpha < \Omega$ we have, by virtue of the

[1]Sierpiński (1926a).

property (v), $x \in S^{\alpha}_{\mu_1}$. Hence, $x \in U_{\alpha}$ for all $\alpha < \Omega$ and we thus have $S \subset \bigcap_{\alpha < \Omega} U_{\alpha}$.

Suppose that $x \in \bigcap_{\alpha < \Omega} U_{\alpha}$. Then for every $\alpha < \Omega$ there is a natural number m such that $x \in S^{\alpha}_m$. Hence, there exists $\mu_1 \in \mathbb{N}$ for which the set $\{\alpha < \Omega : x \in S^{\alpha}_{\mu_1}\}$ is uncountable. But since $S^{\beta}_{\mu_1} \subset S^{\alpha}_{\mu_1}$ for all $\beta > \alpha$, we must have $x \in S^{\alpha}_{\mu_1}$ for every ordinal number $\alpha < \Omega$. From the equality

$$S^{\alpha+1}_{\mu_1} = S^{\alpha}_{\mu_1} \cap \left(\bigcup_{m=1}^{\infty} S^{\alpha}_{\mu_1 m} \right)$$

we see that for each $\alpha < \Omega$ there is a natural number m such that $x \in S^{\alpha}_{\mu_1 m}$. Consequently, there exists $\mu_2 \in \mathbb{N}$ such that $x \in S^{\alpha}_{\mu_1 \mu_2}$ for uncountably many ordinal numbers $\alpha < \Omega$ and, in fact, $x \in S^{\alpha}_{\mu_1 \mu_2}$ for all $\alpha < \Omega$. Similarly, using the equality

$$S^{\alpha+1}_{\mu_1 \mu_2} = S^{\alpha}_{\mu_1 \mu_2} \cap \left(\bigcup_{m=1}^{\infty} S^{\alpha}_{\mu_1 \mu_2 m} \right)$$

we obtain $\mu_3 \in \mathbb{N}$ such that $x \in S^{\alpha}_{\mu_1 \mu_2 \mu_3}$ for all $\alpha < \Omega$. Continuation of this procedure leads to a sequence $\mu = \langle \mu_1, \mu_2, \ldots \rangle \in Z$ such that $x \in S^{\alpha}_{\mu|n}$ for all $\alpha < \Omega$ and all $n \in \mathbb{N}$. In particular, if we take $\alpha = 1$, then we have $x \in S_{\mu|n}$ for every $n \in \mathbb{N}$, so $x \in S$. We conclude that

$$S = \bigcap_{\alpha < \Omega} U_{\alpha}$$

Turning now to the second assertion, suppose that $x \in S$. Then there exists a sequence $\mu = \langle \mu_1, \mu_2, \ldots \rangle \in Z$ such that $x \in \bigcap_{n=1}^{\infty} S_{\mu|n}$. From (v) we know that $x \in S^{\alpha}_{\mu|k}$ for all $\alpha < \Omega$ and all $k \in \mathbb{N}$. Consequently, $x \in U_{\alpha}$ for all $\alpha < \Omega$. The conclusion $x \in T_{\alpha} = U_{\alpha} - V_{\alpha}$ for some $\alpha < \Omega$ will follow once we show that $\bigcap_{\alpha < \Omega} V_{\alpha} = \varnothing$.

Assume, to the contrary, that there is a point $y \in \bigcap_{\alpha < \Omega} V_{\alpha}$. To each ordinal number $\alpha < \Omega$ there then corresponds an element $\sigma \in \mathbb{N}^F$ such that $y \in S^{\alpha}_{\sigma} - S^{\alpha+1}_{\sigma}$. The set $\mathbb{N}^F$ being countable, there exist at least two ordinal numbers α, β, with $\alpha < \beta < \Omega$, whose corresponding elements σ are the same, so that $y \in (S^{\alpha}_{\sigma} - S^{\alpha+1}_{\sigma}) \cap (S^{\beta}_{\sigma} - S^{\beta+1}_{\sigma})$. From $y \in S^{\beta}_{\sigma}$ we obtain, as a consequence of property (iv), $y \in S^{\alpha+1}_{\sigma}$. But this contradicts the fact that $y \in S^{\alpha}_{\sigma} - S^{\alpha+1}_{\sigma}$. We thus have $\bigcap_{\alpha < \Omega} V_{\alpha} = \varnothing$. Therefore, $S \subset \bigcup_{\alpha < \Omega} T_{\alpha}$.

Conversely, suppose that $x \in \bigcup_{\alpha < \Omega} T_{\alpha}$. Then there is an ordinal number $\alpha < \Omega$ such that $x \in U_{\alpha}$ and $x \notin S^{\alpha}_{\sigma} - S^{\alpha+1}_{\sigma}$ for all elements $\sigma \in \mathbb{N}^F$. Since $x \in U_{\alpha}$, there exists $\mu_1 \in \mathbb{N}$ such that $x \in S^{\alpha}_{\mu_1}$. As $x \notin S^{\alpha}_{\mu_1} - S^{\alpha+1}_{\mu_1}$, we must have $x \in S^{\alpha+1}_{\mu_1}$. By property (ii), there exists $\mu_2 \in \mathbb{N}$ such that $x \in S^{\alpha}_{\mu_1 \mu_2}$. As $x \notin S^{\alpha}_{\mu_1 \mu_2} - S^{\alpha+1}_{\mu_1 \mu_2}$, we must have $x \in S^{\alpha+1}_{\mu_1 \mu_2}$. Continuing inductively, we determine a sequence $\mu = \langle \mu_1, \mu_2, \ldots \rangle \in Z$ such that $x \in \bigcap_{n=1}^{\infty} S^{\alpha}_{\mu|n}$. By (iv) and (i) we have

$x \in \bigcap_{n=1}^{\infty} S_{\mu|n}$, so $x \in S$. We conclude that

$$S = \bigcup_{\alpha < \Omega} T_\alpha$$

In the representation of S as the union of the sets T_α, these sets are not necessarily disjoint. Placing

$$Q_1 = T_1$$

and

$$Q_\alpha = T_\alpha - \bigcup_{\xi < \alpha} T_\xi$$

for $1 < \alpha < \Omega$, we obtain

$$T_\alpha = \bigcup_{\xi \leq \alpha} Q_\xi$$

for $\alpha < \Omega$ and

$$S = \bigcup_{\alpha < \Omega} Q_\alpha$$

The disjoint sets Q_α thus defined are called the constituents of the set S with respect to the determinant system $\{S_\sigma : \sigma \in \mathbb{N}^F\}$.[1]

The representation of S as the intersection of the sets U_α yields the representation of the complement of S:

$$X - S = \bigcup_{\alpha < \Omega} (X - U_\alpha)$$

Setting

$$R_1 = X - U_1$$

and

$$R_\alpha = (X - U_\alpha) - \bigcup_{\xi < \alpha} U_\xi$$

for $1 < \alpha < \Omega$, we obtain the representation

$$X - S = \bigcup_{\alpha < \Omega} R_\alpha$$

of the complement of S as the union of disjoint sets R_α. These sets R_α are called the constituents of the complement of S with respect to the determinant system $\{S_\sigma : \sigma \in \mathbb{N}^F\}$.

We next establish a property that is valid when all the sets in the determinant system $\{S_\sigma : \sigma \in \mathbb{N}^F\}$ with nucleus S are Baire sets.[2]

[1] Sierpiński (1933f, pp. 32–33).
[2] Sierpiński (1933f) and Szpilrajn (1933).

THEOREM 23 If $\mathscr{C}$ satisfjes CCC and all the set S_σ are Baire sets, then there exists an ordinal number $\gamma < \Omega$ such that $S—T_\gamma$ is a meager set and consequently $S \approx T_\gamma$.

Proof. For each $\sigma \in \mathsf{N}^F$ the transfinite sequence $\langle S_\sigma^\alpha \rangle_{\alpha < \Omega}$ is a descending transfinite sequence of Baire sets. By Theorem 18, there exists an ordinal number $\gamma < \Omega$ such that $S_\sigma^\gamma—S_\sigma^\xi$ is a meager set for each ordinal number $\xi > \gamma$. In general, the number γ will depend upon the particular element σ in N^F. However, since N^F is countable, there exists an ordinal number $\gamma < \Omega$ such that $S_\sigma^\gamma—S_\sigma^\xi$ is a meager set for all ordinal numbers $\xi > \gamma$ and all elements $\sigma \in \mathsf{N}^F$. Consequently, the set V_γ will also be meager. From the inclusion $S \subset U_\gamma$ we obtain

$$S—T_\gamma = S—(U_\gamma—V_\gamma) \subset U_\gamma—(U_\gamma—V_\gamma) = U_\gamma \cap V_\gamma \subset V_\gamma$$

Therefore, $S—T_\gamma$ is a meager set.

COROLLARY 24 If $\mathscr{C}$ satisfies CCC and all the sets S_σ are Baire sets, then there exists an ordinal number $\gamma < \Omega$ such that the set $\bigcup_{\gamma < \alpha < \Omega} Q_\alpha$ is meager.[1]

In particular, this corollary, as well as Theorem 5, implies that at most countably many constituents of S can be abundant sets. This corollary can be rephrased as follows:

COROLLARY 25 If $\mathscr{C}$ satisfies CCC and all the sets S_σ are Baire sets, then the transfinite sequence $\langle Q_\alpha \rangle_{\alpha < \Omega}$ of constituents of S is a null sequence.

Corollary 24 also implies

COROLLARY 26 If $\mathscr{C}$ satisfies CCC and all the sets S_σ are Baire sets, then a necessary and sufficient condition that S be a meager set is that all its constituents be meager sets.[2]

N. Invariant sets

Let Φ be a set of one-to-one mappings of X onto itself. If $\mathscr{S}$ is a family of subsets of X and $\phi \in \Phi$, then we denote

$$\phi(\mathscr{S}) = \{\phi(S) : S \in \mathscr{S}\}$$

DEFINITION A family $\mathscr{S}$ of subsets of X is invariant under Φ, or Φ-invariant, if $\phi(\mathscr{S}) = \mathscr{S}$ for all $\phi \in \Phi$. A set S is Φ-invariant if the family $\mathscr{S} = \{S\}$ is Φ-invariant; i.e., $\phi(S) = S$ for all $\phi \in \Phi$. When a single mapping ϕ

---

[1]Sélivanowski (1933) and Sierpiński (1933f); see also Ljapunov (1947; 1953; 1973, Sec. 2).

[2]Cf. Sélivanowski (1933, Sec. 5).

is being considered, then we use the terminology "ϕ-invariant" or "invariant under ϕ."

DEFINITION A set is called essentially invariant under Φ if $\phi(S) \approx S$ for all $\phi \in \Phi$.

By a group of one-to-one mappings of X onto itself we understand a set Φ of such mappings that is closed under the operation of composition of mappings and the operation of formation of inverse mappings.

In a straightforward manner, one can verify that

THEOREM 27 If ϕ is a one-to-one mapping of X onto itself and $\mathscr{C}$ is invariant under ϕ, then
  (i) Each of the families $\mathfrak{S}(\mathscr{C})$, $\mathfrak{M}(\mathscr{C})$, and $\mathfrak{B}(\mathscr{C})$ is ϕ-invariant.
  (ii) A set S is abundant everywhere in a region A if and only if $\phi(S)$ is abundant everywhere in the region $\phi(A)$.
  (iii) A set S has a property locally at a point x if and only if $\phi(S)$ has that property locally at the point $\phi(x)$.

In probability theory there occur various theorems whose conclusion is that certain sets (or events) must have probability 0 or 1. Such theorems are known as zero-one laws. Following is a generalization of one of these laws.[1]

THEOREM 28 Assume that Φ is a group of one-to-one mappings of X onto itself, $\mathscr{C}$ is Φ-invariant, and the following condition holds:
  (*) For any two regions A and B, there is a mapping $\phi \in \Phi$ such that $\phi(A) \cap B$ contains a region.
If S is a Baire set that is essentially invariant under Φ, then either S or $X{-}S$ is a meager set.

Proof. Assume that S is an abundant set. We first show that S is abundant everywhere.

Let A be a region in which S is abundant everywhere, let B be any region, let $\phi \in \Phi$ satisfy the condition (*), and let C be a region contained in $\phi(A) \cap B$. According to Theorem 27, the set $\phi(S)$ is abundant everywhere in $\phi(A)$ and hence $C \cap \phi(S)$ is an abundant set. The set $T = S \triangle \phi(S)$ being meager, $C \cap T$ is meager. Therefore, $[C \cap \phi(S)]{-}(C \cap T)$ is an abundant set. From the inclusions

$$[C \cap \phi(S)]{-}(C \cap T) \subset C \cap S \subset B \cap S$$

it follows that $B \cap S$ is an abundant set, for every region B.

[1]Morgan (1977b); see also Sierpiński (1932b, p. 24), Oxtoby (1937, 1961), Gottschalk and Hedlund (1955, p. 55), Hewitt and Savage (1955), Horn and Schach (1970), Christensen (1971, 1972), Bhaskara Ras and Bhaskara Rao (1974), Kuratowski (1974), White (1974b), Stout (1977), Bhaskara Rao and Pol (1978), and Sendler (1978).

Having thus seen that S is abundant everywhere, we use the fact that S is a Baire set to conclude that $X—S$ is a meager set.

IV. BAIRE FUNCTIONS

A. Real-valued Baire functions

For sets X and Y we denote by $f: X \to Y$ a function whose domain is X and whose range is a subset of Y.

DEFINITION   Let $(X,\mathscr{C})$ be a category base. A function $f: X \to \mathbb{R}$ is called a Baire function if for every element $a \in \mathbb{R}$ the set $\{x \in X : f(x) \leqslant a\}$ is a Baire set.

Every function $f: X \to \mathbb{R}$ whose range consists of a single value, called a constant function, is a Baire function.

In the case that $X = \mathbb{R}^n$ and $\mathscr{C}$ is the family of all closed rectangles, the Baire functions are referred to as the functions having the Baire property. When $X = \mathbb{R}^n$ and $\mathscr{C}$ is the family of all compact sets of positive Lebesgue measure, the Baire functions are called Lebesgue measurable functions. We note that both of these classes of functions include the real-valued continuous functions.[1] The functions that have the Baire property constitute the Baire category analogue of the Lebesgue measurable functions.

We leave to the reader the proof of the following theorem.

THEOREM 1   Each of the following conditions is necessary and sufficient for a function $f: X \to \mathbb{R}$ to be a Baire function:
   (i) $\{x \in X : f(x) \leqslant a\} \in \mathfrak{B}(\mathscr{C})$ for every $a \in \mathbb{R}$.
   (ii) $\{x \in X : f(x) < a\} \in \mathfrak{B}(\mathscr{C})$ for every $a \in \mathbb{R}$.
   (iii) $\{x \in X : f(x) \geqslant a\} \in \mathfrak{B}(\mathscr{C})$ for every $a \in \mathbb{R}$.
   (iv) $\{x \in X : f(x) > a\} \in \mathfrak{B}(\mathscr{C})$ for every $a \in \mathbb{R}$.
   (v) $\{x \in X : a < f(x) < b\} \in \mathfrak{B}(\mathscr{C})$ for all $a,b \in \mathbb{R}$ with $a < b$.
   (vi) $\{x \in X : a \leqslant f(x) \leqslant b\} \in \mathfrak{B}(\mathscr{C})$ for all $a,b \in \mathbb{R}$ with $a < b$.
   (vii) $\{x \in X : a \leqslant f(x) < b\} \in \mathfrak{B}(\mathscr{C})$ for all $a,b \in \mathbb{R}$ with $a < b$.
   (viii) $\{x \in X : a < f(x) \leqslant b\} \in \mathfrak{B}(\mathscr{C})$ for all $a,b \in \mathbb{R}$ with $a < b$.
   (ix) $f^{-1}(G) \in \mathfrak{B}(\mathscr{C})$ for every open set $G \subset \mathbb{R}$.
   (x) $f^{-1}(F) \in \mathfrak{B}(\mathscr{C})$ for every closed set $F \subset \mathbb{R}$.

COROLLARY 2   If f is a Baire function, then for every $a \in \mathbb{R}$ the set $f^{-1}(a)$ is a Baire set.

[1] See Theorem 1 of Chapter 5, Section III.

NOTE   In view of the equality

$$\{x \in X : f(x) < a\} = \bigcup \{\{x \in X : f(x) < r\} : r \in \mathbb{Q} \text{ and } r < a\}$$

valid for each $a \in \mathbb{R}$, it suffices to consider only rational numbers a in condition (ii). Similarly, in conditions (iv) and (v) we need only consider rational numbers a,b.

DEFINITION   The characteristic function of a set $S \subset X$ is the function $\chi_S : X \to \mathbb{B}$ defined by

$$\chi_S(x) = \begin{cases} 1 & \text{if } x \in S \\ 0 & \text{if } x \notin S \end{cases}$$

It is a simple matter to prove

THEOREM 3   A set $S \subset X$ is a Baire set if and only if its characteristic function is a Baire function.

The existence of a non-Baire set thus implies the existence of a function $f : X \to \mathbb{R}$ that is not a Baire function.

In order to prove that the sum and product of two real-valued Baire functions is also a Baire function, we first establish a basic fact.

THEOREM 4   If $f : X \to \mathbb{R}$ and $g : X \to \mathbb{R}$ are Baire functions, then the set $\{x \in X : f(x) \leqslant g(x)\}$ is a Baire set.

Proof. From the equality

$$\{x \in X : f(x) > g(x)\} = \bigcup_{r \in \mathbb{Q}} (\{x \in X : f(x) > r\} \cap \{x \in X : g(x) < r\})$$

it is seen that the set $\{x \in X : f(x) > g(x)\}$ is a Baire set and hence so is its complement $\{x \in X : f(x) \leqslant g(x)\}$.

COROLLARY 5   If $f : X \to \mathbb{R}$ and $g : X \to \mathbb{R}$ are Baire functions, then the set $\{x \in X : f(x) = g(x)\}$ is a Baire set.

THEOREM 6   If $f : X \to \mathbb{R}$ and $g : X \to \mathbb{R}$ are Baire functions and $c \in \mathbb{R}$, then each of the functions $f + c$, cf, $f + g$, fg is a Baire function.

Proof. That $f + c$ and cf are Baire functions is an easy consequence of condition (i) of Theorem 1. It follows that the function $h = a - g$ is a Baire function for every $a \in \mathbb{R}$. From Theorem 4 and the equality

$$\{x \in X : f(x) + g(x) \leqslant a\} = \{x \in X : f(x) \leqslant a - g(x)\}$$

we then see $f + g$ is a Baire function.

For $a \in \mathbb{R}$, the set $\{x \in X : [f(x)]^2 \leqslant a\}$ is equal to $\emptyset$ if $a < 0$, is equal to $\{x \in X : f(x) = 0\}$ if $a = 0$, and is equal to $\{x \in X : -a \leqslant f(x) \leqslant a\}$ if $a > 0$.

The latter three sets are Baire sets. Hence the function f^2 is a Baire function. The equality

$$fg = \frac{1}{2}[(f + g)^2 - f^2 - g^2]$$

then yields the fact that the product fg is a Baire function.

B. Extended real-valued Baire functions

In measure theory it is necessary to admit infinite values for measures and functions. The set $\mathbb{R}$ of all real numbers is then augmented by adjoining the two elements $-\infty$ and $+\infty$ to form the set

$$\mathbb{R}^* = \mathbb{R} \cup \{-\infty, +\infty\}$$

constituting the extended real number system. The ordering of $\mathbb{R}$ is extended to an ordering of $\mathbb{R}^*$ by further stipulating $-\infty < +\infty$ and $-\infty < r < +\infty$ for every element $r \in \mathbb{R}$. Any set S of extended real numbers has both an infimum and a supremum, denoted by $\inf S$ and $\sup S$, respectively. For $S = \varnothing$ we have $\inf S = +\infty$ and $\sup S = -\infty$.

We say that a sequence $\langle x_n \rangle_{n \in \mathbb{N}}$ in $\mathbb{R}^*$ has limit ∞ (resp., $-\infty$) and express this symbolically by $\lim_{n \to \infty} x_n = \infty$ (resp., $\lim_{n \to \infty} x_n = -\infty$) if for each element $a \in \mathbb{R}$ the set $\{n \in \mathbb{N} : x_n < a\}$ is finite (resp., $\{n \in \mathbb{N} : x_n > a\}$ is finite). A sequence $\langle x_n \rangle_{n \in \mathbb{N}}$ in $\mathbb{R}^*$ is called

1. monotone increasing if $x_n \leqslant x_{n+1}$ for every $n \in \mathbb{N}$.
2. increasing if $x_n < x_{n+1}$ for every $n \in \mathbb{N}$.
3. monotone decreasing if $x_{n+1} \leqslant x_n$ for every $n \in \mathbb{N}$.
4. decreasing if $x_{n+1} < x_n$ for every $n \in \mathbb{N}$.

We note that any monotone increasing sequence in $\mathbb{R}^*$ and any monotone decreasing sequence in $\mathbb{R}^*$ has a limit in $\mathbb{R}^*$.

Every sequence $\langle x_n \rangle_{n \in \mathbb{N}}$ in $\mathbb{R}^*$ has a limit inferior and limit superior in $\mathbb{R}^*$, denoted by $\liminf_n x_n$ and $\limsup_n x_n$, with

$$\liminf_n x_n = \lim_{n \to \infty} \left(\inf_{k \geqslant n} x_k \right) = \lim_{n \to \infty} (\inf\{x_k : k \in \mathbb{N} \text{ and } k \geqslant n\})$$

$$\limsup_n x_n = \lim_{n \to \infty} \left(\sup_{k \geqslant n} x_k \right) = \lim_{n \to \infty} (\sup\{x_k : k \in \mathbb{N} \text{ and } k \geqslant n\})$$

If for each $n \in \mathbb{N}$ we place $y_n = \inf_{k \geqslant n} x_k$ and $z_n = \sup_{k \geqslant n} x_k$, then $\langle y_n \rangle_{n \in \mathbb{N}}$ is a monotone increasing sequence, $\langle z_n \rangle_{n \in \mathbb{N}}$ is a monotone decreasing sequence, and $y_n \leqslant z_n$ for every n. This implies that

$$\liminf_n x_n = \sup_n \left(\inf_{k \geqslant n} x_k \right) \quad \text{and} \quad \limsup_n x_n = \inf_n \left(\sup_{k \geqslant n} x_k \right)$$

and

$$\liminf_n x_n \leqslant \limsup_n x_n$$

We note the following facts for sequences $\langle x_n \rangle_{n \in \mathbb{N}}$ in $\mathbb{R}^*$ and $x \in \mathbb{R}^*$:

(i) $x = \liminf_n x_n$ if and only if for every $c < x$ the set $\{n \in \mathbb{N} : x_n < c\}$ is finite and for every $d > x$ the set $\{n \in \mathbb{N} : x_n < d\}$ is infinite.

(ii) $x = \limsup_n x_n$ if and only if for every $c < x$ the set $\{n \in \mathbb{N} : x_n > c\}$ is infinite and for every $d > x$ the set $\{n \in \mathbb{N} : x_n > d\}$ is finite.

(iii) If $x = \liminf_n x_n$, then there exists a monotone decreasing subsequence $\langle x_{n_j} \rangle_{j \in \mathbb{N}}$ of $\langle x_n \rangle_{n \in \mathbb{N}}$ such that $\lim_{j \to \infty} x_{n_j} = x$.

(iv) If $x = \limsup_n x_n$, then there exists a monotone increasing subsequence $\langle x_{n_j} \rangle_{j \in \mathbb{N}}$ of $\langle x_n \rangle_{n \in \mathbb{N}}$ such that $\lim_{j \to \infty} x_{n_j} = x$.

(v) $\lim_{n \to \infty} x_n$ exists in $\mathbb{R}^*$ if and only if $\liminf_n x_n = \limsup_n x_n$, in which case $\lim_{n \to \infty} x_n = \liminf_n x_n = \limsup_n x_n$.

The addition and multiplication operations between elements of $\mathbb{R}$, and their basic properties, are extended to $\mathbb{R}^*$ with the exception that $(-\infty) + (+\infty)$ is undefined. We define

$$
\begin{aligned}
(-\infty) + a &= -\infty = a + (-\infty) && \text{for } a \in \mathbb{R}^* \!-\! \{+\infty\} \\
(+\infty) + a &= +\infty = a + (+\infty) && \text{for } a \in \mathbb{R}^* \!-\! \{-\infty\} \\
(\pm\infty) \cdot 0 &= \quad 0 = 0 \cdot (\pm\infty) && \\
(\pm\infty) \cdot a &= \pm\infty = a \cdot (\pm\infty) && \text{for } a \in \mathbb{R}^* \text{ and } a > 0 \\
(\pm\infty) \cdot a &= \mp\infty = a \cdot (\pm\infty) && \text{for } a \in \mathbb{R}^* \text{ and } a < 0
\end{aligned}
$$

The absolute value operation is extended to $\mathbb{R}^*$ by further stipulating

$$|-\infty| = +\infty \qquad |+\infty| = +\infty$$

DEFINITION Let $(X, \mathscr{C})$ be a category base. An extended real-valued function $f : X \to \mathbb{R}^*$ is called a Baire function if for every element $a \in \mathbb{R}^*$ we have $\{x \in X : f(x) \leqslant a\} \in \mathscr{B}(\mathscr{C})$.

THEOREM 7 A function $f : X \to \mathbb{R}^*$ is a Baire function if and only if either one of the conditions (i)–(iv) of Theorem 1 is satisfied or one of the conditions (v)–(x) is satisfied and both of the sets $f^{-1}(-\infty)$, $f^{-1}(+\infty)$ are Baire sets.

Using the same reasoning as in the preceding section, we obtain

THEOREM 8 If $f : X \to \mathbb{R}^*$ and $g : X \to \mathbb{R}^*$ are Baire functions, then the set $\{x \in X : f(x) \leqslant g(x)\}$ is a Baire set.

THEOREM 9   If $f: X \to \mathbb{R}^*$ is a Baire function and $c \in \mathbb{R}$, then the functions $f + c$ and cf are Baire functions.

Suppose that $f: X \to \mathbb{R}^*$, $g: X \to \mathbb{R}^*$, and let U denote the subset of X on which $f + g$ is undefined; i.e.,

$$U = \{x \in X : f(x) = -\infty \text{ and } g(x) = +\infty\}$$
$$\cup \{x \in X : f(x) = +\infty \text{ and } g(x) = -\infty\}$$

Placing $D = X - U$, we have

THEOREM 10   If $f: X \to \mathbb{R}^*$ and $g: X \to \mathbb{R}^*$ are Baire functions and $c \in \mathbb{R}^*$, then the function

$$h = (f + g)\chi_D + c\chi_U$$

is a Baire function.

Proof. If $a \in \mathbb{R}$, then

$$\{x \in X : h(x) \leqslant a\} = \{x \in D : h(x) \leqslant a\} \cup \{x \in U : h(x) \leqslant a\}$$
$$= \{x \in D : f(x) + g(x) \leqslant a\} \cup \{x \in U : c \leqslant a\}$$
$$= (D \cap \{x \in X : f(x) \leqslant a - g(x)\}) \cup (U \cap \{x \in X : c \leqslant a\})$$

Hence, $\{x \in X : h(x) \leqslant a\}$ is a Baire set for every $a \in \mathbb{R}$.

For any function $f: X \to \mathbb{R}^*$ we define the positive and negative parts f^+ and f^- of f by

$$f^+(x) = \max\{0, f(x)\}$$
$$f^-(x) = \min\{0, f(x)\} = \max\{0, -f(x)\}$$

for each $x \in X$. These functions are nonnegative and we have

$$f = f^+ - f^-$$
$$|f| = f^+ + f^-$$

THEOREM 11   If $f: X \to \mathbb{R}^*$ and $g: X \to \mathbb{R}^*$ are Baire functions, then each of the functions $\max\{f,g\}$, $\min\{f,g\}$, f^+, f^-, $|f|$ are Baire functions.

Proof. That $\max\{f,g\}$ and $\min\{f,g\}$ are Baire functions follows from the equalities

$$\{x \in X : \max\{f(x), g(x)\} \leqslant a\} = \{x \in X : f(x) \leqslant a\} \cap \{x \in X : g(x) \leqslant a\}$$
$$\{x \in X : \min\{f(x), g(x)\} \leqslant a\} = \{x \in X : f(x) \leqslant a\} \cup \{x \in X : g(x) \leqslant a\}$$

which are valid for every $a \in \mathbb{R}$. Clearly, then, f^+, f^-, and $|f|$ are also Baire functions.

C. Sequences of functions

For a sequence $\langle f_n \rangle_{n \in \mathbb{N}}$ of extended real-valued functions defined on X we define the supremum, infimum, limit superior, and limit inferior, denoted by $\sup_n f_n$, $\inf_n f_n$, $\limsup_n f_n$, and $\liminf_n f_n$, respectively, in the following pointwise manner:

$$\left(\sup_n f_n \right)(x) = \sup_n f_n(x) = \sup\{f_n(x) : n \in \mathbb{N}\}$$

$$\left(\inf_n f_n \right)(x) = \inf_n f_n(x) = \inf\{f_n(x) : n \in \mathbb{N}\}$$

$$\left(\limsup_n f_n \right)(x) = \lim_{n \to \infty} \sup_{k \geq n} f_k(x) = \lim_{n \to \infty} (\sup\{f_k(x) : k \in \mathbb{N} \text{ and } k \geq n\})$$

$$\left(\liminf_n f_n \right)(x) = \lim_{n \to \infty} \inf_{k \geq n} f_k(x) = \lim_{n \to \infty} (\inf\{f_k(x) : k \in \mathbb{N} \text{ and } k \geq n\})$$

for all $x \in X$.

We note that

$$\limsup_n f_n = \inf_n \left(\sup_{k \geq n} f_k \right) \qquad \liminf_n f_n = \sup_n \left(\inf_{k \geq n} f_k \right)$$

and

$$\liminf_n f_n \leq \limsup_n f_n$$

We note also that $\langle f_n \rangle_{n \in \mathbb{N}}$ converges pointwise to an extended real-valued function f if and only if $\limsup_n f_n = \liminf_n f_n$, in which case $f = \lim_{n \to \infty} f_n = \limsup_n f_n = \liminf_n f_n$.

THEOREM 12 If $\langle f_n \rangle_{n \in \mathbb{N}}$ is a sequence of extended real-valued Baire functions, then each of the extended real-valued functions $\sup_n f_n$, $\inf_n f_n$, $\limsup_n f_n$, $\liminf_n f_n$ is a Baire function.

Proof. For every $a \in \mathbb{R}$ we have the equality

$$\left\{ x \in X : \sup_n f_n(x) \leq a \right\} = \bigcap_{n=1}^{\infty} \{x \in X : f_n(x) \leq a\}$$

which reveals that $\sup_n f_n$ is a Baire function. One can similarly show that $\inf_n f_n$ is a Baire function or this fact may be deduced from the equality

$$\inf_n f_n = -\sup_n (-f_n)$$

If $\langle g_n \rangle_{n \in \mathbb{N}}$ is a monotone decreasing sequence of extended real-valued Baire functions, then

$$\lim_{n \to \infty} g_n = \inf_n g_n$$

is a Baire function. The sequence $\langle g_n \rangle_{n \in \mathbb{N}}$ defined by $g_n = \sup_{k \geq n} f_k$ being monotone decreasing, the function

$$\limsup_n f_n = \lim_{n \to \infty} g_n$$

is thus a Baire function. Hence,

$$\liminf_n f_n = -\limsup_n (-f_n)$$

is also a Baire function.

THEOREM 13 If a sequence of extended real-valued Baire functions converges pointwise to an extended real-valued function f, then f is a Baire function.[1]

Proof. This is a consequence of the preceding theorem and the fact that if $\langle f_n \rangle_{n \in \mathbb{N}}$ converges pointwise to f, then

$$f = \lim_{n \to \infty} f_n = \limsup_n f_n$$

THEOREM 14 If $\langle f_n \rangle_{n \in \mathbb{N}}$ is a sequence of extended real-valued Baire functions and $\langle S_n \rangle_{n \in \mathbb{N}}$ is a sequence of disjoint Baire sets with $X = \bigcup_{n=1}^{\infty} S_n$, then the function

$$h = \sum_{n=1}^{\infty} f_n \chi_{S_n}$$

is a Baire function.

Proof. Use the fact that, for every $a \in \mathbb{R}$,

$$\{x \in X : h(x) \leq a\} = \bigcup_{n=1}^{\infty} \{x \in S_n : h(x) \leq a\}$$

$$= \bigcup_{n=1}^{\infty} (S_n \cap \{x \in X : f_n(x) \leq a\})$$

[1]Cf. Borel (1903), Lebesgue (1903), Kuratowski (1924, Théorème III), Sierpiński (1935c), and Szpilrajn (1935, Sec. 4.2).

D. Approximate relations

Throughout this section all functions are assumed to be extended real-valued functions defined on X.

DEFINITION Let S be a set and let π be a property of points. We say that the property π holds essentially everywhere in S if the set of points in S for which π does not hold is a meager set. If π holds essentially everywhere in X, then we say simply that π holds essentially everywhere.

REMARK In the case of Lebesgue measure the term "almost everywhere" is customarily used in lieu of "essentially everywhere."

NOTATION For functions f, g we write $f \prec g$ or $g \succ f$ if $f(x) \leqslant g(x)$ essentially everywhere. We say that f, g are equal essentially everywhere, and write $f \approx g$, if $f(x) = g(x)$ essentially everywhere.

THEOREM 15 If $f \approx g$ and f is a Baire function, then g is a Baire function.

Proof. Suppose that $a \in \mathbb{R}$. Let $S = \{x \in X : f(x) \leqslant a\}$ and $T = \{x \in X : g(x) \leqslant a\}$. The set $S \Delta T$ is a subset of the set $\{x \in X : f(x) \neq g(x)\}$ which is meager by hypothesis. Hence, $S \Delta T$ is a meager set. By property $(\gamma 1)$ of Section III.I, T is a Baire set. We conclude that g is a Baire function.

THEOREM 16 If a sequence $\langle f_n \rangle_{n \in \mathbb{N}}$ of Baire functions converges essentially everywhere to a function f, then f is a Baire function.

Proof. According to Theorem 12, the function $\limsup_n f_n$ is a Baire function. Our hypothesis implies that the set $\{x \in X : f(x) \neq \limsup_n f_n(x)\}$ is meager. The conclusion now follows from Theorem 15.

THEOREM 17 If $\langle f_n \rangle_{n \in \mathbb{N}}$, $\langle g_n \rangle_{n \in \mathbb{N}}$ converge essentially everywhere to f, g, respectively, and $f_n \prec g_n$ for all $n \in \mathbb{N}$, then $f \prec g$.

Proof. The set $\{x \in X : f(x) > g(x)\}$ is contained in the union of the three meager sets

$$R = \{x \in X : \langle f_n(x) \rangle_{n \in \mathbb{N}} \text{ does not converge to } f(x)\}$$

$$S = \{x \in X : \langle g_n(x) \rangle_{n \in \mathbb{N}} \text{ does not converge to } g(x)\}$$

$$T = \bigcup_{n=1}^{\infty} \{x \in X : f_n(x) > g_n(x)\}$$

and thus is a meager set.

In particular, this theorem implies the essential uniqueness of limits of sequences of functions.

THEOREM 18 If a sequence of functions converges essentially everywhere to a function f and also to a function g, then $f \approx g$.

DEFINITION A sequence of functions $\langle f_n \rangle_{n \in \mathbb{N}}$ is called essentially monotone increasing (resp., decreasing) if $f_n \prec f_{n+1}$ (resp., $f_n \succ f_{n+1}$) for all $n \in \mathbb{N}$.

THEOREM 19 Every essentially monotone increasing (resp., decreasing) sequence of Baire functions converges essentially everywhere to a Baire function.

Proof. Assume that $\langle f_n \rangle_{n \in \mathbb{N}}$ is essentially monotone increasing and let $S_n = \{x \in X : f_n(x) > f_{n+1}(x)\}$ for each $n \in \mathbb{N}$. Then $S = \bigcup_{n=1}^{\infty} S_n$ is a meager set. For every $x \in X - S$ the sequence $\langle f_n(x) \rangle_{n \in \mathbb{N}}$ is monotone increasing and converges to $\sup_n f_n(x)$. Hence $\langle f_n \rangle_{n \in \mathbb{N}}$ converges essentially everywhere to the function $\sup_n f_n$, which, according to Theorem 12, is a Baire function.

The situation where $\langle f_n \rangle_{n \in \mathbb{N}}$ is essentially monotone decreasing is treated similarly.

DEFINITION A function g is called an essential supremum for a family $\mathscr{R}$ of functions if the following conditions are satisfied:
 (a) g is a Baire function.
 (b) $f \prec g$ for all $f \in \mathscr{R}$.
 (c) If h is a Baire function and $f \prec h$ for all $f \in \mathscr{R}$, then $g \prec h$.
A function g is called an essential infimum for a family $\mathscr{R}$ of functions if conditions (a)–(c) are satisfied with $\prec$ replaced by $\succ$.

THEOREM 20 Every nonempty family of Baire functions has an essential supremum and an essential infimum.[1]

Proof. In view of the fact that an essential infimum of a family of functions is the negative of an essential supremum of the family of negatives of the functions, it suffices to show that every nonempty family $\mathscr{R} = \{f_\alpha : \alpha \in I\}$ of Baire functions has an essential supremum.

For any function f and any $r \in \mathbb{R}$, let

$$S(f,r) = \{x \in X : f(x) > r\}$$

We note that

(*) $f \prec g$ if and only if $S(f,r) \prec S(g,r)$ for all $r \in \mathbb{Q}$

[1]Cf. Birkhoff (1948, p. 241), Goffman and Waterman (1960, Lemma 1), Goffman and Zink (1960, Lemma 1), and Wagner (1978, Lemma 1; 1981, Lemma 1). The full generality of this theorem and the proof given are due to K. Schilling (personal communication, 1988).

This is a consequence of the equality

$$\{x \in X : f(x) > g(x)\} = \bigcup_{r \in Q} [S(f,r) - S(g,r)]$$

For each $r \in Q$, we apply Theorem 15 of Section III to obtain an essential hull E_r for the family $\{S(f_\alpha,r) : \alpha \in I\}$. If $r, s \in Q$ with $r > s$, then $S(f_\alpha,r) \subset S(f_\alpha,s)$ for all $\alpha \in I$ and consequently $E_r - E_s$ is a meager set. This implies that the set

$$Y = X - \bigcup \{E_r - E_s : r, s \in Q \text{ and } r > s\}$$

is a comeager set.

Define $g: X \to \mathbb{R}^*$ by

$$g(x) = \sup\{r \in Q : x \in E_r\}$$

We show that g is an essential supremum for $\mathcal{R}$.

For each $a \in Q$ we have

$$\{x \in Y : g(x) > a\} = Y \cap \left(\bigcup_{n \in N} E_{a + 1/n} \right)$$

From the equality

$$\{x \in X : g(x) > a\} = \{x \in Y : g(x) > a\} \cup \{x \in X - Y : g(x) > a\}$$

it then follows that the set on the left-hand side is a Baire set for every $a \in Q$. This means that g is a Baire function.

Suppose that $\alpha \in I$. For each $r \in Q$, we have

$$S(f_\alpha,r) = \bigcup_{n \in N} S\left(f_\alpha, r + \frac{1}{n}\right) \prec \bigcup_{n \in N} E_{r + 1/n} \approx S(g,r)$$

Hence, by (*), we have $f_\alpha \prec g$ for every $\alpha \in I$.

Finally, suppose that h is any Baire function and $f_\alpha \prec h$ for all $\alpha \in I$. For any $q \in Q$ we have $S(f_\alpha,q) \prec S(h,q)$, which implies that $E_q \prec S(h,q)$. Consequently, for any $r \in Q$,

$$S(g,r) \approx \bigcup_{n \in N} E_{r + 1/n} \prec \bigcup_{n \in N} S\left(h, r + \frac{1}{n}\right) = S(h,r)$$

Therefore, $g \prec h$.

REMARK We have derived Theorem 20 from Theorem 15 of Section III. Conversely, Theorem 15 of Section III can be derived from Theorem 20.[1]

---

[1]Cf. Wagner (1978, Lemma 1; 1981, Lemma 1).

E. Categorical limits

All functions considered in this section are assumed to be at most extended real-valued functions defined on X.

DEFINITION   A sequence of functions $\langle f_n \rangle_{n \in \mathbb{N}}$ converges categorically to a function f if every subsequence $\langle f_{n_k} \rangle_{k \in \mathbb{N}}$ of $\langle f_n \rangle_{n \in \mathbb{N}}$ has a subsequence $\langle f_{n_{k_j}} \rangle_{j \in \mathbb{N}}$ that converges essentially everywhere to f.[1]

For the category base of all compact sets of positive Lebesgue measure contained in the unit interval [0,1], categorical convergence coincides with the measure-theoretic notion of "convergence in measure."[2]

In this section we establish a version of Wagner's generalization of a theorem of Goffman and Waterman concerning lower and upper categorical limits and catorical convergence.[3]

DEFINITION   A sequence of functions $\langle g_n \rangle_{n \in \mathbb{N}}$ is said to be eventually a subsequence of a sequence of functions $\langle f_n \rangle_{n \in \mathbb{N}}$ if there is an index m such that the sequence $\langle g_{m+n} \rangle_{n \in \mathbb{N}}$ is a subsequence of $\langle f_n \rangle_{n \in \mathbb{N}}$.

THEOREM 21   Assume that $\mathscr{C}$ satisfies CCC. If a sequence $\langle f_n \rangle_{n \in \mathbb{N}}$ of Baire functions does not converge categorically to 0, then at least one of the following conditions is satisfied:
   (i) There exists a subsequence $\langle f_{n_k} \rangle_{k \in \mathbb{N}}$ of $\langle f_n \rangle_{n \in \mathbb{N}}$, an abundant Baire set S, and a natural number m such that for every subsequence $\langle f_{n_{k_j}} \rangle_{j \in \mathbb{N}}$ of $\langle f_{n_k} \rangle_{k \in \mathbb{N}}$ we have $\limsup_j f_{n_{k_j}} \geq 1/m$ essentially everywhere in S.
   (ii) There exists a subsequence $\langle f_{n_k} \rangle_{k \in \mathbb{N}}$ of $\langle f_n \rangle_{n \in \mathbb{N}}$, an abundant Baire set S, and a natural number m such that for every subsequence $\langle f_{n_{k_j}} \rangle_{j \in \mathbb{N}}$ of $\langle f_{n_k} \rangle_{k \in \mathbb{N}}$ we have $\liminf_j f_{n_{k_j}} \leq -1/m$ essentially everywhere in S.

Proof. If X is a meager set, then every sequence of Baire functions converges categorically to 0. We thus assume that X is abundant.

Assume that $\langle f_n \rangle_{n \in \mathbb{N}}$ is a sequence of Baire functions and suppose that the condition (i) does not hold. Then for every subsequence $\langle f_{n_k} \rangle_{k \in \mathbb{N}}$ of $\langle f_n \rangle_{n \in \mathbb{N}}$, for every abundant Baire set S, and for every natural number m, there exists a subsequence $\langle f_{n_{k_j}} \rangle_{j \in \mathbb{N}}$ of $\langle f_{n_k} \rangle_{k \in \mathbb{N}}$ such that the set

$$T = \left\{ x \in S : \limsup_j f_{n_{k_j}}(x) < \frac{1}{m} \right\}$$

is abundant. The functions $f_{n_{k_j}}$ being Baire functions, it is clear that T is a Baire set.

[1] Cf. Wagner (1978, 1981) and Wagner and Wilczyński (1980).
[2] Cf. Royden (1968, Chap. 4, Sec. 5) and Munroe (1971, Sec. 31).
[3] Goffman and Waterman (1960) and Wagner (1981).

Let $\langle f_{m_k} \rangle_{k \in \mathbb{N}}$ be any subsequence of $\langle f_n \rangle_{n \in \mathbb{N}}$. We shall establish the existence of a subsequence $\langle f_{n'_k} \rangle_{k \in \mathbb{N}}$ of $\langle f_{m_k} \rangle_{k \in \mathbb{N}}$ such that $\limsup_k f_{n'_k} \leqslant 0$ essentially everywhere. Utilizing transfinite induction, we first show there exists a subsequence $\langle f_{m_k,1} \rangle_{k \in \mathbb{N}}$ of $\langle f_{m_k} \rangle_{k \in \mathbb{N}}$ such that $\limsup_k f_{m_k,1} < 1$ essentially everywhere.

Placing $S = X$ and $m = 1$ in the negation of condition (i) given above we see that there is a subsequence $\langle f_{n_k}^{(1)} \rangle_{k \in \mathbb{N}}$ of $\langle f_{m_k} \rangle_{k \in \mathbb{N}}$ such that the set

$$T_1 = \left\{ x \in X : \limsup_k f_{n_k}^{(1)}(x) < 1 \right\}$$

is abundant. We set $E_1 = T_1$.

Assume that $1 < \alpha < \Omega$ and for every ordinal number β, with $1 \leqslant \beta < \alpha$, we have already determined a subsequence $\langle f_{n_k}^{(\beta)} \rangle_{k \in \mathbb{N}}$ of $\langle f_{m_k} \rangle_{k \in \mathbb{N}}$ and an abundant Baire set E_β such that if $1 \leqslant \gamma < \beta$, then $\langle f_{n_k}^{(\beta)} \rangle_{k \in \mathbb{N}}$ is eventually a subsequence of $\langle f_{n_k}^{(\gamma)} \rangle_{k \in \mathbb{N}}$, $E_\gamma \subset E_\beta$, and $\limsup_k f_{n_k}^{(\beta)}(x) < 1$ for all $x \in E_\beta$. We consider separately the cases where α is a successor ordinal and a limit ordinal.

Suppose that $\alpha = \beta + 1$ for some ordinal number β. If $X - E_\beta$ is a meager set, then we define $f_{n_k}^{(\alpha)} = f_{n_k}^{(\beta)}$ for all $k \in \mathbb{N}$ and set $E_\alpha = E_\beta$. If $X - E_\beta$ is abundant, then, using the negation of condition (i) with $S = X - E_\beta$ and $m = 1$, we determine a subsequence $\langle f_{n_k}^{(\alpha)} \rangle_{k \in \mathbb{N}}$ of $\langle f_{n_k}^{(\beta)} \rangle_{k \in \mathbb{N}}$ such that the set

$$T_\alpha = \left\{ x \in X - E : \limsup_k f_{n_k}^{(\alpha)}(x) < 1 \right\}$$

is abundant and set $E_\alpha = E_\beta \cup T_\alpha$. Clearly, $\limsup_k f_{n_k}^{(\alpha)}(x) < 1$ for all $x \in E_\alpha$.

Suppose that α is a limit ordinal. If $X - E_\beta$ is a meager set for some ordinal number $\beta < \alpha$, then we define $f_{n_k}^{(\alpha)} = f_{n_k}^{(\beta_0)}$ for all $k \in \mathbb{N}$, where β_0 is the smallest such number β, and set $E_\alpha = E_{\beta_0}$. Assume, on the other hand, that $X - E_\beta$ is abundant for all ordinal numbers $\beta < \alpha$. Let $\langle \beta_i \rangle_{i \in \mathbb{N}}$ be an increasing sequence of ordinal numbers with $\sup_i \beta_i = \alpha$. Define $\langle f_{n_k}^{(\alpha)} \rangle_{k \in \mathbb{N}}$ to be the diagonal sequence of the matrix $\langle f_{n_k}^{(\beta_i)} : i \in \mathbb{N}, k \in \mathbb{N} \rangle$; that is, the ith term of the sequence $\langle f_{n_k}^{(\alpha)} \rangle_{k \in \mathbb{N}}$ is equal to the ith term of the sequence $\langle f_{n_k}^{(\beta_i)} \rangle_{k \in \mathbb{N}}$ for each $i \in \mathbb{N}$. The set $E_\alpha = \bigcup_{\beta < \alpha} E_\beta = \bigcup_{i=1}^{\infty} E_{\beta_i}$ is an abundant Baire set and $\limsup_k f_{n_k}^{(\alpha)}(x) < 1$ for all $x \in E_\alpha$.

By this procedure we determine an ascending transfinite sequence $\langle E_\alpha \rangle_{\alpha < \Omega}$ of abundant Baire sets. According to Theorem 18 of Section III, this sequence is essentially stationary; i.e., there exists a smallest ordinal number $\mu_1 < \Omega$ such that $E_\alpha \approx E_{\mu_1}$ for all ordinal numbers $\alpha \geqslant \mu_1$.

From the definition of the set E_{μ_1+1} and the fact that $E_{\mu_1+1} - E_{\mu_1}$ is meager it follows that $X - E_{\mu_1}$ is a meager set. Accordingly, $\langle f_{n_k}^{(\mu_1)} \rangle_{k \in \mathbb{N}}$ is a subsequence of $\langle f_{m_k} \rangle_{k \in \mathbb{N}}$ with $\limsup_k f_{n_k}^{(\mu_1)}(x) < 1$ for all $x \in E_{\mu_1}$, that is, essentially everywhere. We denote the sequence $\langle f_{n_k}^{(\mu_1)} \rangle_{k \in \mathbb{N}}$ by $\langle f_{m_k,1} \rangle_{k \in \mathbb{N}}$.

We now repeat the foregoing reasoning starting with $\langle f_{n_k,1}\rangle_{k\in N}$ and $m = 2$ to obtain a subsequence $\langle f_{n_k,1}^{(\mu_2)}\rangle_{k\in N}$ of $\langle f_{n_k,1}\rangle_{k\in N}$ with $\limsup_k f_{n_k,1}^{(\mu_2)} < 1/2$ essentially everywhere. We denote the sequence $\langle f_{n_k,1}^{(\mu_2)}\rangle_{k\in N}$ by $\langle f_{n_k,2}\rangle_{k\in N}$. Continuing for $m = 3,4,\ldots$, we determine sequences $\langle f_{n_k,1}\rangle_{k\in N}$, $\langle f_{n_k,2}\rangle_{k\in N}$, $\langle f_{n_k,3}\rangle_{k\in N}$, $\langle f_{n_k,4}\rangle_{k\in N}$, ..., each a subsequence of those sequences that precede it, with $\limsup_k f_{n_k,m} < 1/m$ essentially everywhere for each $m \in N$. Now defining $\langle f_{n_k'}\rangle_{k\in N}$ to be the diagonal sequence of the matrix $\langle f_{n_k,m} : k \in N, \ m \in N\rangle$, $\langle f_{n_k'}\rangle_{k\in N}$ is a subsequence of $\langle f_{n_k}\rangle_{k\in N}$ with $\limsup_k f_{n_k'} \leqslant 0$ essentially everywhere.

Suppose that condition (ii) also fails to hold. Then by a similar argument we determine a subsequence $\langle f_{n_k''}\rangle_{k\in N}$ of $\langle f_{n_k'}\rangle_{k\in N}$ such that $\liminf_k f_{n_k''} \geqslant 0$ essentially everywhere. Denoting the sequence $\langle f_{n_k''}\rangle_{k\in N}$ by $\langle f_{n_{k_j}}\rangle_{j\in N}$, we have

$$0 \leqslant \liminf_j f_{n_{k_j}} \leqslant \limsup_j f_{n_{k_j}} \leqslant 0$$

essentially everywhere.

Summarizing, if neither condition (i) nor condition (ii) holds, then every subsequence $\langle f_{n_k}\rangle_{k\in N}$ of $\langle f_n\rangle_{n\in N}$ contains a subsequence $\langle f_{n_{k_j}}\rangle_{j\in N}$ that converges to 0 essentially everywhere; i.e., $\langle f_n\rangle_{n\in N}$ converges categorically to 0.

DEFINITION Two sequences of functions $\langle f_n\rangle_{n\in N}$ and $\langle g_n\rangle_{n\in N}$ are said to be categorically equivalent if the sequence $\langle f_n - g_n\rangle_{n\in N}$ converges categorically to 0.

The notion of categorical equivalence is readily seen to be an equivalence relation between sequences of functions.

DEFINITION Let $\langle f_n\rangle_{n\in N}$ be a given sequence of real-valued Baire functions and let Γ denote the class of all sequences $\langle g_n\rangle_{n\in N}$ of real-valued Baire functions that are categorically equivalent to $\langle f_n\rangle_{n\in N}$. An extended real-valued Baire function U is called an upper categorical limit for $\langle f_n\rangle_{n\in N}$ if U is an essential infimum for the family of functions

$$\left\{\limsup_n g_n : \langle g_n\rangle_{n\in N} \in \Gamma\right\}$$

An extended real-valued Baire function L is called a lower categorical limit for $\langle f_n\rangle_{n\in N}$ if L is an essential supremum for the family of functions

$$\left\{\liminf_n g_n : \langle g_n\rangle_{n\in N} \in \Gamma\right\}$$

The existence of upper and lower categorical limits is guaranteed via Theorems 12 and 20.

THEOREM 22 If L and U are lower and upper categorical limits for a sequence $\langle f_n \rangle_{n \in \mathbb{N}}$ of real-valued Baire functions, then $L \prec U$.

Proof. Assume to the contrary that $U(x) < L(x)$ for all x belonging to some abundant set. Then there exists a sequence $\langle g_n \rangle_{n \in \mathbb{N}}$ in Γ such that $\limsup_n g_n(x) < L(x)$ for all x belonging to some abundant set and there exists a sequence $\langle h_n \rangle_{n \in \mathbb{N}}$ in Γ such that $\limsup_n g_n(x) < \liminf_n h_n(x)$ for all x belonging to an abundant set S. We have

$$S \subset \bigcup_{r \in \mathbb{Q}^+} \left\{ x \in X : \limsup_n g_n(x) + r < \liminf_n h_n(x) \right\}$$

Hence, there is a positive rational number r_0 for which the set

$$\left\{ x \in X : \limsup_n g_n(x) + r_0 < \liminf_n h_n(x) \right\}$$

is abundant. This set is expressible in the form

$$\bigcup_{n=1}^{\infty} \left\{ x \in X : \sup_{k \geq n} g_k(x) + r_0 < \inf_{k \geq n} h_k(x) \right\}$$

so that there also exists a natural number n_0 for which the set

$$T = \left\{ x \in X : \sup_{k \geq n_0} g_k(x) + r_0 < \inf_{k \geq n_0} h_k(x) \right\}$$

is abundant. For each $x \in T$ we have $h_n(x) - g_n(x) > r_0 > 0$ for all $n \geq n_0$. This implies that the sequence $\langle h_n - g_n \rangle_{n \in \mathbb{N}}$ does not converge categorically to 0, contradicting the fact that both $\langle g_n \rangle_{n \in \mathbb{N}}$ and $\langle h_n \rangle_{n \in \mathbb{N}}$ belong to Γ. We must therefore have $L \prec U$.

THEOREM 23 Assume that $\mathscr{C}$ satisfies CCC and let $\langle f_n \rangle_{n \in \mathbb{N}}$ be a sequence of real-valued Baire functions.

    (i) If $\langle f_n \rangle_{n \in \mathbb{N}}$ converges categorically to a real-valued Baire function f, then $L \approx U \approx f$.

    (ii) If the lower and upper categorical limits L, U are real-valued Baire functions and $L \approx U$, then $\langle f_n \rangle_{n \in \mathbb{N}}$ converges categorically to U.

Proof. Suppose that $\langle f_n \rangle_{n \in \mathbb{N}}$ converges categorically to f. Then $\langle f_n \rangle_{n \in \mathbb{N}}$ is categorically equivalent to the sequence $\langle g_n \rangle_{n \in \mathbb{N}}$ in Γ, where $g_n = f$ for all $n \in \mathbb{N}$. Since $\limsup_n g_n = f$, we have $U \prec f$. Since $\liminf_n g_n = f$, we also have $f \prec L$. From $U \prec f \prec L$ and the preceding theorem, we obtain $f \approx U \approx L$.

    Now suppose that $U \approx L$. Without loss of generality we may assume that $U \approx 0$. Suppose that $\langle f_n \rangle_{n \in \mathbb{N}}$ does not converge categorically to 0. Then one of the conditions (i), (ii) of Theorem 21 is satisfied. We treat only the case that condition (i) is satisfied; the other case is treated in a similar manner.

Let $\langle g_n \rangle_{n\in N}$ be any sequence in Γ that is categorically equivalent to $\langle f_n \rangle_{n\in N}$. Setting $h_n = g_n - f_n$ for each $n \in N$, we obtain a sequence $\langle h_n \rangle_{n\in N}$ of real-valued Baire functions that converges categorically to 0. Let $\langle f_{n_k} \rangle_{k\in N}$ be a subsequence of $\langle f_n \rangle_{n\in N}$, let S be an abundant Baire set, and let m be a natural number as specified in condition (i) of Theorem 21.

Because $\langle h_n \rangle_{n\in N}$ converges categorically to 0, there is a subsequence $\langle h_{n_{k_j}} \rangle_{j\in N}$ of $\langle h_{n_k} \rangle_{k\in N}$ and a meager set M such that $\langle h_{n_{k_j}} \rangle_{j\in N}$ converges to 0 for all $x \in X - M$. If $x \in S - M$, then

$$\limsup_n g_n(x) \geq \limsup_j g_{n_{k_j}}(x) = \limsup_j [f_{n_{k_j}}(x) + h_{n_{k_j}}(x)] \geq \frac{1}{m}$$

Due to the arbitrariness of the sequence $\langle g_n \rangle_{n\in N}$, it follows from the definition of U that $U \succ 1/m$ essentially everywhere in S. This contradicts $U \approx 0$.

F.   On Fréchet's theorem

If $\langle f_{m,n} \rangle_{m,n\in N}$ is a double sequence of real-valued continuous functions defined on $\mathbb{R}$ whose iterated limit

(1) $$f(x) = \lim_{m \to \infty} \left(\lim_{n \to \infty} f_{m,n}(x) \right)$$

exists for every $x \in \mathbb{R}$, then it is not necessarily true that there exists increasing sequences $\langle m_k \rangle_{k\in N}$ and $\langle n_k \rangle_{k\in N}$ of natural numbers such that

(2) $$f(x) = \lim_{k \to \infty} f_{m_k, n_k}(x)$$

for every x. For example, if

$$f_{m,n}(x) = [\cos(m!\,\pi x)]^{2n}$$

and f is the characteristic function of the set $\mathbb{Q}$ of rational numbers, then (1) holds for every $x \in \mathbb{R}$, but there do not exist sequences $\langle m_k \rangle_{k\in N}$ and $\langle n_k \rangle_{k\in N}$ such that (2) will be valid for every $x \in \mathbb{R}$.[1] However, the following theorem of Fréchet[2] is valid:

If $\langle f_{m,n} \rangle_{m,n\in N}$ is a double sequence of real-valued Lebesgue measurable functions defined on $\mathbb{R}^n$ whose iterated limit (1) exists for every $x \in \mathbb{R}^n$, then there exists a set $E \subset \mathbb{R}^n$ of Lebesgue measure zero and there exist increasing sequences $\langle m_k \rangle_{k\in N}$, $\langle n_k \rangle_{k\in N}$ of natural numbers such that (2) is valid for all $x \in \mathbb{R}^n - E$.

[1] See Kuratowski (1966, Sec. 31.X, Theorem 1).
[2] Fréchet (1906, pp. 15–16).

Briefly stated: If the iterated limit of a double sequence of Lebesgue measurable functions converges pointwise to a function f for every x, then we can always extract a single sequence from the double sequence that will converge pointwise to f for all x not belonging to a certain set of Lebesgue measure zero.

We shall discuss the Baire category analogue of Fréchet's theorem later.[1] In the present section we give a characterization for the validity of a general converse to Fréchet's theorem.

DEFINITION The limit superior of a sequence $\langle S_n \rangle_{n \in \mathbb{N}}$ of sets, denoted by $\limsup_n S_n$, is defined by

$$\limsup_n S_n = \bigcap_{k=1}^{\infty} \bigcup_{n=k}^{\infty} S_n$$

This set consists of all points that belong to infinitely many of the terms S_n of the sequence.

DEFINITION A category base $(X, \mathscr{C})$ satisfies the condition (F) if for every abundant Baire set S and every double sequence $\langle S_{p,r} \rangle_{p,r \in \mathbb{N}}$ of Baire sets satisfying

(a) $S_{p,r} \subset S_{p,r+1}$ for all $p, r \in \mathbb{N}$

(b) $S = \bigcup_{r=1}^{\infty} S_{p,r}$ for all $p \in \mathbb{N}$

(c) $S_{p_1, r_1} \supset S_{p_2, r_2}$ if $p_1 < p_2$ and $p_1 + r_1 = p_2 + r_2$

there exists a sequence $\langle r_p \rangle_{p \in \mathbb{N}}$ of natural numbers for which $\limsup_p S_{p,r_p}$ is an abundant set.[2]

The category base of all closed rectangles in $\mathbb{R}^n$ and the category base of all compact sets of positive Lebesgue measure in $\mathbb{R}^n$ satisfy the condition (F).[3]

We utilize the following lemma.[4]

LEMMA 24 If $\langle S_{p,r} \rangle_{p,r \in \mathbb{N}}$ is a double sequence of Baire sets having the properties (a) and (c) above, then there exists a double sequence $\langle T_{m,n} \rangle_{m,n \in \mathbb{N}}$ of Baire sets such that

$$S_{p,r} = \bigcup_{m=p}^{p+r-1} \bigcup_{n=p}^{p+r-1} T_{m,n} \qquad \text{for all } p, r \in \mathbb{N}$$

[1]See Chapter 5, Section III.G; see also Chapter 3, Section I.B.
[2]Wagner (1981, Definition 4).
[3]See Theorem 4 of Chapter 3, Section I, and for Lebesgue measure, Wagner (1981, p. 100).
[4]Wagner (1981, Lemma 5).

Proof. Define

$$T_{m,n} = \begin{cases} S_{n,1} & \text{for } m = n \\ S_{m,n-m+1}-(S_{m,n-m} \cup S_{m+1,n-m}) & \text{for } m < n \\ \varnothing & \text{for } m > n \end{cases}$$

for all $m,n \in \mathbb{N}$. To establish the asserted equality, we proceed by induction on r with p being arbitrary.

The equality is obviously satisfied when $r = 1$. Assume that the equality is satisfied for a given natural number r. Then

$$\bigcup_{m=p}^{p+r} \bigcup_{n=p}^{p+r} T_{m,n} = \left(\bigcup_{m=p}^{p+r-1} \bigcup_{n=p}^{p+r-1} T_{m,n} \right) \cup T_{p,p+r} \cup T_{p+1,p+r} \cup$$

$$\cdots \cup T_{p+r-1,p+r} \cup T_{p+r,p+r}$$

$$= S_{p,r} \cup [S_{p,r+1}-(S_{p,r} \cup S_{p+1,r})]$$

$$\cup [S_{p+1,r}-(S_{p+1,r-1} \cup S_{p+2,r-1})]$$

$$\cdots \cup [S_{p+r-1,2}-(S_{p+r-1,1} \cup S_{p+r,1})] \cup S_{p+r,1}$$

Now, for all $m,n \in \mathbb{N}$ it follows from the inclusions $S_{m,n} \subset S_{m,n+1}$ and $S_{m+1,n} \subset S_{m,n+1}$ that

$$S_{m,n+1} = S_{m,n} \cup [S_{m,n+1}-(S_{m,n} \cup S_{m+1,n})] \cup S_{m+1,n}$$

In particular, this equality yields

$$S_{p,r+1} = S_{p,r} \cup [S_{p,r+1}-(S_{p,r} \cup S_{p+1,r})] \cup S_{p+1,r}$$

$$S_{p+1,r} = S_{p+1,r-1} \cup [S_{p+1,r}-(S_{p+1,r-1} \cup S_{p+2,r-1})] \cup S_{p+2,r-1}$$

$$\vdots$$

$$S_{p+r-1,2} = S_{p+r-1,1} \cup [S_{p+r-1,2}-(S_{p+r-1,1} \cup S_{p+r,1})] \cup S_{p+r,1}$$

Putting these equations together and using the facts that $S_{p+1,r-1} \subset S_{p,r}, \ldots, S_{p+r-1,1} \subset S_{p,r}$, we obtain from the initial equality of this paragraph that

$$\bigcup_{m=p}^{p+r} \bigcup_{n=p}^{p+r} T_{m,n} = S_{p,r+1}$$

THEOREM 25 Let $\langle f_{m,n} \rangle_{m,n \in \mathbb{N}}$ be a double sequence of real-valued Baire functions defined on X. The convergence essentially everywhere to a function $f: X \to \mathbb{R}$ by all subsequences $\langle f_{m_k,n_k} \rangle_{k \in \mathbb{N}}$, where $\langle m_k \rangle_{k \in \mathbb{N}}$, $\langle n_k \rangle_{k \in \mathbb{N}}$ are sequences of natural numbers with $\lim_{k \to \infty} m_k = \infty$, $\lim_{k \to \infty} n_k = \infty$, implies the convergence essentially everywhere to f of the double sequence $\langle f_{m,n}\rangle_{m,n\in\mathbb{N}}$ if and only if $(X,\mathscr{C})$ satisfies the condition (F).[1]

Proof. Sufficiency: Without loss of generality, we can assume that $f = 0$ [otherwise, replace $\langle f_{m,n}\rangle_{m,n\in\mathbb{N}}$ with the double sequence $\langle f'_{m,n}\rangle_{m,n\in\mathbb{N}}$, where $f'_{m,n} = f_{m,n} - f$].

Assume that condition (F) is satisfied and suppose that the set

$$E = \{x\in X : \langle f_{m,n}(x)\rangle_{m,n\in\mathbb{N}} \text{ does not converge to } 0\}$$

is an abundant set. For each $j\in\mathbb{N}$, let E_j denote the set of all points $x\in E$ for which there exist sequences $\langle m_k\rangle_{k\in\mathbb{N}}$, $\langle n_k\rangle_{k\in\mathbb{N}}$ of natural numbers, with $\lim_{k\to\infty} m_k = \infty$, $\lim_{k\to\infty} n_k = \infty$, such that $|f_{m_k,n_k}(x)| \geq 1/j$ for every $k\in\mathbb{N}$. Then

$$E = \bigcup_{j=1}^{\infty} E_j$$

Accordingly, there exists an index j_0 such that E_{j_0} is an abundant set. Denote this set E_{j_0} by S. For $m,n\in\mathbb{N}$ define

$$T_{m,n} = \left\{x\in S : |f_{m,n}(x)| \geq \frac{1}{j_0}\right\}$$

It is readily seen that

$$S = \bigcup_{m=p}^{\infty}\bigcup_{n=p}^{\infty} T_{m,n}$$

for each $p\in\mathbb{N}$. Placing

$$S_{p,r} = \bigcup_{m=p}^{p+r-1}\bigcup_{n=p}^{p+r-1} T_{m,n}$$

we obtain a double sequence $\langle S_{p,r}\rangle_{p,r\in\mathbb{N}}$ of Baire sets having the properties (a)–(c) of the preceding definition. Since $(X,\mathscr{C})$ satisfies the condition (F), there exists a sequence $\langle r_p\rangle_{p\in\mathbb{N}}$ of natural numbers for which the set

$$R = \limsup_p S_{p,r_p}$$

is abundant. We determine a specific sequence $\langle (m_k,n_k)\rangle_{k\in\mathbb{N}}$ of pairs of natural numbers by defining (m_k,n_k) to be the kth element in the enumeration

$$(1,1),\ldots,(1, 1 + r_1 - 1), (2,1),\ldots,(2, 1 + r_1 - 1),\ldots,$$

$$(1 + r_1 - 1, 1),\ldots,(1 + r_1 - 1, 1 + r_1 - 1),$$

[1]Cf. Sierpiński (1950, Théorème 1) and Wagner (1981, Theorem 3).

$$(2,2), \dots, (2, 2 + r_2 - 1), (3,2), \dots, (3, 2 + r_2 - 1), \dots,$$

$$(2 + r_2 - 1, 2), \dots, (2 + r_2 - 1, 2 + r_2 - 1)$$

$$\vdots$$

$$(p,p), \dots, (p + r_p - 1), (p + 1, p), \dots, (p + 1, p + r_p - 1), \dots,$$

$$(p + r_p - 1, p), \dots, (p + r_p - 1, p + r_p - 1)$$

$$\vdots$$

Then $\lim_{k \to \infty} m_k = \infty$ and $\lim_{k \to \infty} n_k = \infty$. If $x \in R$, then x belongs to infinitely many of the sets S_{p,r_p} and consequently the sequence $\langle f_{m_k,n_k}(x) \rangle_{k \in \mathbb{N}}$ contains infinitely many terms whose absolute values are greater than $1/j$. Hence, the sequence $\langle f_{m_k,n_k} \rangle_{k \in \mathbb{N}}$ does not converge to 0 essentially everywhere. We have thus arrived at a contradiction.

Necessity: Assume that condition (F) is not satisfied. Then there exists an abundant Baire set S and a double sequence $\langle S_{p,r} \rangle_{p,r \in \mathbb{N}}$ of Baire sets having the foregoing properties (a)–(c) such that $\limsup_p S_{p,r_p}$ is a meager set for every sequence $\langle r_p \rangle_{p \in \mathbb{N}}$ of natural numbers. Let $\langle T_{m,n} \rangle_{m,n \in \mathbb{N}}$ be a double sequence of sets as given in Lemma 24 and let $f_{m,n} = \chi_{T_{m,n}}$ be the characteristic function of the set $T_{m,n}$ for all $m,n \in \mathbb{N}$. We shall show that for any given sequences $\langle m_k \rangle_{k \in \mathbb{N}}$, $\langle n_k \rangle_{k \in \mathbb{N}}$ of natural numbers with $\lim_{k \to \infty} m_k = \infty$, $\lim_{k \to \infty} n_k = \infty$, the sequence $\langle f_{m_k,n_k} \rangle_{k \in \mathbb{N}}$ converges essentially everywhere to 0. Since for each $x \in X$ the sequence $\langle f_{m_k,n_k}(x) \rangle_{k \in \mathbb{N}}$ does not converge to 0 if and only if $x \in \limsup_k T_{m_k,n_k}$, we have only to show $\limsup_k T_{m_k,n_k}$ is a meager set.

Let

$$M = \{m_k : k \in \mathbb{N}\} \quad \text{and} \quad N = \{n_k : k \in \mathbb{N}\}$$

We determine a sequence $\langle r_p \rangle_{p \in \mathbb{N}}$ of natural numbers by defining $r_p = 1$ if $p \notin M \cup N$, and if $p \in M \cup N$ we define

$$r_p = \max \left\{ \max_j \{m_j : n_j = p\}, \ \max_j \{n_j : m_j = p\} \right\}$$

(taking 0 as the maximum for an empty set).

Suppose now that k is any given natural number and let $p = \min\{m_k, n_k\}$. Assume that $p = m_k$. Then $p \in M \cup N$ and we have

$$p = m_k \leqslant n_k \leqslant \max_j \{n_j : m_j = p\} \leqslant r_p \leqslant p + r_p - 1$$

According to Lemma 24, this implies the inclusion

$$T_{m_k,n_k} \subset S_{p,r_p}$$

The same inclusion results if we assume that $p = n_k$. This implies that

$$\limsup_k T_{m_k,n_k} \subset \limsup_p S_{p,r_p}$$

The set $\limsup_p S_{p,r_p}$ being meager, so also is $\limsup_k T_{m_k,n_k}$. Thus, the sequence $\langle f_{m_k,n_k} \rangle_{k \in \mathbb{N}}$ converges essentially everywhere to 0.

On the other hand, the double sequence $\langle f_{m,n}(x) \rangle_{m,n \in \mathbb{N}}$ does not converge to 0 for every $x \in S$. For, suppose that $x \in S$ and let p_0 be any given natural number. From property (b), there is a natural number r_0 such that $x \in S_{p_0,r_0}$. By Lemma 24, there exist $m,n \in \mathbb{N}$ such that $m \geqslant p_0$, $n \geqslant p_0$, and $x \in T_{m,n}$. Since we can thus find arbitrarily large natural numbers m,n such that $x \in T_{m,n}$ and consequently $f_{m,n}(x) = 1$, the sequence $\langle f_{m,n}(x) \rangle_{m,n \in \mathbb{N}}$ does not converge to 0. We thereby see that the stated implication in the theorem fails to hold.

REMARK Assuming CH, one can show that the assumption that the functions $f_{m,n}$ are Baire functions is essential.[1]

---

[1] See Sierpiński (1950, Théorème 2).

2

Point-Meager and Baire Bases

I. DEFINITIONS AND BASIC PROPERTIES

A. Point-meager bases

DEFINITION A category base is called point-meager if every set consisting of a single point is meager.

This defining condition can also be given in either of the following two equivalent conditions:

(i) Every finite set is singular.
(ii) Every countable set is meager.

The family of all closed rectangles in $\mathbb{R}^n$ is a point-meager base, as is the family of all compact sets in $\mathbb{R}^n$ of positive Lebesgue measure. The family of all closed sets in $\mathbb{R}^n$ of positive Hausdorff measure μ^h for $h \in \mathcal{H}_c$ and the family of all closed sets in $\mathbb{R}^n$ of positive Hausdorff dimension furnish additional examples.

The following special result will be utilized below.

THEOREM 1 (Assume CH.) If $(X, \mathscr{C})$ is a point-meager base satisfying CCC that has power at most $2^{\aleph_0}$ and X is an abundant set, then the family of all meager $\mathcal{K}_{\delta\sigma}$-sets has power $2^{\aleph_0}$.

Proof. Because $\mathscr{C}$ has power at most $2^{\aleph_0}$, the family of all meager $\mathscr{K}_{b\sigma}$-sets also has power at most $2^{\aleph_0}$. On the other hand, Theorem 5 of Chapter 1, Section II implies that this family has power at least $\aleph_1 = 2^{\aleph_0}$.

B. Baire bases

DEFINITION A category base is a Baire base if every region is an abundant set.

In many instances[1] the proof that a given category base is a Baire base is the consequence of a general set-theoretical result.

Let $\mathscr{T}$ be an arbitrary nonempty family of nonempty subsets of a nonempty set X. A subset S of X is called singular if every set in $\mathscr{T}$ has a subset in $\mathscr{T}$ which is disjoint from S.

DEFINITION $(X,\mathscr{T})$, or briefly $\mathscr{T}$, is a complete family if there exists a sequence $\langle \phi_n \rangle_{n \in \mathbb{N}}$ of mappings from $\mathscr{T}$ to $\mathscr{T}$ having the properties
   (a) $\phi_n(A) \subset A$ for every $A \in \mathscr{T}$ and every $n \in \mathbb{N}$.
   (b) Every sequence $\langle A_n \rangle_{n \in \mathbb{N}}$ of sets in $\mathscr{T}$, for which the sequence $\langle \phi_n(A_n) \rangle_{n \in \mathbb{N}}$ is descending, has a nonempty intersection.

NOTE If $(X,\mathscr{T})$ and $(Y,\mathscr{U})$ are complete families, then the cartesian product $(X \times Y, \mathscr{T} \times \mathscr{U})$, where

$$\mathscr{T} \times \mathscr{U} = \{A \times B : A \in \mathscr{T} \text{ and } B \in \mathscr{U}\}$$

is also a complete family. We note also that a one-to-one image of a complete family is a complete family.

PROPOSITION 2 If $(X,\mathscr{T})$ is complete, then X cannot be represented as a countable union of singular sets.

Proof. Let $\langle S_n \rangle_{n \in \mathbb{N}}$ be any sequence of singular sets. Start with a given set $A \in \mathscr{T}$. Choose $A_1 \in \mathscr{T}$ so that $A_1 \subset A$ and $A_1 \cap S_1 = \varnothing$. Continuing inductively, we define $A_n \in \mathscr{T}$ so that $A_n \subset \phi_{n-1}(A_{n-1})$ and $A_n \cap S_n = \varnothing$ for all natural numbers $n > 1$. From property (a) we see that the sequence $\langle \phi_n(A_n) \rangle_{n \in \mathbb{N}}$ is descending. Hence, by property (b), there is an element of X that belongs to none of the sets S_n. We conclude that $X \neq \bigcup_{n=1}^{\infty} S_n$.

Suppose that $X = \mathbb{R}$ and $\mathscr{T}$ is the family of all closed intervals. Relative to a fixed well-ordering of $\mathscr{T}$, for each set $A \in \mathscr{T}$ and each $n \in \mathbb{N}$ define $\phi_n(A)$ to be the first subset of A that belongs to $\mathscr{T}$ and has a length $\leqslant 1/n$. Then the

---

[1]Including various notions of "completeness' discussed in Aarts and Lutzer (1974) and Haworth and McCoy (1977).

preceding proposition yields the so-called Baire Category Theorem:[1]

The real line is a set of the second category.

If we take $\mathscr{T}$ to be the family of all closed rectangles contained in a fixed rectangle and use diameter in place of length, then it follows that:

The family of all closed rectangles in $\mathbb{R}^n$ is a Baire base.

In addition, the family of all compact sets in $\mathbb{R}^n$ of positive Lebesgue measure, of all closed sets in $\mathbb{R}^n$ of positive Hausdorff measure μ^h for $h \in \mathscr{H}_c$, and of all closed sets in $\mathbb{R}^n$ of positive Hausdorff dimension are Baire bases.

For Baire bases, the converse of Theorem 5 of Chapter 1, Section III is valid.

THEOREM 3 Assuming that $(X, \mathscr{C})$ is a Baire base, the following statements are equivalent:

(i) $\mathscr{C}$ satisfies CCC.

(ii) Every family of essentially disjoint regions is countable.

(iii) Every family of essentially disjoint, abundant Baire sets is countable.

(iv) The family of all abundant Baire sets satisfies CCC.

Proof. Denoting by $\neg$ the logical negation symbol, we shall establish the implications

$$\neg \,(iv) \Rightarrow \neg \,(iii) \Rightarrow \neg \,(ii) \Rightarrow \neg \,(i) \Rightarrow \neg \,(iv)$$

Obviously, $\neg \,(iv) \Rightarrow \neg \,(iii)$.

Assume that (iii) does not hold. Then there is a transfinite sequence

$$S_1, S_2, \ldots, S_\alpha, \ldots \qquad (\alpha < \Omega)$$

of essentially disjoint, abundant Baire sets. Let

$$B_1, B_2, \ldots, B_\alpha \ldots \qquad (\alpha < \Omega)$$

be a family of regions such that S_α is abundant everywhere in B_α, for each $\alpha < \Omega$. These regions must be essentially disjoint. For, suppose that $\alpha \neq \beta$ and $B_\alpha \cap B_\beta$ is an abundant set. Let C be a region contained in $B_\alpha \cap B_\beta$. As both S_α and S_β are abundant everywhere in C, the set $(C - S_\alpha) \cup (C - S_\beta)$ is meager. But

$$C = [C \cap (S_\alpha \cap S_\beta)] \cup [C - (S_\alpha \cap S_\beta)]$$

is an abundant set. Hence, $S_\alpha \cap S_\beta$ is an abundant set, contradicting the

[1] Although commonly attributed to Baire, this theorem was given in equivalent forms prior to Baire's enunciation. First stated by P. du Bois Reymond, his "proof" was based upon the mistaken belief that all nowhere dense sets are countable. See du Bois Reymond (1882, Secs. 50, 53), Osgood (1896, p. 290; 1897, p. 173; 1900, p. 462), Baire (1899a, p. 65; 1899b, p. 948), and Young (1926, pp. 426, 428). The basic method used to prove this theorem is found in Hankel [1870 (1882, p. 90)].

assumption that S_α and S_β are essentially disjoint. The regions B_α are therefore essentially disjoint and (ii) does not hold.

Assume that (ii) does not hold. Then there exists a transfinite sequence

$$B_1, B_2, \ldots, B_\alpha, \ldots \qquad (\alpha < \Omega)$$

of essentially disjoint regions. Proceeding by transfinite induction, we define a transfinite sequence

$$A_1, A_2, \ldots, A_\alpha, \ldots \qquad (\alpha < \Omega)$$

of disjoint regions. Set $A_1 = B_1$. Assume that $1 < \alpha < \Omega$ and for all ordinal numbers $\beta < \alpha$ we have defined a region $A_\beta \subset B_\beta$ such that $\mathscr{D}_\alpha = \{A_\beta : \beta < \alpha\}$ is a disjoint family of regions. If $B_\alpha \cap (\bigcup \mathscr{D}_\alpha)$ contains a region, then, by Axiom 2a, for some ordinal number $\beta < \alpha$ the set $B_\alpha \cap A_\beta$ contains a region and consequently is an abundant set. However, this contradicts the assumption that B_α and B_β are essentially disjoint. Thus, $B_\alpha \cap (\bigcup \mathscr{D}_\alpha)$ contains no region. We now apply Axiom 2b to define A_α to be a region contained in $B_\alpha - \bigcup \mathscr{D}_\alpha$. From the existence of the transfinite sequence $\langle A_\alpha \rangle_{\alpha < \Omega}$ so determined, we conclude that (i) does not hold.

Finally, the implication $\neg$ (i) $\Rightarrow \neg$ (iv) is an immediate consequence of the fact that each region is an abundant Baire set.

Either of the conditions given in the next theorem serves as an appropriate generalization of the notion of a set in $\mathbb{R}^n$ having Lebesgue inner measure zero.

THEOREM 4 Assuming that $(X, \mathscr{C})$ is a Baire base, the following conditions are equivalent for a set S:

   (i) S contains no abundant Baire set.
   (ii) The complement of S is abundant everywhere.

Proof. Assume that condition (ii) holds and suppose T is any abundant subset of S. Then T is abundant everywhere in some region A. Since $X-S$ is abundant everywhere in A, so is $X-T$. Therefore, T is not a Baire set. Condition (i) is thus seen to hold.

Assume now that condition (ii) does not hold. Then there exists a region A such that $A \cap (X-S)$ is a meager set. From Theorem 6 of Chapter 1, Section III and the equality

$$A = (A \cap S) \cup [A \cap (X-S)]$$

it follows that $A \cap S$ is an abundant Baire set. Thus, condition (i) also fails to hold.

NOTE Only one further theorem in this chapter utilizes the notion of a Baire base in its hypothesis: Theorem 16 of Section II.

II. GENERAL PROPERTIES

A. Transfinite matrices of sets

We first establish an important result of Ulam.[1]

PROPOSITION 1 If S is a set of power $\aleph_1$, then there exists a transfinite matrix $\langle S_{n,\beta} : 1 \leqslant n < \omega, 1 \leqslant \beta < \Omega \rangle$ of subsets of S

$$S_{11}, S_{12}, \ldots, S_{1\beta}, \ldots$$

$$S_{21}, S_{22}, \ldots, S_{2\beta}, \ldots$$

$$\cdots\cdots\cdots\cdots\cdots\cdots\cdots$$

$$S_{n1}, S_{n2}, \ldots, S_{n\beta}, \ldots$$

$$\cdots\cdots\cdots\cdots\cdots\cdots\cdots$$

$$\cdots\cdots\cdots\cdots\cdots\cdots\cdots$$

with $\aleph_0$ rows and $\aleph_1$ columns satisfying the following conditions:
  (i) The sets in any given row are disjoint.
  (ii) The union of the sets in any given row is equal to S.
  (iii) The sets in any given column are disjoint.
  (iv) The union of the sets in any given column differs from S by at most a countable set; i.e., for every ordinal number $\beta < \Omega$, the set $S - \bigcup_{n=1}^{\infty} S_{n,\beta}$ is countable.

Proof. Well-order all points of the set S into a transfinite sequence

$$x_1, x_2, \ldots, x_\beta, \ldots \qquad (\beta < \Omega)$$

We consider the sets $S_{n,\beta}$ initially to be empty boxes into which we shall place the points x in such a way that each point will appear once and only once in each row and at most once in each column.

First, place each of the points $x_1, x_2, \ldots, x_n, \ldots (n < \omega)$ in all the boxes $S_{11}, S_{22}, \ldots, S_{nn}, \ldots (n < \omega)$. For $\omega \leqslant \beta < \Omega$, place the point x_β once in each row and once in each column up to the column whose index is β. This is possible because β is a denumerable ordinal number and there is a one-to one correspondence between the denumerably many rows and the denumerably many columns up to and including the βth column.

Since each point appears just once in each row, the sets in each row are disjoint and the union of each row yields S. Since each point appears at most once in each column, the sets in each column are also disjoint. Moreover, the union of each column yields the set S with the exception of at most countably many points, because all points with index larger than β will lie in the βth column.

[1] Ulam (1930, 1964b).

PROPOSITION 2 If S is a set of power $\aleph_1$ and $\mathcal{M}$ is a family of subsets of S such that for every sequence $\langle S_n \rangle_{n \in \mathbb{N}}$ of sets in $\mathcal{M}$, the difference $S - \bigcup_{n=1}^{\infty} S_n$ is uncountable, then there exists an uncountable family of disjoint subsets of S, none of which belongs to $\mathcal{M}$.

Proof. This is a consequence of the existence of the matrix given in Proposition 1 for a set of power $\aleph_1$. In each column there must be a set that does not belong to $\mathcal{M}$. Therefore, there is at least one row in which there are uncountably many sets that do not belong to $\mathcal{M}$.

Proposition 2 can also be formulated in the following manner.[1]

PROPOSITION 3 If all subsets of a set S of power $\aleph_1$ are partitioned into two classes $\mathcal{M}$ and $\mathcal{N}$, with $\mathcal{N}$ satisfying CCC, then there exists a sequence $\langle S_n \rangle_{n \in \mathbb{N}}$ of sets in $\mathcal{M}$ such that $S - \bigcup_{n=1}^{\infty} S_n$ is countable.

THEOREM 4 Assume that $(X, \mathscr{C})$ is a point-meager base. If $\mathscr{C}$ satisfies CCC and $\langle S_{n,\beta} : 1 \leq n < \omega, 1 \leq \beta < \Omega \rangle$ is a transfinite matrix of Baire subsets of an abundant set S such that the sets in each row n are disjoint, then there exists an ordinal number $\beta < \Omega$ such that $S - \bigcup_{n=1}^{\infty} S_{n,\beta}$ is uncountable.

Proof. Assume to the contrary that for every ordinal number $\beta < \Omega$, the set $S - \bigcup_{n=1}^{\infty} S_{n,\beta}$ is countable. The set S being abundant, there exists at least one abundant set in each column β. Hence, in some row n there are uncountably many, disjoint abundant Baire sets. This contradicts Theorem 5 of Chapter 1, Section III.

In connection with Theorem 4, we note that if $\mathscr{C}$ satisfies CCC and $\langle S_{\alpha,\beta} : 1 \leq \alpha < \Omega, 1 \leq \beta < \Omega \rangle$ is a transfinite matrix of Baire subsets of an abundant set S such that the sets in each row α are disjoint, then it is not necessarily true that there exists an index $\beta < \Omega$ for which the set $S - \bigcup_{1 \leq \alpha < \Omega} S_{\alpha,\beta}$ is uncountable.

For example, let X be the set of all countable ordinal numbers and let $\mathscr{C}$ be the cocountable base for X. Then $(X, \mathscr{C})$ is a point-meager category base satisfying CCC. Define by transfinite induction singleton sets $S_{\alpha,\beta} = \{x_{\alpha,\beta}\}$ such that each point of X occurs exactly once in each row and exactly once in each column. First, define $x_{\alpha,1} = \alpha$ and $x_{1,\beta} = \beta$ for all countable ordinal numbers α, β. For $1 < \alpha < \Omega$ and $1 < \beta < \Omega$, define $x_{\alpha,\beta}$ to be the smallest ordinal number that differs from all the numbers $x_{\alpha,\gamma}$ with $\gamma < \beta$ and from all the numbers $x_{\xi,\beta}$ with $\xi < \alpha$. The set $S = X$ is an abundant set for which $S - \bigcup_{1 \leq \alpha < \Omega} S_{\alpha,\beta} = \varnothing$ for every ordinal number $\beta < \Omega$.

---

[1]Ulam (1930), Sierpiński (1933e), and Tarski (1938).

On the other hand, we have the following result.[1]

THEOREM 5 Assume that $(X,\mathscr{C})$ is a point-meager base. If $\mathscr{C}$ satisfies CCC and $\langle S_{\alpha,\beta}: 1 \leqslant \alpha < \Omega, 1 \leqslant \beta < \Omega \rangle$ is a transfinite matrix of Baire subsets of an abundant set S such that the sets in each row are disjoint, then there exists an increasing function ϕ mapping the set of countable ordinal numbers into itself such that the set $S - \bigcup_{1 \leqslant \alpha < \Omega} S_{\alpha,\phi(\alpha)}$ is uncountable.

Proof. We define by transfinite induction the function ϕ and a transfinite sequence $\langle x_{\alpha} \rangle_{\alpha < \Omega}$ of distinct points of S.

As the sets $S_{1,\beta}$, $1 \leqslant \beta < \Omega$, in the first row are disjoint, there is an index β for which $S_{1,\beta}$ is a meager set. Define $\phi(1)$ to be the smallest such index β and let x_1 be any point of $S - S_{1,\phi(1)}$. Assume that $1 < \alpha < \Omega$ and we have already defined $\phi(\xi)$ and x_{ξ} for all ordinal numbers $\xi < \alpha$, so that ϕ is an increasing function, the sets $S_{\xi,\phi(\xi)}$ are meager, the points x_{ξ} are distinct elements of S, and $x_{\xi} \notin \bigcup_{\gamma < \xi} S_{\gamma,\phi(\gamma)}$. In the αth row there are uncountably many indices β for which $S_{\alpha,\beta}$ is a meager set and $x_{\xi} \notin S_{\alpha,\beta}$ for all ordinal numbers $\xi < \alpha$. We then define $\phi(\alpha)$ to be the smallest such index β which is larger than all the ordinal numbers $\phi(\xi)$ with $\xi < \alpha$. Subsequently, we define x_{α} to be any point of the abundant set

$$ S - \left[\left(\bigcup_{\xi \leqslant \alpha} S_{\xi,\phi(\xi)} \right) \cup \{ x_{\xi} : \xi < \alpha \} \right] $$

The set $T = \{ x_{\alpha} : 1 \leqslant \alpha < \Omega \}$ will then be an uncountable set with $T \cap (\bigcup_{1 \leqslant \alpha < \Omega} S_{\alpha,\phi(\alpha)}) = \emptyset$, so that $T \subset S - \bigcup_{1 \leqslant \alpha < \Omega} S_{\alpha,\phi(\alpha)}$.

REMARK Assuming CH, it has been shown that Theorem 5 fails to hold if the assumption that the sets $S_{\alpha,\beta}$ are Baire sets is deleted.[2]

B. Existence of non-Baire sets

Combining Proposition 1 and Theorem 4, we obtain the following existence theorem, which, as we shall demonstrate, is also a consequence of Proposition 2.

THEOREM 6 Assume that $(X,\mathscr{C})$ is a point-meager base. If $\mathscr{C}$ satisfies CCC, then every abundant set of power $\aleph_1$ contains a set that is not a Baire set.[3]

---

[1]Kuratowski (1966, p. 416).

[2]Cf. Kuratowski (1966, p. 417), Braun and Sierpiński [1932, Proposition (Q)], and Sierpiński (1956, Chap. I).

[3]Ulam (1933).

Proof. Let S be an abundant subset of X of power $\aleph_1$ and let $\mathscr{M}$ be the family of all subsets of S that are either meager or non-Baire sets. If every subset of S is a Baire set, then $\mathscr{M}$ consists exclusively of meager sets and applying Proposition 2, there must exist uncountably many disjoint, abundant Baire sets contained in S. However, by virtue of Theorem 5 of Chapter 1, Section III, $\mathscr{C}$ cannot satisfy CCC. Therefore, there exists a subset of S that is not a Baire set.

COROLLARY 7 Assume $(X,\mathscr{C})$ is a point-meager base. If $\mathscr{C}$ satisfies CCC, then a necessary and sufficient condition that a set S of power $\aleph_1$ be a meager set is that every subset of S be a Baire set.[1]

We note that one can also obtain from Theorem 6 the solution of Banach and Kuratowski to the generalized measure problem:[2] If X is any set of power $\aleph_1$, then there does not exist any real-valued function $\mu(S)$ defined for all sets $S \subset X$ with the following properties:

(i) $\mu(X) \neq 0$.
(ii) For any sequence $\langle S_n \rangle_{n \in \mathbb{N}}$ of disjoint subsets of X,

$$\mu\left(\bigcup_{n=1}^{\infty} S_n \right) = \sum_{n=1}^{\infty} \mu(S_n)$$

(iii) $\mu(\{x\}) = 0$ for every $x \in X$.

C. Restriction

DEFINITION A set R is a restricted Baire set relative to a set P if $R \cap P = E \cap P$ for some Baire set E.[3]

THEOREM 8 Assume that $(X,\mathscr{C})$ is a point-meager base. If $\mathscr{C}$ satisfies CCC, then every abundant set S of power $\aleph_1$ contains a set that is not a restricted Baire set relative to S.[4]

Proof. Apply Theorem 6 to the category base $(S,\mathscr{C}^*)$ given in Theorem 12 in Chapter 1, Section 3.

COROLLARY 9 Assume that $(X,\mathscr{C})$ is a point-meager base. If $\mathscr{C}$ satisfies CCC, then a necessary and sufficient condition that a set S of power $\aleph_1$ be meager is that every subset of S be a restricted Baire set relative to S.

[1]See also Theorem 13 of Chapter 5, Section II.
[2]See Banach and Kuratowski (1929), Sierpiński (1929c; 1956, pp. 107–109), Braun and Sierpiński (1932), and Sierpiński and Szpilrajn (1936b).
[3]Cf. Hausdorff (1914, p. 415).
[4]See Eilenberg (1932) and Sierpiński (1934c, pp. 133–134; 1956, Chap. IV, Sec. 3). See also Theorem 9 of Chapter 3, Section I.

DEFINITION   For $S \subset X$, a function $f: S \to \mathbb{R}^*$ is called a restricted Baire function relative to S if for every element $a \in \mathbb{R}$ the set $\{x \in S: f(x) \leqslant a\}$ is a restricted Baire set relative to S.

Other equivalent conditions utilized below are obtainable from Theorems 1 and 7 of Chapter 1, Section IV.

It is easy to see that the restriction f_S of a Baire function $f: X \to \mathbb{R}^*$ is a restricted Baire function relative to S. In fact, the restricted Baire functions relative to a set S coincide with the restrictions of Baire functions defined on X. This is established by showing any restricted Baire function can be extended to a Baire function.[1]

THEOREM 10   If $S \subset X$ and $f: S \to \mathbb{R}^*$ is a restricted Baire function relative to S, then there exists a Baire function $\hat{f}: X \to \mathbb{R}^*$ such that $\hat{f}(x) = f(x)$ for every $x \in S$.

Proof. The function f being a restricted Baire function relative to S, for each $r \in \mathbb{R}$ there exists a Baire set E_r such that

$$\{x \in S: f(x) \geqslant r\} = S \cap E_r$$

Define the function $\hat{f}$ for each $x \in X$ by

$$\hat{f}(x) = \sup\{r \in \mathbb{Q}: x \in E_r\}$$

For $x \in S$ we have $x \in E_r$ if and only if $f(x) \geqslant r$, whence

$$\hat{f}(x) = \sup\{r \in \mathbb{Q}: f(x) \geqslant r\} = f(x)$$

For each $x \in \mathbb{R}$ we have

$$\{x \in X: \hat{f}(x) > a\} = \bigcup \{E_r: r \in \mathbb{Q} \text{ and } r > a\}$$

The sets E_r being Baire sets, this implies that $\hat{f}$ is a Baire function.

THEOREM 11   The following properties are equivalent for a set S:[2]
  (i) Every subset of S is a restricted Baire set relative to S.
  (ii) Every function $f: S \to \mathbb{R}$ can be extended to a Baire function $\hat{f}: X \to \mathbb{R}^*$.

Proof. (i) $\Rightarrow$ (ii). If every subset of S is a restricted Baire set relative to S, then every function $f: S \to \mathbb{R}$ is a restricted Baire function relative to S, so (ii) is a consequence of Theorem 10.

---

[1]Eilenberg (1932, Théorème 1).
[2]See Eilenberg (1932, Théorème 2).

(ii) $\Rightarrow$ (i). Let T be any subset of S. Define $f: S \to \mathbb{R}$ by

$$f(x) = \begin{cases} 1 & \text{if } x \in T \\ 0 & \text{if } x \in S-T \end{cases}$$

and let $\hat{f}: X \to \mathbb{R}^*$ be a Baire function such that $\hat{f}(x) = f(x)$ for all $x \in S$. Then $E = \{x \in X : \hat{f}(x) > 0\}$ is a Baire set. By virtue of the definitions of f and $\hat{f}$ we have $T = E \cap S$. Thus, T is a restricted Baire set relative to S.

Combining Theorems 8 and 11, we obtain[1]

THEOREM 12 Assume that $(X, \mathscr{C})$ is a point-meager base. If $\mathscr{C}$ satisfies CCC, then for every abundant set S of power $\aleph_1$ there exists a function $f: S \to \mathbb{R}$ that cannot be extended to a Baire function defined on X.

D. Decomposition theorems

Utilizing Proposition 2, we first derive some decomposition theorems of Ulam[2] for sets of power $\aleph_1$.

THEOREM 13 Assume that $(X, \mathscr{C})$ is a point-meager base. Every abundant set of power $\aleph_1$ can be decomposed into an uncountable family of disjoint abundant sets.

Proof. Let S be an abundant set of power $\aleph_1$ and let $\mathscr{M}$ be the family of all meager subsets of S. By Proposition 2, there exists an uncountable family of disjoint subsets of S each of which is abundant. Upon adjoining to one of these subsets all points of S that belong to none of these subsets, the desired decomposition is obtained.

THEOREM 14 Assume that $(X, \mathscr{C})$ is a point-meager base. Every set of power $\aleph_1$ that is not a Baire set can be decomposed into an uncountable family of sets none of which is a Baire set.

Proof. Let S be a non-Baire set of power $\aleph_1$ and let $\mathscr{M}$ be the family of all Baire subsets of S. By Proposition 2, there exists an uncountable family of disjoint non-Baire subsets of S. If the set T of all points of S that belong to none of these subsets is not a Baire set, then we already have the desired decomposition. Otherwise, we adjoin all the points of T to one of the given subsets of S to obtain the decomposition.

[1]Cf. Eilenberg (1932) and Sierpiński (1956, Chap. IV, Sec. 3).
[2]Cf. Ulam (1931, 1933).

THEOREM 15 Assume that $(X, \mathscr{C})$ is a point-meager base. If $\mathscr{C}$ satisfies CCC, then every abundant set of power $\aleph_1$ can be decomposed into an uncountable family of disjoint sets, none of which is a Baire set.

Proof. Let S be an abundant set of power $\aleph_1$. If S is not a Baire set, then the conclusion is an immediate consequence of Theorem 14. Thus, we assume that S is a Baire set.

By Theorem 6, there exists a non-Baire subset T of S. Applying Theorem 14, we decompose T into an uncountable family of disjoint sets none of which is a Baire set. In view of the equality $T = S—(S—T)$ and Theorem 6 of Chapter 1, Section III, the set $S—T$ is not a Baire set. Adjoining this set to the given family, we obtain the desired decomposition.

We now give Grzegorek's unification of the following two theorems of Sierpiński:[1]

(τ) If $2^{\aleph_0} = \aleph_1$, then every subset of $\mathbb{R}$ which is of the second category in every interval contains an uncountable family of disjoint sets each of which is of the second category in every interval.

(μ) If $2^{\aleph_0} = \aleph_1$, then every subset of $\mathbb{R}$ with positive Lebesgue outer measure contains an uncountable family of disjoint sets each of which has the same measure as the given set.

THEOREM 16 Assume that $(X, \mathscr{C})$ is a point-meager, Baire base. If $\mathscr{C}$ satisfies CCC and every region has power $\aleph_1$, then every abundant set S can be decomposed into an uncountable family of disjoint sets each of which is abundant in every region in which S is abundant.

Proof. According to Theorem 12 of Chapter 1, Section III,

$$\mathscr{C}^* = [\mathfrak{B}(\mathscr{C}) \cap S]—[\mathfrak{M}(\mathscr{C}) \cap S]$$

is a category base satisfying CCC with

$$\mathfrak{B}(\mathscr{C}^*) = \mathfrak{B}(\mathscr{C}) \cap S \qquad \mathfrak{M}(\mathscr{C}^*) = \mathfrak{M}(\mathscr{C}) \cap S$$

We first prove that every $\mathscr{C}^*$-region A contains a $\mathscr{C}^*$-region B satisfying the condition

(δ) There exists an uncountable family $\{S_{B,\alpha} : \alpha < \Omega\}$ of disjoint subsets of B each of which is $\mathscr{C}^*$-abundant everywhere in B.

By Theorem 13, the $\mathscr{C}^*$-region A can be decomposed into an uncountable family $\mathscr{R} = \{Q_\xi : \xi < \Omega\}$ consisting of $\mathscr{C}^*$-abundant sets. According to Theorem 19 of Chapter 1, Section III, there is an asymptotic hull E for $\langle Q_\xi \rangle_{\xi < \Omega}$. From the definition of an asymptotic hull, it is a simple matter to

[1]See Sierpiński (1934a, 1934c), Poprużenko (1957), Itzkowitz (1974), and Grzegorek (1979). See also Chapter 3, Section I.D.

determine a collection $\{\mathscr{P}_\alpha : \alpha < \Omega\}$ consisting of disjoint, denumerable subfamilies $\mathscr{P}_\alpha$ of $\mathscr{R}$ such that E is essentially a $\mathscr{C}^*$-hull for the set $P_\alpha = \bigcup \mathscr{P}_\alpha$ for every $\alpha < \Omega$.

Being $\mathscr{C}^*$-abundant in A, the set E is $\mathscr{C}^*$-abundant everywhere in some $\mathscr{C}^*$-region $B \subset A$. Setting

$$S_{B,\alpha} = B \cap P_\alpha$$

for each $\alpha < \Omega$, we obtain a family of disjoint subsets of B. It remains only to show each of the sets $S_{B,\alpha}$ is $\mathscr{C}^*$-abundant everywhere in B.

Suppose that C is any $\mathscr{C}^*$-subregion of B. Assume that $C \cap S_{B,\alpha}$ is a $\mathscr{C}^*$-meager set. From the equality

$$S_{B,\alpha} - (E - C) = (S_{B,\alpha} - E) \cup (C \cap S_{B,\alpha})$$

and the fact that $B \cap E$ is essentially a $\mathscr{C}^*$-hull for $S_{B,\alpha}$ it follows that the set $S_{B,\alpha} - (E - C)$ is $\mathscr{C}^*$-meager. The set $F = E - C$ is an element of $\mathfrak{B}(\mathscr{C}^*)$ such that $S_{B,\alpha} - F$ is $\mathscr{C}^*$-meager. From the definition of an essential $\mathscr{C}^*$-hull for $S_{B,\alpha}$ we see that $E - F$ must be a $\mathscr{C}^*$-meager set. But this is impossible, since E is $\mathscr{C}^*$-abundant in C and

$$E - F = E - (E - C) = E \cap C$$

Thus, $C \cap S_{B,\alpha}$ must be a $\mathscr{C}^*$-abundant set for every $\mathscr{C}^*$-region $C \subset B$ and for every $\alpha < \Omega$.

Having thus established that each $\mathscr{C}^*$-region A contains a $\mathscr{C}^*$-region B satisfying the condition (δ), we apply Lemmas 3 and 4 of Chapter 1, Section II, to obtain a countable family $\mathscr{M}$ of disjoint $\mathscr{C}^*$-regions B satisfying condition (δ) such that $S - \bigcup \mathscr{M}$ is $\mathscr{C}^*$-meager. Defining

$$S_\alpha = \bigcup \{S_{B,\alpha} : B \in \mathscr{M}\}$$

for each $\alpha < \Omega$, we obtain an uncountable family of disjoint subsets S_α of S each of which is $\mathscr{C}^*$-abundant everywhere and hence $\mathscr{C}$-abundant in every region in which S is abundant. Upon adjoining the set $S - \bigcup \mathscr{M}$ to one of the sets S_α, we obtain the desired decomposition of S.

Theorems 6, 8, 12, 13, 14, 15, and 16 are all based on Proposition 2, or its equivalent form, Proposition 3. The latter proposition has been shown by Sierpiński to hold for any set of power $\mathfrak{m} > \aleph_0$, provided that there exists no (weakly) inaccessible number less than or equal to $\mathfrak{m}$. Accordingly, these theorems extend to sets of such powers.[1] In addition, Theorem 16 can be strengthened.[2]

[1]See Ulam (1933), Sierpiński (1933e; 1934a; 1934c; 1956, Chap. V), Tarski (1938), Popruzenko (1957), and Grzegorek (1979).
[2]See Grzegorek (1979).

E. Quotient algebras

THEOREM 17 Assume CH. Also, assume that $(X,\mathscr{C})$ is a point-meager base and X is an abundant set of power $2^{\aleph_0}$. If $\mathscr{C}$ satisfies CCC and has power at most $2^{\aleph_0}$, then the quotient algebra of all subsets of X modulo the ideal of meager sets is not complete.[1]

Proof. According to Theorem 13, X can be decomposed into a family $\mathscr{S} = \{S_\alpha : \alpha \in I\}$ consisting of $2^{\aleph_0}$ disjoint abundant sets. We show that $\mathscr{S}$ has no least upper bound with respect to the relation $\prec$. Assuming that E is any set such that $S_\alpha \prec E$ for every $\alpha \in I$, we establish the existence of a set F such that $S_\alpha \prec F$ for all $\alpha \in I$, $F \prec E$, and $F \not\approx E$.

By virtue of Theorem 1 of Section I, the family of all meager $\mathscr{K}_{\delta\sigma}$-sets can be written in the form $\{R_\alpha : \alpha \in I\}$. The sets $S_\alpha - R_\alpha$ being abundant and $S_\alpha - E$ meager, the sets

$$T_\alpha = (S_\alpha - R_\alpha) - (S_\alpha - E)$$

are nonempty. These sets T_α are also disjoint. Choose a single point $x_\alpha \in T_\alpha$ for each $\alpha \in I$ and place

$$Q = \{x_\alpha : \alpha \in I\}$$
$$F = E - Q$$

From the equality

$$S_\alpha - F = (S_\alpha - E) \cup \{x_\alpha\}$$

we obtain $S_\alpha \prec F$ for every $\alpha \in I$. We obviously have $F \prec E$. However, $Q = E - F$ is not a meager set. For, if Q were meager, then, by Theorem 5 of Chapter 1, Section II, $Q \subset R_\alpha$ for some $\alpha \in I$. But this is impossible, since $x_\alpha \in Q - R_\alpha$. Therefore, $E \not\approx F$.

III. RARE SETS

A. Definition and existence

DEFINITION A set is a rare set if every uncountable subset of the set is abundant.

Equivalently, a set is a rare set if it has at most countably many points in common with each meager set.

---

[1]Cf. Sikorski (1949; 1964, pp. 78–79) and Halmos (1963, Sec. 25). See also Theorem 27 of Chapter 5, Section II.

It is clear that the family of all rare sets forms a σ-ideal that contains all countable sets. If $\mathscr{C}$ is the cocountable base for an uncountable set X, then the meager sets coincide with the countable sets and, accordingly, every subset of X is a rare set.

Concerning the general existence of uncountable rare sets, we have[1]

THEOREM 1 The following condition is equivalent to the Continuum Hypothesis:

(R) If $(X,\mathscr{C})$ is a point-meager category base satisfying CCC and $\mathscr{C}$ has power at most $2^{\aleph_0}$, then every abundant set contains a rare set of power $2^{\aleph_0}$.

Proof. $(CH) \Rightarrow (R)$.

Assume that S is an abundant set and let

(1) $$x_1, x_2, \ldots, x_\alpha, \ldots \qquad (\alpha < \Theta)$$

be a well-ordering of S, where Θ is the smallest ordinal number whose power is the same as the power of S. In addition, let

(2) $$R_1, R_2, \ldots, R_\alpha, \ldots \qquad (\alpha < \Omega)$$

be a well-ordering of all meager $\mathscr{X}_{\delta\sigma}$-sets (cf. Theorem 1 of Section I). We define an uncountable rare subset T of S by transfinite induction.

Define $y_1 = x_1$. Assume that $1 < \alpha < \Omega$ and the points y_β have already been defined for all ordinal numbers $\beta < \alpha$. The set

$$S_\alpha = \left(\bigcup_{\beta < \alpha} R_\beta \right) \cup \{ y_\beta : \beta < \alpha \}$$

being a meager set, $S - S_\alpha$ is nonempty. We then define y_α to be the first element of the sequence (1) that does not belong to S_α. The set

$$T = \{ y_\alpha : 1 \leqslant \alpha < \Omega \}$$

will then be a subset of S with power $\aleph_1$.

Suppose that U is any meager subset of T. By virtue of Theorem 5 of Chapter I, Section II, U is a subset of some meager $\mathscr{X}_{\delta\sigma}$-set R. We have $R = R_\gamma$, where R_γ is a set occurring in the enumeration (2). For all ordinal numbers $\alpha > \gamma$, the point y_α does not belong to R_γ and hence does not belong to U. Therefore, U must be a countable set. It follows that every uncountable subset of T is an abundant set.

$(R) \Rightarrow (CH)$.

---

[1] Cf. Sierpiński (1933a; 1956, Propositions P_8 and $P_{8'}$) and van Douwen, Tall, and Weiss (1977).

Assume that the condition (R) holds, but that $2^{\aleph_0} > \aleph_1$. Then $\aleph_2 \leqslant 2^{\aleph_0}$. Let X denote the set of all ordinal numbers less than the first ordinal number of power $\aleph_2$ and let $\mathscr{C}$ be the family of all subsets of X of the form $\{x \in X : x \geqslant x_0\}$ for some element $x_0 \in X$. Being closed under finite intersections, $(X, \mathscr{C})$ is a category base. As is easily verified, $\mathscr{C}$ satisfies CCC and has power $\aleph_2 \leqslant 2^{\aleph_0}$. For this category base the singular sets coincide with the subsets of X that are bounded above by some element of X and are the same as the meager sets. This implies that $(X, \mathscr{C})$ is a point-meager base.

According to condition (R), there exists a subset S of X, with power $2^{\aleph_0}$, every uncountable subset of which is abundant. But any subset of S of power $\aleph_1$ is bounded above in X and must thus be a meager set. Hence, we cannot have $2^{\aleph_0} > \aleph_1$. We conclude that $2^{\aleph_0} = \aleph_1$.

The following consequence of Theorem 1 is useful for applications.[1]

THEOREM 2   Let X be a set of power $2^{\aleph_0}$ and let $\mathscr{E}$ be a family of subsets of X satisfying
  (i) X is the union of the entire family $\mathscr{E}$.
  (ii) X is not the union of any countable subfamily of $\mathscr{E}$.
  (iii) $\mathscr{E}$ has power $2^{\aleph_0}$.
If $2^{\aleph_0} = \aleph_1$, then there exists an uncountable subset of X which has at most countably many points in common with each set in $\mathscr{E}$.

Proof. Let $\mathscr{C}$ be the family of all subsets of X whose complement is an $\mathscr{E}_\sigma$-set. Then $(X, \mathscr{C})$ is a category base that is closed under finite intersections. For this base the meager sets coincide with the sets that are subsets of $\mathscr{E}_\sigma$-sets. By condition (i), every countable set is a subset of an $\mathscr{E}_\sigma$-set, whence $\mathscr{C}$ is a point-meager base. From condition (ii) we see that every set in $\mathscr{C}$ is nonempty and, being closed under finite intersections, this implies that $\mathscr{C}$ satisfies CCC. Condition (ii) also implies that X is an abundant set. From condition (iii) we see that the family $\mathscr{C}$ has power $2^{\aleph_0}$. The conclusion now follows from Theorem 1.

B.   Mahlo-Luzin sets

In the case that $\mathscr{C}$ consists of all closed rectangles in $\mathbb{R}^n$, Theorem 1 yields

THEOREM 3   If $2^{\aleph_0} = \aleph_1$, then there exists an uncountable set in $\mathbb{R}^n$ that has at most countably many points in common with each set of the first category.

[1]Tumarkin (1971).

In view of the fact that every set in $\mathbb{R}^n$ of the first category is contained in a countable union of nowhere dense, closed sets, Theorem 3 is equivalent to the statement that the Continuum Hypothesis implies the existence of an uncountable set $S \subset \mathbb{R}^n$ having the property

(L) Each nowhere dense, closed set has at most countably many points in common with S.

The existence of an uncountable set having this property was first established, using CH, by Mahlo and shortly thereafter by Luzin. Accordingly, uncountable rare sets for this category base are called Mahlo-Luzin sets.[1]

We now state three theorems, which give properties that are equivalent to the existence of a Mahlo-Luzin set having the power of the continuum, whose proofs are found in the cited references.

DEFINITION A set $M \subset \mathbb{N}^F$ is called a complete set if for every $n \in \mathbb{N}$ and every $\sigma \in \mathbb{N}^n$ there exists a natural number $p \geq n$ and there exists $\tau \in \mathbb{N}^p$ such that $\tau \in M$ and $\tau | n = \sigma$.

Briefly stated, $M \subset \mathbb{N}^F$ is complete if every $\sigma \in \mathbb{N}^F$ is the initial part of some $\tau \in M$.

THEOREM 4 The existence of a Mahlo-Luzin set in $\mathbb{R}$ having the power of the continuum is equivalent to the existence of a set S of power of the continuum and a double sequence $\langle S_{m,n} \rangle_{m,n \in \mathbb{N}}$ of subsets of S with the following properties:
    (i) For each $m \in \mathbb{N}$,

$$S = \bigcup_{n=1}^{\infty} S_{m,n}$$

   (ii) For all $m,n,n' \in \mathbb{N}$ with $n \neq n'$,

$$S_{m,n} \cap S_{m,n'} = \varnothing$$

   (iii) For every complete set M, the set

$$S - \bigcup_{\langle \mu_1, \ldots, \mu_n \rangle \in M} \bigcup_{m=1}^{n} S_{m,\mu_m}$$

is countable.[2]

THEOREM 5 The existence of a Mahlo-Luzin set in $\mathbb{R}$ of power of the continuum is equivalent to the existence of a denumerable field $\mathscr{A}$ of sets with

---

[1] Mahlo (1913, Aufgabe 5, pp. 294–295) and Luzin (1914, Théorème I).
[2] Kuratowski (1934) and Sierpiński (1956, Chap. II, Sec. 8).

the following properties:
  (i) There is a family $\mathcal{M}$ of power of the continuum consisting of disjoint sets belonging to the family $\mathcal{A}_\delta$.
  (ii) Every set in the family $\mathcal{A}_\delta$ differs from some set in the family $\mathcal{A}_\sigma$ by at most a countable set.[1]

THEOREM 6 The existence of a Mahlo-Luzin set in $\mathbb{R}$ having the power of the continuum is equivalent to the existence of a determinant system $\{S_\sigma : \sigma \in \mathbb{N}^F\}$ of nonempty sets having the following properties:
  (i) For each sequence $\mu \in \mathbb{Z}$ and each $n \in \mathbb{N}$,

$$S_{\mu|n+1} \subset S_{\mu|n}$$

  (ii) If $n \in \mathbb{N}$, $\sigma = \langle \mu_1, \ldots, \mu_n \rangle \in \mathbb{N}^n$, $\mu'_n \in \mathbb{N}$, and $\mu'_n \neq \mu_n$, then

$$S_{\mu_1,\ldots,\mu_n} \cap S_{\mu_1,\ldots,\mu'_n} = \varnothing$$

  (iii) For every sequence $\mu \in \mathbb{Z}$ the set $\bigcap_{n=1}^\infty S_{\mu|n}$ contains at most one point.
  (iv) The nucleus

$$S = \bigcup_{\mu \in \mathbb{Z}} \bigcap_{n=1}^\infty S_{\mu|n}$$

  of the determinant system $\{S_\sigma : \sigma \in \mathbb{N}^F\}$ has the power of the continuum.
  (v) If T is a union of sets of the determinant system that has at least one point in common with each set of the determinant system, then the set $S - T$ is countable.[2]

C. Sierpiński sets

When $X = \mathbb{R}^n$ and $\mathscr{C}$ consists of all compact sets of positive Lebesgue measure, Theorem 1 yields the following result of Sierpiński.[3]

THEOREM 7 If $2^{\aleph_0} = \aleph_1$, then there exists an uncountable set in $\mathbb{R}^n$ that has at most countably many points in common with each set of Lebesgue measure zero.

An uncountable rare set for this category base is called a Sierpiński set. A statement equivalent to the existence of a Sierpiński set having the power of the continuum is given in the next theorem.[4]

[1]Sierpiński (1937d).
[2]Sierpiński (1937e).
[3]Sierpiński (1924b; 1956, Proposition C_{26}). We note that Theorem 7 is also a consequence of Theorem 2.
[4]Sierpiński (1937f).

THEOREM 8   The existence of a Sierpiński set in $\mathbb{R}$ of power of the continuum is equivalent to the existence of a set S having the power of the continuum and a family $\{S_\sigma : \sigma \in \mathbb{B}^F\}$ of subsets of S having the properties:

(i) For each $n \in \mathbb{N}$,

$$S = \bigcup_{\sigma \in \mathbb{B}^n} S_\sigma$$

(ii) For each $n \in \mathbb{N}$ and each $\sigma \in \mathbb{B}^n$,

$$S_{\sigma 0} \cup S_{\sigma 1} \subset S_\sigma$$

(iii) If $n \in \mathbb{N}$, $\sigma, \tau \in \mathbb{B}^n$, and $\sigma \neq \tau$, then

$$S_\sigma \cap S_\tau = \varnothing$$

(iv) For each $\mu \in \mathbb{B}^\infty$ the set $\bigcap_{n=1}^\infty S_{\mu|n}$ contains at most one point.

(v) If $T \subset S$ and there exists for every number $\varepsilon > 0$ a sequence $\langle \sigma_k \rangle_{k \in \mathbb{N}}$, with $\sigma_k = \langle \mu_1^k, \ldots, \mu_{n_k}^k \rangle \in \mathbb{B}^F$ for every $k \in \mathbb{N}$, such that

$$T \subset \bigcup_{k=1}^\infty S_{\sigma_k} \quad \text{and} \quad \sum_{k=1}^\infty \frac{1}{2^{n_k}} < \varepsilon$$

then T is a countable set.

NOTE   The simultaneous existence of a Mahlo-Luzin set of power of the continuum and a Sierpiński set of the power of the continuum is equivalent to the Continuum Hypothesis.[1]

D.   Concentrated sets

A set $S \subset \mathbb{R}^n$ is said to be concentrated on a set $D \subset \mathbb{R}^n$ if for every open set G containing D, the set $S-G$ is countable. For an everywhere dense countable set D we have the following result:[2]

THEOREM 9   If $2^{\aleph_0} = \aleph_1$, then there exists an uncountable set S that is concentrated on D.

Proof.[3]   Take $X = \mathbb{R}^n - D$, take $\mathscr{E}$ to be the family of all nowhere dense closed sets disjoint from D, and let S denote the uncountable set whose existence is established by Theorem 2. If G is any open set containing D, then the complement of G is a nowhere dense closed set disjoint from D and consequently $S-G$ is a countable set.

[1] Rothberger (1938); see also Sierpiński (1956) and Bagemihl (1959b).
[2] Besicovitch (1934).
[3] Proof of Dieudonné given in Rogers (1970, pp. 73–77).

THEOREM 10 The existence of a linear set having the power of the continuum that is concentrated on a denumerable set is equivalent to the existence of a pointwise convergent sequence of functions of a real variable that does not converge uniformly on any uncountable set.[1]

E. Invariant rare sets

THEOREM 11 Assume CH. Also assume that $(X,\mathscr{C})$ is a point-meager base and Φ is a countable group of one-to-one mappings of X onto itself. If $\mathscr{C}$ satisfies CCC, has power at most $2^{\aleph_0}$, and is Φ-invariant, then every Φ-invariant abundant set contains a Φ-invariant rare set of power of the continuum.

Proof. Let S be any Φ-invariant abundant set and let

(1) $$x_1, x_2, \ldots, x_\alpha, \ldots \qquad (\alpha < \Theta)$$

(2) $$R_1, R_2, \ldots, R_\alpha, \ldots \qquad (\alpha < \Omega)$$

be well-orderings of S and the family of all meager $\mathscr{K}_{\delta\sigma}$-sets, respectively, where the latter well-ordering is based on Theorem 1 in Section I. We first determine a transfinite sequence

(3) $$y_1, y_2, \ldots, y_\alpha, \ldots \qquad (\alpha < \Omega)$$

of distinct points of S.

Place $y_1 = x_1$. Assume that $1 < \alpha < \Omega$ and the points y_β have already been defined for all ordinal numbers $\beta < \alpha$. The set

$$S_\alpha = \bigcup_{\beta < \alpha} \bigcup_{\phi \in \Phi} \phi(R_\beta)$$

being a meager set, $S - S_\alpha$ is an uncountable set. Accordingly, we define y_α to be the first term of the sequence (1) which does not belong to $S_\alpha \cup \{y_\beta : \beta < \alpha\}$.

The sequence (3) being thus determined, we define

$$T = \{\phi(y_\alpha) : \alpha < \Omega \text{ and } \phi \in \Phi\}$$

Clearly, T is an uncountable Φ-invariant subset of S.

Suppose that U is any meager subset of T. By Theorem 5 of Chapter 1, Section 2, there is an ordinal number $\gamma < \Omega$ such that $U \subset R_\gamma$. For any ordinal number $\alpha > \gamma$ and any mapping $\phi \in \Phi$, the point $\phi(y_\alpha)$ does not belong to the set $S_{\gamma+1}$ and hence does not belong to R_γ. We thus have

$$U = U \cap R_\gamma \subset \{\phi(y_\alpha) : \alpha \leqslant \gamma \text{ and } \phi \in \Phi\}$$

which implies that U is a countable set. Therefore, T is a rare set.

[1]Sierpiński (1939b); see also Sierpiński (1938c).

If S is a countable set of real numbers, then one can always find a suitable translation of S that will be disjoint from S; e.g., use a translation through a length t, where t is any number that does not belong to the countable set $\{x - y : x, y \in S\}$ of all differences of elements of S. As we next show, this is not true of all rare sets, if we assume CH.[1]

THEOREM 12   Assume CH. Also assume that $(X, \mathscr{C})$ is a point-meager base, X is abundant, and Φ is a group of one-to-one mappings of X onto itself which has power at most $2^{\aleph_0}$. If $\mathscr{C}$ satisfies CCC, has power at most $2^{\aleph_0}$, and is Φ-invariant, then there exists an uncountable rare set S such that $S \triangle \phi(S)$ is countable for each mapping $\phi \in \Phi$.

Proof. Let

(1) $$x_1, x_2, \ldots, x_\alpha, \ldots \qquad (\alpha < \Theta)$$

(2) $$R_1, R_2, \ldots, R_\alpha, \ldots \qquad (\alpha < \Omega)$$

(3) $$\phi_1, \phi_2, \ldots, \phi_\alpha, \ldots \qquad (\alpha < \Omega)$$

be well-orderings of X, the family of all meager $\mathscr{K}_{\delta\sigma}$-sets, and all mappings in Φ (where ϕ_1 is the identity mapping). [If Φ is countable, we repeat one element of Φ an uncountable number of times to obtain a transfinite sequence (3) of type Ω.] We first define a transfinite sequence

(4) $$y_1, y_2, \ldots, y_\alpha, \ldots \qquad (\alpha < \Omega)$$

of distinct points of X.

Place $y_1 = x_1$. Assume that $1 < \alpha < \Omega$ and the points y_β have already been defined for all ordinal numbers $\beta < \alpha$. The set $P_\alpha = \{y_\beta : \beta < \alpha\}$ is then a countable set. Let T_α be the union of all the sets

(5) $$\phi_{\beta_n}^{-1} \phi_{\beta_{n-1}}^{-1} \cdots \phi_{\beta_1}^{-1}(R_\beta)$$

where $\langle \beta, \beta_1, \beta_2, \ldots, \beta_n \rangle$ is any $(n + 1)$-tuple of ordinal numbers less than α. Because $\alpha < \Omega$ the set of all such finite sequence is countable. Since the sets R_β are meager and $\mathscr{C}$ is Φ-invariant, the set T_α is meager, as is also the set $P_\alpha \cup T_\alpha$. We define y_α to be the first element occurring in the enumeration (1) that does not belong to $P_\alpha \cup T_\alpha$. We thus determine the sequence (4).

Let S be the set consisting of y_1 and all elements of the form

(6) $$\phi_{\beta_1} \phi_{\beta_2} \cdots \phi_{\beta_n}(y_\alpha)$$

where $1 < \alpha < \Omega$, $n \in \mathbb{N}$, and $\langle \beta_1, \beta_2, \ldots, \beta_n \rangle$ is an arbitrary n-tuple of ordinal numbers less than α. Because ϕ_1 is the identity mapping, S contains all

---

[1] Cf. Sierpiński (1932b, Théorème I; 1935a; 1935h).

elements of the sequence (4). Consequently, S is an uncountable set. We show that $S \cap R_y$ is a countable set for each ordinal number $\gamma < \Omega$.

Let y be any element of $S \cap R_\gamma$ with $y \neq y_1$. Then y is representable in the form (6). We claim that $\alpha \leqslant \gamma$. Assume to the contrary that $\alpha > \gamma$. According to the definition of the set T_α, we have, for $\alpha > \gamma$,

$$(7) \qquad\qquad \phi_{\beta_n}^{-1}\phi_{\beta_{n-1}}^{-1} \cdots \phi_{\beta_1}^{1}(R_\gamma) \subset T_\alpha$$

Since $y \in R_\gamma$ and y is of the form (6) we have $\phi_{\beta_1}\phi_{\beta_2} \cdots \phi_{\beta_n}(y_\alpha) \in R_\gamma$ and hence $y_\alpha \in \phi_{\beta_n}^{-1}\phi_{\beta_{n-1}}^{-1} \cdots \phi_{\beta_1}^{-1}(R_\gamma)$. In view of (7), we have $y_\alpha \in T_\alpha$, contrary to the definition of y_α. Therefore, if $y \in S \cap R_\gamma$ and $y \neq y_1$, then y is of the form (6) with $\alpha \leqslant \gamma$. The set of all elements of the form (6) with $\alpha \leqslant \gamma$ being a countable set, it follows that $S \cap R_\gamma$ is countable. Now, since every meager subset of S is contained in some $\mathscr{K}_{\delta\sigma}$-set R_γ in (2), every meager subset of S is countable. Therefore, S is a rare set.

Suppose now that ϕ is any mapping in Φ; say that $\phi = \phi_\beta$ in the enumeration (3). If y is any element of $\phi_\beta(S)$—S, then, from the definition of S,

$$(8) \qquad\qquad y = \phi_\beta\phi_{\beta_1}\phi_{\beta_2} \cdots \phi_{\beta_n}(y_\alpha)$$

where $1 \leqslant \alpha < \Omega$, $n \in \mathbb{N}$, and $\beta_1, \beta_2, \ldots, \beta_n$ are ordinal numbers less than α. We must have $\alpha \leqslant \beta$, since the contrary situation implies that $y \in S$. Each element y of $\phi_\beta(S)$—S is thus of the form (8) for some ordinal numbers $\alpha, \beta_1, \beta_2, \ldots, \beta_n$, all of which are less than or equal to β. There being only countably many elements of this sort, the set $\phi_\beta(S)$—$S = \phi(S)$—S is countable.

In a similar manner it is seen that $\phi^{-1}(S)$—S is also a countable set and the same is true of S—$\phi(S)$, its image under the mapping ϕ. We conclude that $S \triangle \phi(S)$ is a countable set for each $\phi \in \Phi$.

F. Some applications of rare sets

We first derive a simple consequence of Theorem 1.

THEOREM 13 Assume CH. Also assume that $(X, \mathscr{C})$ is a point-meager base and X has power $2^{\aleph_0}$. If $\mathscr{C}$ satisfies CCC and has power at most $2^{\aleph_0}$, then there exists a one-to-one mapping of X into itself under which each uncountable subset of X corresponds to an abundant set.

Proof. Let f be a one-to-one mapping of X onto a rare set of power $2^{\aleph_0}$ contained in X.

Particular instances of this result include:[1]

(τ) If $2^{\aleph_0} = \aleph_1$, then there exists a one-to-one function mapping $\mathbb{R}^n$ onto a

[1]Sierpiński (1929a).

subset of itself which transforms each uncountable set into a set of the second category.

(μ) If $2^{\aleph_0} = \aleph_1$, then there exists a one-to-one function mapping $\mathbb{R}^n$ onto a subset of itself which transforms each uncountable set into a set of positive Lebesgue outer measure.

We next establish a generalized version of the result of Sierpiński:[1] If $2^{\aleph_0} = \aleph_1$, then any interval in $\mathbb{R}$ is representable as the union of $2^{2^{\aleph_0}}$ sets, none of which is Lebesgue measurable, such that the intersection of any two different sets is countable.

We shall utilize a set-theoretical fact.[2]

PROPOSITION 14 If $2^{\aleph_0} = \aleph_1$, then every set of power $\aleph_1$ is representable as the union of $2^{\aleph_1}$ uncountable sets, the intersection of any two different sets of which is countable.

Proof. For each countable ordinal number α, let E_α denote the set of all sequences $\langle x_\beta \rangle_{\beta < \alpha}$ of type α, with $x_\beta \in \mathbb{B}$ for each β. Each set E_α has power at most $2^{\aleph_0} = \aleph_1$ and the set

$$E = \bigcup_{\alpha < \Omega} E_\alpha$$

has power $\aleph_1 \cdot \aleph_1 = \aleph_1$.

Let T denote the set of all transfinite sequences $\langle x_\beta \rangle_{\beta < \Omega}$ of type Ω with $x_\beta \in \mathbb{B}$ for each β. Clearly, T has power $2^{\aleph_1}$.

With each transfinite sequence $x = \langle x_\beta \rangle_{\beta < \Omega}$ in T we associate the set

$$S_x = \{\langle x_\beta \rangle_{\beta < \alpha} : \alpha < \Omega\}$$

consisting of all initial sections of x. Each set S_x is a subset of E and has power $\aleph_1$. In addition, if $x, y \in T$ and $x \neq y$, then $S_x \cap S_y$ is a countable set. The set

$$S = \bigcup_{x \in T} S_x$$

is therefore a set of power $\aleph_1$ having a representation of the prescribed form.

For an arbitrary set of power $\aleph_1$ the desired decomposition is obtained by means of a one-to-one correspondence between the given set and the set S.

THEOREM 15 Assume CH. Also assume that $(X, \mathscr{C})$ is a point-meager base. If $\mathscr{C}$ satisfies CCC and has power at most $2^{\aleph_0}$, then every abundant set is representable as the union of $2^{2^{\aleph_0}}$ non-Baire sets such that the intersection of any two different sets is countable.

[1]Sierpiński (1928b; 1956, Chap. IV, Sec. 5); see also Tarski (1928, Corollaire 20).
[2]Cf. Sierpiński (1928b).

Proof. Let Q be any abundant set. According to Theorem 1 and Theorem 6 of Section II, Q contains a rare set R of power $\aleph_1$ which is not a Baire set. Hence, there is a region A such that the set $S = R \cap A$ and its complement are both abundant everywhere in A.

Applying Proposition 14, we obtain a representation of S as the union of $2^{2^{\aleph_0}}$ uncountable sets S_x, the intersection of any two different sets of which is countable. Each set S_x, as an uncountable subset of the rare set R, is abundant. Moreover, because the complement of S is abundant everywhere in A, none of the sets S_x can be a Baire set.

If $Q-S$ is not a Baire set, then this set, together with the sets S_x, yield the desired representation. If $Q-S$ is a Baire set, then we adjoin it to one of the sets S_x to obtain the representation sought.

G.   Further results

If S is a Sierpiński set in $\mathbb{R}^2$, then it can be shown[1] that S is nonmeasurable with respect to planar Lebesgue measure and, furthermore, S may be used to establish the validity of the following fact:

(M) If $2^{\aleph_0} = \aleph_1$, then there exists a set $S \subset \mathbb{R}^2$ having infinite Carathéodory linear measure such that every subset of S is measurable with respect to Carathéodory linear measure of plane sets.

This fact, first discovered by Marczewski,[2] can be established in a more direct manner using the following general result for Hausdorff measures μ^h with $h \in \mathscr{H}_c$.

THEOREM 16   If $2^{\aleph_0} = \aleph_1$, then for any non-σ-finite Hausdorff measure in $\mathbb{R}^n$, there exists an uncountable set of infinite measure, every uncountable subset of which has infinite measure.

This theorem is a consequence of Theorem 2 with $\mathscr{E}$ taken to be the family of all $\mathscr{G}_\delta$-sets in $\mathbb{R}^n$ of finite measure. From this theorem one can derive.

COROLLARY 17   If $2^{\aleph_0} = \aleph_1$, then for any non-σ-finite Hausdorff measure in $\mathbb{R}^n$, there exists a set of infinite measure every subset of which is Hausdorff measurable with measure 0 or ∞.

Such a set, although having infinite measure, can be viewed as a "negligible set" in mathematical deliberations, since it is a singular set with

---

[1]See Theorem 15 of Chapter 5, Section II.
[2]Cf. Szpilrajn (1931), Besicovitch (1934), Sierpiński (1956, Proposition C_{39}), and Choquet (1946); see also Eggleston (1954, 1958) and Ostaszewski (1975).

respect to the category base of all closed sets in $\mathbb{R}^n$ of positive Hausdorff measure.

For an arbitrary Hausdorff measure μ^h with $h \in \mathcal{H}_c$ one can establish a general result that encompasses Theorem 7.

THEOREM 18 If $2^{\aleph_0} = \aleph_1$, then for any Hausdorff measure in $\mathbb{R}^n$, there exists an uncountable set in $\mathbb{R}^n$ which has at most countably many points in common with each set of measure zero.

An analogous result for Hausdorff dimension is obtained from Theorem 2 with $\mathscr{E}$ taken to be the family of all $\mathscr{G}_\delta$-sets in $\mathbb{R}^n$ which have Hausdorff dimension $\leqslant p$, where $0 \leqslant p < n$ is given.

THEOREM 19 Assume that $0 \leqslant p < n$. If $2^{\aleph_0} = \aleph_1$, then there exists an uncountable set in $\mathbb{R}^n$ which has at most countably many points in common with each set of Hausdorff dimension $\leqslant p$.

For $0 \leqslant p \leqslant n$, a set $S \subset \mathbb{R}^n$ is said to be essentially at least p-dimensional if it cannot be represented as a countable union of sets each of which has dimension $\leqslant p$.

Let $\mu^{(r)}$ denote Hausdorff r-dimensional measure in $\mathbb{R}^n$. Utilizing Theorem 2, one can prove two further results.[1]

THEOREM 20 Assume that $0 \leqslant p \leqslant n$. If $2^{\aleph_0} = \aleph_1$, then every essentially at least p-dimensional set in $\mathbb{R}^n$ has an essentially at least p-dimensional subset which is $\mu^{(q)}$-measurable for every $q < p$.

THEOREM 21 Assume that $0 \leqslant p \leqslant n$. If $2^{\aleph_0} = \aleph_1$, then every essentially at least p-dimensional set in $\mathbb{R}^n$, which is not σ-finite with respect to $\mu^{(p)}$, has an essentially at least p-dimensional subset which is $\mu^{(q)}$-measurable for every $q \leqslant p$.

In the case of topological dimension, Hurewicz has proved the following proposition to be equivalent to the Continuum Hypothesis.[2]

(H) There exists in Hilbert space an uncountable set of infinite topological dimension which contains no uncountable set of finite topological dimension.[3]

[1] Davies (1979).
[2] See Hurewicz (1932), Sierpiński (1956, Proposition P_{10}), and Kuratowski (1966, esp. p. 298, IV, Theorem 1, and pp. 316–318).
[3] See also Tumarkin (1971) for additional examples in topological dimension theory.

IV. THE DUALITY PRINCIPLE

A. Decomposition into meager sets

We establish here a fundamental decomposition theorem.[1]

THEOREM 1   Assume that $(X, \mathscr{C})$ is a point-meager base satisfying CCC for which the following conditions hold:
   (a) The family of all meager $\mathscr{K}_{\delta\sigma}$-sets has power $\aleph_1$.
   (b) The complement of any meager set contains an uncountable meager set.
Then there exists a decomposition

$$X = \bigcup_{\alpha < \Omega} X_\alpha$$

of X into $\aleph_1$ disjoint, uncountable meager sets X_α having the property that a set S is a meager set if and only if S is contained in some countable union of sets X_α.

Proof. We note that condition (b) implies that X is an abundant set.
   Let

(1) $R_1, R_2, \dots, R_\gamma, \dots$ $(\gamma < \Omega)$

be a well-ordering of all meager $\mathscr{K}_{\delta\sigma}$-sets. Place $S_1 = \varnothing$ and

$$S_\gamma = \bigcup_{\beta < \gamma} R_\beta$$

for $1 < \gamma < \Omega$. Then

(2) $S_1, S_2, \dots, S_\gamma, \dots$ $(\gamma < \Omega)$

is an ascending transfinite sequence of meager sets. We first show that for every ordinal number $\mu < \Omega$ there is an ordinal number ν, with $\mu < \nu < \Omega$, such that $S_\nu - \bigcup_{\beta < \nu} S_\beta$ is an uncountable set.
   By condition (b), the set $X - \bigcup_{\beta \leqslant \mu} S_\beta$ contains an uncountable meager set which, according to Theorem 5 of Chapter 1, Section II, is contained in a meager $\mathscr{K}_{\delta\sigma}$-set and hence is contained in some set S_γ. Consequently, there is a smallest ordinal number $\nu > \mu$ for which the set $S_\nu - \bigcup_{\beta \leqslant \mu} S_\beta$ is uncountable. If $\mu < \gamma < \nu$, then $S_\gamma - \bigcup_{\beta \leqslant \mu} S_\beta$ must be a countable set and hence so also is the set $S_\gamma - \bigcup_{\beta < \gamma} S_\beta$. From the uncountability of the left-hand side

[1]Cf. Sierpiński (1934b; 1956, Proposition C_{25}), Erdös (1943), Marczewski (1946), Dressler and Kirk (1972), and Oxtoby (1980, Theorem 19.5).

of the inclusion

$$S_\gamma - \bigcup_{\beta \leq \mu} S_\beta \subset \bigcup_{\mu < \gamma \leq \nu} \left(S_\gamma - \bigcup_{\beta < \gamma} S_\beta \right)$$

we conclude $S_\nu - \bigcup_{\beta < \nu} S_\beta$ is an uncountable set.

Having thus shown that there are uncountably many ordinal numbers $\gamma < \Omega$ such that $S_\gamma - \bigcup_{\beta < \gamma} S_\beta$ is an uncountable set, we proceed by transfinite induction to select a subsequence

(3) $S_{\gamma_1}, S_{\gamma_2}, \ldots, S_{\gamma_\alpha}, \ldots$ $(\alpha < \Omega)$

of the sequence (2) with the property that each of the sets

$$X_\alpha = S_{\gamma_\alpha} - \bigcup_{\xi < \alpha} S_{\gamma_\xi}$$

for $\alpha < \Omega$, is uncountable. Obviously, the sets X_α are disjoint.

Suppose that S is a subset of X that is contained in a countable union of sets X_α. These sets being meager, S must be a meager set. On the other hand, suppose that S is any meager set. Then S is contained in some meager $\mathscr{X}_{\delta\sigma}$-set R that occurs among the terms of the sequence (1), say $R = R_\gamma$. The sequence (3) being cofinal with the sequence (2), there is an ordinal number $\mu < \Omega$ such that

$$R_\gamma \subset S_{\gamma_\mu} \subset \bigcup_{\alpha \leq \mu} X_\alpha$$

Therefore, S is contained in a countable union of sets X_α. In particular, this implies each point of X must belong to one of the sets X_α, so we have

$$X = \bigcup_{\alpha < \Omega} X_\alpha$$

We note that the category bases of all closed rectangles in $\mathbb{R}^n$ and all compact sets of positive Lebesgue measure in $\mathbb{R}^n$ satisfy the hypothesis of Theorem 1 assuming CH [to verify the condition (a)].

As a simple consequence of Theorem 1, we have[1]

COROLLARY 2   Assume that $(X, \mathscr{C})$ is a point-meager base satisfying CCC for which the conditions (a) and (b) hold. Then there exists a decomposition

$$X = \bigcup_{\alpha < \Omega} X_\alpha$$

of X into $\aleph_1$ disjoint, uncountable meager sets X_α having the property that an uncountable set S is a rare set if and only if each of the sets $S \cap X_\alpha$ is countable.

[1]Cf. Dressler and Kirk (1972).

If we select a single point from each of the sets X_α in the decomposition of X given in Theorem 1, then, according to this corollary, the set of points selected is an uncountable rare set.

B. A mapping of Sierpiński

By 1930 it was quite apparent that there are theorems about subsets of $\mathbb{R}$ involving only the notion of a set of the first category and abstract set-theoretic concepts in their statements which remain true when the phrase "set of the first category" is replaced everywhere therein by the phrase "set of Lebesgue measure zero." In order to elucidate this duality between first category and Lebesgue measure zero, Sierpiński proved:[1]

If $2^{\aleph_0} = \aleph_1$, then there exists a one-to-one mapping ϕ of $\mathbb{R}$ onto itself which transforms each set of the first category into a set of Lebesgue measure zero and whose inverse mapping ϕ^{-1} transforms each set of Lebesgue measure zero into a set of the first category. It follows that ϕ transforms each set of the second category into a set of positive Lebesgue outer measure and ϕ^{-1} transforms each set of positive Lebesgue outer measure into a set of the second category.

Suppose, for instance, that one has proved

(τ_1) If $2^{\aleph_0} = \aleph_1$, then there exists a Mahlo-Luzin set $S \subset \mathbb{R}$ (i.e., an uncountable set that contains no uncountable set of the first category).

One can then apply Sierpiński's mapping ϕ to the Mahlo-Luzin set S immediately to obtain [with $T = \phi(S)$]

(μ_1) If $2^{\aleph_0} = \aleph_1$, then there exists a Sierpiński set $T \subset \mathbb{R}$ (i.e., an uncountable set that contains no uncountable set of Lebesgue measure zero).

We note that some loss of generality may occur when invoking Sierpiński's result. For example, the Lebesgue measure analogue of

(τ_2) $\mathbb{R}$ can be decomposed into continuum many disjoint sets of the second category.

reads

(μ_2) $\mathbb{R}$ can be decomposed into continuum many disjoint set of positive Lebesgue outer measure.

However, the derivation of one of these statements from the other using

[1] Sierpiński (1934b).

Sierpiński's result requires CH, although the original statements do not require this hypothesis.[1]

We give here a general version of Sierpiński's theorem.

THEOREM 3   Assume that $(X,\mathscr{C})$ and $(Y,\mathscr{D})$ are point-meager bases satisfying CCC for which the conditions (a) and (b) of Theorem 1 hold and both X and Y have power $\aleph_1$. Then there exists a one-to-one mapping ϕ of X onto Y such that a set $S \subset X$ is $\mathscr{C}$-meager if and only if $\phi(S)$ is $\mathscr{D}$-meager.

Proof. Let $X = \bigcup_{\alpha<\Omega} X_\alpha$ and $Y = \bigcup_{\alpha<\Omega} Y_\alpha$ be decompositions of X and Y as given in Theorem 1. The sets X_α and Y_α having the same power for each $\alpha < \Omega$, we can define a one-to-one mapping ϕ of X onto Y such that $\phi(X_\alpha) = Y_\alpha$ for each $\alpha < \Omega$. It is a simple matter to verify ϕ has the desired property.

We now proceed to establish the existence of such a mapping ϕ for other category bases.[2] First, we prove

THEOREM 4   Assume that $(X,\mathscr{C})$ is a point-meager base and X is abundant. Then one and only one of the following conditions holds:
  (α) Every meager set is countable.
  (β) X is representable as the union of two disjoint, uncountable sets P and Q, where P is a rare set and Q is a meager set.
  (γ) The complement of every meager set contains an uncountable meager set.

Proof. Assume that conditions (α) and (β) are not satisfied. Suppose that S is any meager set. Because condition (α) is not satisfied, there exists an uncountable meager set T. Then $Q = S \cup T$ is an uncountable meager set. As X is abundant and every countable set is meager, the set $P = X{-}Q$ must be uncountable. Since condition (β) is not satisfied, P cannot be a rare set. This implies that $X{-}S$ contains an uncountable meager set. Thus, condition (γ) is satisfied.

Clearly, no two of these three conditions are compatible.

The condition (α) holds for the cofinite and cocountable bases for an uncountable set. The condition (γ) holds for the category bases of all closed rectangles in $\mathbb{R}^n$ and all compact sets in $\mathbb{R}^n$ of positive Lebesgue measure. Following are examples for which condition (β) holds.

EXAMPLE   Let X be the union of two disjoint, uncountable sets P, Q and let

$$\mathscr{C} = \{A : P{-}A \text{ is finite}\}$$

[1]These two statements are consequence of Theorem 27 of Chapter 5, Section I.
[2]Cf. Cholewa (1982).

Then $(X,\mathscr{C})$ is a category base closed under finite intersections that is a point-meager, Baire base satisfying CCC. The singular sets are the sets S representable in the form $S = K \cup L$, where K is a finite subset of P and L is any subset of Q.[1]

EXAMPLE Let X be the set of all countable ordinal numbers, let P be the set consisting of 1 and all limit ordinal numbers, and let $\mathscr{C}$ consist of $\varnothing$ and all sets of the form $\{x \in X : a \leqslant x < b\}$, where $a \in P$, $b \in X$, and $a < b$. Then $(X,\mathscr{C})$ is a category base closed under finite intersections which is a Baire base that is not point-meager and does not satisfy CCC. The singular sets coincide with the subsets of $X - P$.

THEOREM 5 Assume that $(X,\mathscr{C})$, $(Y,\mathscr{D})$ are point-meager and that X, Y both have power $\aleph_1$. If both bases satisfy the condition (α) or both bases satisfy the condition (β), then there exists a one-to-one mapping ϕ of X onto Y such that a set $S \subset X$ is $\mathscr{C}$-meager if and only if $\phi(S)$ is $\mathscr{D}$-meager.

Proof. If both bases satisfy condition (α), then we may take ϕ to be any one-to-one mapping of X onto Y.

Assume that both bases satisfy condition (β). Then there exist four uncountable sets P_X, Q_X, P_Y, Q_Y such that $X = P_X \cup Q_X$, $P_X \cap Q_X = \varnothing$, $Y = P_Y \cup Q_Y$, $P_Y \cap Q_Y = \varnothing$, P_X and P_Y are rare sets, and Q_X, Q_Y are meager sets. Since both X and Y have power $\aleph_1$, all four of these sets also have power $\aleph_1$. Let f be any one-to-one mapping of P_X onto P_Y and let g be any one-to-one mapping of Q_X onto Q_Y. The mapping ϕ defined by

$$\phi(x) = \begin{cases} f(x), & x \in P_X \\ g(x), & x \in Q_X \end{cases}$$

is a one-to-one mapping of X onto Y with the asserted property.

C. The Sierpiński-Erdös Duality Principle

After establishing the result noted in the preceding section, Sierpiński posed the question whether there exists a one-to-one mapping of $\mathbb{R}$ onto itself which transforms each set of the first category into a set of Lebesgue measure zero and simultaneously transforms each set of Lebesgue measure zero into a set of the first category. Utilizing the complementary nature of the two underlying category bases Erdös proved that such a mapping does exist, assuming CH.

[1]Cf. Cholewa (1982).

DEFINITION Two category bases $(X,\mathscr{C})$ and $(X,\mathscr{C}')$ are called complementary if X is representable as the union of two disjoint sets M and M', where M is $\mathscr{C}$-meager and M' is $\mathscr{C}'$-meager.

THEOREM 6 For $X = \mathbb{R}^n$, the category bases of all closed rectangles and of all compact sets of poisitive Lebesgue measure are complementary. More precisely, $\mathbb{R}^n$ is representable as the disjoint union of a $\mathscr{G}_\delta$-set of Lebesgue measure zero and an $\mathscr{F}_\sigma$-set of the first category.[1]

Proof. Let $p_1, p_2, \ldots, p_m \ldots$ ($m \in \mathbb{N}$) be an enumeration of all rational points in $\mathbb{R}^n$. For all $m,n \in \mathbb{N}$, let $E_{m,n}$ be an open rectangle containing the point p_m having volume $1/2^{m+n}$. Define $G_n = \bigcup_{m=1}^{\infty} E_{m,n}$ for each $n \in \mathbb{N}$ and let $M = \bigcap_{n=1}^{\infty} G_n$. Each set G_n is an open set whose complement is a nowhere dense, closed set and $\mu(G_n) \leqslant \Sigma_{m=1}^{\infty} 1/2^{m+n} = 1/2^n$. Hence, M is a $\mathscr{G}_\delta$-set of Lebesgue measure zero and its complement M' is an $\mathscr{F}_\sigma$-set of the first category.

REMARK In the case that $X = \mathbb{R}$, a decomposition of the asserted type is provided by the set of all Liouville numbers and its complement.[2]

THEOREM 7 Assume that $(X,\mathscr{C})$ and $(X,\mathscr{C}')$ are complementary, point-meager bases satisfying CCC for which the conditions (a) and (b) of Theorem 1 hold and that X has power $\aleph_1$. Then there exists a one-to-one mapping ϕ of X onto itself, with $\phi = \phi^{-1}$, such that a set S is $\mathscr{C}$-meager if and only if $\phi(S)$ is $\mathscr{C}'$-meager. [It follows from these properties that a set S is $\mathscr{C}'$-meager if and only if $\phi(S)$ is $\mathscr{C}$-meager.][3]

Proof. Let X be representable as the union of disjoint sets M and M', where M is $\mathscr{C}$-meager and M' is $\mathscr{C}'$-meager.

By Theorem 5 of Chapter 1, Section II, M is contained in a meager $\mathscr{K}_{\delta\sigma}$-set R. We may assume without loss of generality that $R = R_1$ in the enumeration (1) in the proof of Theorem 1. The set R_1 is then clearly uncountable and consequently we may assume that $X_1 = R$ in the decomposition $X = \bigcup_{\alpha < \Omega} X_\alpha$ of X into $\aleph_1$ disjoint, uncountable $\mathscr{C}$-meager sets X_α.

The set $R' = X - R$ is an uncountable $\mathscr{C}'$-meager subset of M', but we do not know whether or not R' is a $\mathscr{K}_{\delta\sigma}'$-set. Accordingly, in applying the proof of Theorem 1 to $\mathscr{C}'$, we modify the enumeration (1) by taking $R_1' = R'$ and taking

$$R_2', R_3', \ldots, R_\gamma', \ldots \qquad (1 < \gamma < \Omega)$$

[1]Extensions of this theorem are given in Szpilrajn (1934), Marczewski and Sikorski (1949), Kuczma (1973), and Oxtoby (1980, Chap. 16).

[2]See Oxtoby (1980, Chap. 2).

[3]Cf. Erdös (1943) and Oxtoby (1980, Theorem 19.6); see also Mendez (1976, 1978).

to be an enumeration of all meager $\mathscr{K}'_{\delta\sigma}$-sets that are not equal to R'. We may then assume that $X'_1 = R'$ in the decomposition $X = \bigcup_{\alpha<\Omega} X'_\alpha$ of X into $\aleph_1$ disjoint, uncountable $\mathscr{C}'$-meager sets X'_α. We now have

(*) $$X_1 = \bigcup_{1<\alpha<\Omega} X'_\alpha \qquad X'_1 = \bigcup_{1<\alpha<\Omega} X_\alpha$$

For each ordinal number α, with $1 < \alpha < \Omega$, let ϕ_α be a one-to-one mapping of X_α onto X'_α and define a mapping ϕ by

$$\phi(x) = \begin{cases} \phi_\alpha(x) & \text{if } x \in X_\alpha \quad \text{for } 1 < \alpha < \Omega \\ \phi_\alpha^{-1}(x) & \text{if } x \in X'_\alpha \quad \text{for } 1 < \alpha < \Omega \end{cases}$$

Since

$$X = \left(\bigcup_{1<\alpha<\Omega} X_\alpha \right) \cup \left(\bigcup_{1<\alpha<\Omega} X'_\alpha \right)$$

and the sets X_α and X'_β are disjoint for all ordinal numbers α and β larger than 1, the mapping ϕ is a well-defined one-to-one mapping of X onto itself with $\phi(X_\alpha) = X'_\alpha$ for $1 < \alpha < \Omega$. Moreover, from (*) we see that $\phi(X_1) = X'_1$. Hence, $\phi(X_\alpha) = X'_\alpha$ for all $\alpha < \Omega$.

Due to the nature of the decomposition of X obtained from Theorem 1, a set $S \subset X$ is $\mathscr{C}$-meager if and only if $\phi(S)$ is $\mathscr{C}'$-meager. Because $\phi = \phi^{-1}$, a set $S \subset X$ is $\mathscr{C}'$-meager if and only if $\phi(S)$ is $\mathscr{C}$-meager.

COROLLARY 8 If $2^{\aleph_0} = \aleph_1$, then there exists a one-to-one mapping ϕ of $\mathbb{R}^n$ onto itself which transforms each set of the first category into a set of Lebesgue measure zero and simultaneously transforms each set of Lebesgue measure zero into a set of the first category.

This corollary leads to the following general principle:[1]

SIERPIŃSKI-ERDÖS DUALITY PRINCIPLE Let π be any statement about sets in $\mathbb{R}^n$ involving only first category, Lebesgue measure zero, and abstract set-theoretical concepts. Let π' be the statement obtained from π by replacing first category with Lebesgue measure zero and replacing Lebesgue measure zero with first category. Then, assuming CH, the statement π and π' are logically equivalent.

For instance, assuming CH, the following statements are logically equivalent:

(π) A set is of the first category if its intersection with each set of Lebesgue measure zero is countable.

[1] Cf. Oxtoby (1980, Chap. 19).

(π') A set is of Lebesgue measure zero if its intersection with each set of the first category is countable.

These statements, which can be proved using Theorem 6, imply respectively that every Sierpiński set is a set of the first category and every Mahlo-Luzin set has Lebesgue measure zero.

We note that this principle cannot be extended to include the interchanging of the Baire property and Lebesgue measurability.[1] We also note, in connection with Corollary 8, the following result of Marczewski.[2]

THEOREM 9 There does not exist a one-to-one mapping of ℝ onto itself which transforms the family of all sets with the Baire property into the family of all Lebesgue measurable sets.[3]

D. Quotient algebras

We derive here one consequence of the mapping ϕ defined in Theorem 3.

THEOREM 10 If $2^{\aleph_0} = \aleph_1$, then the quotient algebra of all subsets of ℝ^n modulo the ideal of sets of the first category is isomorphic to the quotient algebra of all subsets of ℝ^n modulo the ideal of sets of Lebesgue measure zero.[4]

Proof. Let both $\mathscr{A}$ and $\mathscr{B}$ denote the σ-field of all subsets of ℝ^n, let $\mathscr{I}$ denote the σ-ideal of sets in ℝ^n of the first category, and let $\mathscr{J}$ denote the σ-ideal of sets in ℝ^n of Lebesgue measure zero. For the mapping ϕ given in Theorem 3 we have for subsets S and T of ℝ^n

$$S \Delta T \in \mathscr{I} \quad \text{if and only if} \quad \phi(S) \Delta \phi(T) \in \mathscr{J}$$

or, in an equivalent form,

$$[S]_{\mathscr{I}} = [T]_{\mathscr{I}} \quad \text{if and only if} \quad [\phi(S)]_{\mathscr{J}} = [\phi(T)]_{\mathscr{J}}$$

Hence, the function $\hat{\phi}$ defined by

$$\hat{\phi}([S]_{\mathscr{I}}) = [\phi(S)]_{\mathscr{J}}$$

is a one-to-one mapping of $\mathscr{A}/\mathscr{I}$ onto $\mathscr{B}/\mathscr{J}$. It is readily verified that $\hat{\phi}$ preserves the operations $\wedge$, $\vee$, $\sim$.

[1]Cf. Oxtoby, 1980, Chap. 21. See also Chapter 3, Section I.B.
[2]See Szpilrajn (1934, Sec. 3), Oxtoby (1980, p. 82), and Covington (1986, pp. 36–38).
[3]This theorem is a consequence of Theorem 13 in Chapter 5, Section II, Theorem 5 in Chapter 3, Section I, and the remark following the latter theorem.
[4]Sikorski (1964, pp. 77–78).

3

Separable Bases

I. SEPARABILITY

A. Definitions and basic properties

DEFINITION A family $\mathscr{B}$ of regions having the property that each abundant region is abundant everywhere in at least one region in $\mathscr{B}$ is called a quasi-base.[1]

DEFINITION A category base is called separable if it has a countable quasi-base.

EXAMPLE If $\mathscr{C}$ is the cofinite or cocountable base for an uncountable set X, then $\mathscr{B} = \{X\}$ is a countable quasi-base for $(X,\mathscr{C})$.

EXAMPLE The family of all closed rectangles with rational vertices is a countable quasi-base for the category base of all closed rectangles in $\mathbb{R}^n$.

THEOREM 1 If $(X,\mathscr{C})$ is separable and A is any region in $\mathscr{C}$, then $(A,\mathscr{C}_A)$ is also separable.

---

[1]This definition is given in an equivalent form in Morgan (1982).

Proof. Let $\mathscr{B}$ be a countable quasi-base for $\mathscr{C}$.

If A is $\mathscr{C}$-meager, then every subset of A is $\mathscr{C}_A$-meager and the empty family is a countable quasi-base for $\mathscr{C}_A$. Assume therefore that A is $\mathscr{C}$-abundant.

The family

$$\mathscr{D} = \{B \in \mathscr{B} : A \cap B \text{ is } \mathscr{C}\text{-abundant}\}$$

is then a nonempty countable family of $\mathscr{C}$-regions. Applying Theorem 2 of Chapter 1, Section II, we choose a $\mathscr{C}$-region $E \subset A \cap B$ for each $B \in \mathscr{D}$. The countable family $\mathscr{E}$ of $\mathscr{C}$-regions E thus selected will be quasi-base for $\mathscr{C}_A$.

For if D is any $\mathscr{C}_A$-abundant region in $\mathscr{C}_A$, then D is a $\mathscr{C}$-abundant region in $\mathscr{C}$. Because $\mathscr{B}$ is a quasi-base for $\mathscr{C}$, there is a $\mathscr{C}$-region $B \in \mathscr{B}$ in which D is $\mathscr{C}$-abundant everywhere. The set $A \cap B$, being $\mathscr{C}$-abundant, contains a $\mathscr{C}$-region $E \in \mathscr{E}$. Since D is $\mathscr{C}$-abundant everywhere in B, it is $\mathscr{C}$-abundant everywhere in E. Therefore, D is $\mathscr{C}_A$-abundant everywhere in the $\mathscr{C}_A$-region $E \in \mathscr{E}$.

THEOREM 2   If $(X,\mathscr{C})$ is separable, then the family of all abundant regions satisfies CCC.

Proof. Let $\mathscr{B}$ be a countable quasi-base for $\mathscr{C}$ and assume to the contrary that there are uncountably many disjoint abundant regions. Then there exists a region $B \in \mathscr{B}$ and two disjoint abundant regions A and C, each of which is abundant everywhere in B. According to Theorem 2 of Chapter 1, Section II, there is a region $D \subset A \cap B$. Since C is abundant everywhere in B, the set $C \cap D$ must be abundant. But $C \cap D = \varnothing$ is a meager set. We have thus reached a contradiction.

This theorem implies

THEOREM 3   Every separable, Baire base satisfies CCC.

As is readily seen, every separable, Baire base satisfies the following condition (which also implies CCC):

(k) Every uncountable family of regions contains an uncountable subfamily the intersection of any two regions of which is always nonempty.[1]

It can be shown that the category base of all compact sets in $\mathbb{R}^n$ of positive Lebesgue measure (or, more generally, of all measurable sets of positive measure for any σ-finite measure) also satisfies the condition (k).[2]

---

[1]Condition (k) has been investigated in Knaster (1945), Szpilrajn-Marczewski (1945), and Marczewski (1947); see also Galvin (1980).
[2]See Marczewski (1947) and Erdös (1954).

The category base of all closed rectangles in $\mathbb{R}^n$ further satisfies the condition [which implies condition (k)]

(l) Every uncountable family of regions has an uncountable subfamily whose intersection is nonempty.

On the other hand, assuming CH, there exists an uncountable family of compact sets in $\mathbb{R}^n$ of positive Lebesgue measure the intersection of every uncountable subfamily of which is empty.[1]

B. Double sequences

THEOREM 4   Assume that $(X,\mathscr{C})$ is separable. If S is a Baire set and $\langle S_{k,n}\rangle_{k,n\in\mathbb{N}}$ is a double sequence of Baire sets satisfying the conditions

   (i) $S_{k,n} \subset S_{k,n+1}$ for all $k,n \in \mathbb{N}$
   (ii) $S = \bigcup_{n=1}^{\infty} S_{k,n}$ for each $k \in \mathbb{N}$

then there exists a sequence $\langle n_k\rangle_{k\in\mathbb{N}}$ of natural numbers such that the set $(\limsup_k S_{k,n_k}) \Delta S$ is meager.[2]

Proof. The assertion being obviously true if S is itself a meager set, we assume that S is abundant.

Let $\mathscr{B}$ be a countable quasi-base for $(X,\mathscr{C})$ and let $\langle B_k\rangle_{k\in\mathbb{N}}$ be a sequence comprising all regions in $\mathscr{B}$ in which S is abundant, with each such region occurring infinitely many times among the terms of the sequence. For each $k \in \mathbb{N}$ we define n_k to be the first index such that S_{k,n_k} is abundant in the region B_k. For each $j \in \mathbb{N}$ the set $\bigcup_{k=j}^{\infty} S_{k,n_k}$ will then be abundant in every region in which S is abundant, which implies that the set $S - \bigcup_{k=j}^{\infty} S_{k,n_k}$ is meager. It follows that

$$\bigcup_{j=1}^{\infty} \left(S - \bigcup_{k=j}^{\infty} S_{k,n_k} \right) = S - \bigcap_{j=1}^{\infty} \bigcup_{k=j}^{\infty} S_{k,n_k} = S - \limsup_k S_{k,n_k}$$

is also meager. Therefore, $(\limsup_k S_{k,n_k}) \Delta S$ is a meager set.

REMARK   The Lebesgue measure analogue of this theorem is also valid.

THEOREM 5   Assume that $(X,\mathscr{C})$ is separable. If S is a set and $\langle S_{m,n}\rangle_{m,n\in\mathbb{N}}$ is a double sequence of Baire sets satisfying the condition

(*) $S \subset \bigcup_{m=r}^{\infty} \bigcup_{n=r}^{\infty} S_{m,n}$ for all $r \in \mathbb{N}$

---

[1] See Sierpiński (1934d).
[2] This theorem is a simple generalization of Wagner (1981, Theorem 4).

then there exist increasing sequences $\langle m_k \rangle_{k \in \mathbb{N}}$, $\langle n_k \rangle_{k \in \mathbb{N}}$ of natural numbers such that $S - \limsup_k S_{m_k, n_k}$ is a meager set.[1]

Proof. We assume that S is an abundant set. Let $\mathscr{B}$ be a countable quasi-base for $(X, \mathscr{C})$ and let $\langle B_k \rangle_{k \in \mathbb{N}}$ be a sequence comprising all regions in $\mathscr{B}$ in which S is abundant, with each such region occurring infinitely many times among the terms of the sequence.

The set $\bigcup_{m=1}^{\infty} \bigcup_{n=1}^{\infty} S_{m,n}$ being abundant in B_1, there exist natural numbers m_1, n_1 such that S_{m_1, n_1} is abundant in B_1. Continuing inductively, assume that $S_{m_1, n_1}, \ldots, S_{m_k, n_k}$ have already been defined. Let

$$r_k = \max\{m_1, \ldots, m_k, n_1, \ldots, n_k\} + 1$$

The set $\bigcup_{m=r_k}^{\infty} \bigcup_{n=r_k}^{\infty} S_{m,n}$ is abundant in B_{k+1}. Hence there exist natural numbers m_{k+1}, n_{k+1}, with $m_{k+1} \geqslant r_k$, $n_{k+1} \geqslant r_k$, such that $S_{m_{k+1}, n_{k+1}}$ is abundant in B_{k+1}. In this manner we determine the increasing sequences $\langle m_k \rangle_{k \in \mathbb{N}}$, $\langle n_k \rangle_{k \in \mathbb{N}}$.

For each $j \in \mathbb{N}$, the set $\bigcup_{k=j}^{\infty} S_{m_k, n_k}$ is abundant in every region in which S is abundant, so $S - \bigcup_{k=j}^{\infty} S_{m_k, n_k}$ is meager. Therefore, the set

$$S - \limsup_k S_{m_k, n_k} = \bigcup_{j=1}^{\infty} \left(S - \bigcup_{k=1}^{\infty} S_{m_k, n_k} \right)$$

is meager.

REMARK   The Lebesgue measure analogue of this theorem is not valid.[2] This may be seen by taking $X = \mathbb{R}$, $S = [0,1]$, and

$$S_{m,n} = \left[\frac{n-m}{2^m}, \frac{n-m+1}{2^m} \right] \qquad \text{for all } m, n \in \mathbb{N}$$

We now utilize Theorem 5 to derive a further result[3] concerning double sequences of Baire functions.

THEOREM 6   Assume that $(X, \mathscr{C})$ is separable. If $\langle f_{m,n} \rangle_{m,n \in \mathbb{N}}$ is a double sequence of real-valued Baire functions such that for all increasing sequences $\langle m_k \rangle_{k \in \mathbb{N}}$, $\langle n_k \rangle_{k \in \mathbb{N}}$ of natural numbers, the sequence $\langle f_{m_k, n_k} \rangle_{k \in \mathbb{N}}$ converges essentially everywhere to 0, then the double sequence $\langle f_{m,n} \rangle_{m,n \in \mathbb{N}}$ converges essentially everywhere to 0.

[1] Cf. Wagner (1981) (proof of Theorem 5).
[2] Cf. Sierpiński (1950, Théorème 3).
[3] Cf. Wagner (1981, Theorem 5).

Proof. Assume, to the contrary, that the set E of all points x for which the sequence $\langle f_{m,n}(x) \rangle_{m,n \in \mathbb{N}}$ does not converge to 0 is abundant. Placing

$$S_{m,n,p} = \left\{ x \in X : |f_{m,n}(x)| \geqslant \frac{1}{p} \right\}$$

for all $m,n,p \in \mathbb{N}$, we have

$$E = \bigcup_{p=1}^{\infty} \bigcap_{q=1}^{\infty} \bigcup_{m=q+1}^{\infty} \bigcup_{n=q+1}^{\infty} S_{m,n,p}$$

The set E being abundant, there exists a natural number p_0 such that the set

$$S = \bigcap_{q=1}^{\infty} \bigcup_{m=q+1}^{\infty} \bigcup_{n=q+1}^{\infty} S_{m,n,p_0}$$

is abundant.

Applying Theorem 5 to the double sequence $\langle S_{m,n,p_0} \rangle_{m,n \in \mathbb{N}}$, we obtain increasing sequences $\langle m_k \rangle_{k \in \mathbb{N}}$, $\langle n_k \rangle_{k \in \mathbb{N}}$ of natural numbers such that S— $\limsup_k S_{m_k,n_k,p_0}$ is a meager set. The set $T = \limsup_k S_{m_k,n_k,p_0}$ is therefore abundant. But if $x \in T$, then the sequence $\langle f_{m_k,n_k}(x) \rangle_{k \in \mathbb{N}}$ does not converge to 0. We have thus contradicted the hypothesis of the theorem.

REMARK   The Lebesgue measure analogue of this theorem is not valid.[1]

C.   A decomposition theorem

Kuratowski asked whether or not every set of real numbers that is of the second category in every interval can be decomposed into two disjoint sets of the same nature.[2] An affirmative answer was first given by Sierpiński, assuming CH.[3] We give here a generalized version of Luzin's solution that avoids this hypothesis.[4]

DEFINITION   A family $\mathscr{B}$ of regions having the property that each region contains at least one region in $\mathscr{B}$ is called a pseudo-base.[5]

Clearly, any pseudo-base is a quasi-base. The converse, however, is not true. The cofinite base for an uncountable set has a countable quasi-base, but does not have a countable pseudo-base.

---

[1] See Sierpiński (1950, Theorem 3).
[2] Probléme 21, Fundamenta Mathematicae 4, 1923, 368.
[3] Sierpiński (1924b, pp. 185–186).
[4] Cf. Sierpiński (1956, pp. 172–176).
[5] Oxtoby (1961); see also Morgan (1982).

The theorems in this part are valid under the following restriction.

ASSUMPTION $(X,\mathscr{C})$ is a point-meager category base such that for every abundant region A, the category base $(A,\mathscr{C}_A)$ has a countable pseudo-base.

THEOREM 7   Under the given assumption, a set that is abundant in a given region contains two disjoint subsets both of which are abundant everywhere in a certain subregion of the given region.

Proof. Assume that S is a set that is abundant in a region H. By Theorem 8 of Chapter 1, Section II, S is abundant everywhere in some subregion A of H. We shall show that there is a subregion B of A and two disjoint subsets of S, each of which is abundant everywhere in B.

Let

$$x_1, x_2, \ldots, x_\xi, \ldots \qquad (\xi < \Theta)$$

be a well-ordering of the set $S \cap A$. Since $S \cap A$ is abundant, there exists a smallest ordinal number μ, with $\Omega \leqslant \mu \leqslant \Theta$, for which the set

$$(1) \qquad\qquad\qquad T = \{x_\xi : \xi < \mu\}$$

is abundant. Each of the sets

$$(2) \qquad\qquad\qquad P_\alpha = \{x_\xi : \xi < \alpha\}$$

where $\alpha < \mu$, is thus meager and we have

$$(3) \qquad\qquad\qquad P_\alpha = \bigcup_{n=1}^{\infty} Q_{\alpha,n}$$

where the sets $Q_{\alpha,n}$ are singular.

For each ordinal number $\beta < \mu$ and each $n \in \mathbb{N}$, set

$$(4) \qquad\qquad\qquad U_{\beta,n} = \{x_\alpha \in T : x_\beta \in Q_{\alpha,n}\}$$

As is easily seen, we have for $\beta < \mu$

$$T - (P_\beta \cup \{x_\beta\}) = \{x_\alpha \in T : x_\beta \in P_\alpha\}$$

and consequently, by (3) and (4),

$$(5) \qquad\qquad\qquad T - (P_\beta \cup \{x_\beta\}) = \bigcup_{n=1}^{\infty} U_{\beta,n}$$

for each $\beta < \mu$. The set T being abundant and the sets (2) being meager for $\alpha < \mu$, we conclude the sets (5) are abundant. Hence, by virtue of the equality (5), there exists for every ordinal number $\beta < \mu$ a smallest number $n_\beta \in \mathbb{N}$ for which the set U_{β,n_β} is abundant. Placing

$$(6) \qquad\qquad\qquad T_m = \{x_\beta \in T : n_\beta = m\}$$

for each natural number m, we obviously have

(7) $$T = \bigcup_{m=1}^{\infty} T_m$$

Because T is abundant, there exists a smallest natural number m_0 such that the set T_{m_0} is abundant.

Let

(8) $$\mathscr{B} = \{A_i : i \in I\}$$

be a pseudo-base for $\mathscr{C}_A$, where $I \subset \mathbb{N}$.

Suppose that β is any ordinal number such that $x_\beta \in T_{m_0}$. According to (6) we have $n_\beta = m_0$, and by virtue of the definition of the number n_β, the set U_{β,m_0} is abundant. Consequently, for every ordinal number β for which $x_\beta \in T_{m_0}$, there exists a smallest natural number $i_\beta \in I$ such that the set U_{β,m_0} is abundant everywhere in A_{i_β}. Setting

(9) $$T_{m_0,k} = \{x_\beta \in T_{m_0} : i_\beta = k\}$$

for each natural number k, we clearly have

(10) $$T_{m_0} = \bigcup_{k=1}^{\infty} T_{m_0,k}$$

Since T_{m_0} is abundant, there is a smallest natural number k_0 such that T_{m_0,k_0} is an abundant set and hence is abundant everywhere in some region $C \subset A$. As the sets Q_{α,m_0} are singular, there exists, for each ordinal number $\alpha < \mu$, a smallest natural number $j_\alpha \in I$ such that

(11) $$A_{j_\alpha} \subset C \quad \text{and} \quad A_{j_\alpha} \cap Q_{\alpha,m_0} = \varnothing$$

Setting

(12) $$S_p = \{x_\alpha \in T \cap A_{k_0} : j_\alpha = p\}$$

for each natural number p, we have

(13) $$T \cap A_{k_0} = \bigcup_{p=1}^{\infty} S_p$$

Now, because T_{m_0,k_0} is abundant, it is nonempty. Hence, there exists an ordinal number β such that $x_\beta \in T_{m_0,k_0}$ and, according to (9), $i_\beta = k_0$. By the definition of i_β, the set U_{β,m_0} is abundant everywhere in $A_{i_\beta} = A_{k_0}$. According to (4), the set U_{β,m_0} is contained in T, so T is also abundant everywhere in A_{k_0}. Using (13), we then obtain the existence of a natural number p_0 for which S_{p_0} is abundant. Let B be a subregion of A_{k_0} in which S_{p_0} is abundant everywhere. We proceed to show that the set $S - S_{p_0}$ is also abundant everywhere in B.

The set S_{p_0} being nonempty, there is a point $x_\alpha \in S_{p_0}$. From (12) we obtain $j_\alpha = p_0$, which implies that $A_{p_0} \subset C$. Since T_{m_0, k_0} is abundant everywhere in C, it is abundant everywhere in A_{p_0}. Denoting by x_γ a given point of $T_{m_0, k_0} \cap A_{p_0}$, we have, by virtue of (9), $x_\gamma \in T_{m_0}$ and $i_\gamma = k_0$. From the definition of i_β we see that U_{γ, m_0} is abundant everywhere in $A_{i_\gamma} = A_{k_0}$ and hence also in B, since $B \subset A_{k_0}$. To complete the proof we have only to show that $U_{\gamma, m_0} \subset S - S_{p_0}$ or, equivalently, that

(14) $$U_{\gamma, m_0} \cap S_{p_0} = \varnothing$$

Assume to the contrary that (14) does not hold. Let x_α be a point in $U_{\gamma, m_0} \cap S_{p_0}$. Then $x_\alpha \in U_{\gamma, m_0}$ and, by (4), we have $x_\gamma \in Q_{\alpha, m_0}$. Now (11) gives

(15) $$A_{j_\alpha} \cap Q_{\alpha, m_0} = \varnothing$$

Because $x_\alpha \in S_{p_0}$, we obtain $j_\alpha = p_0$ from (12). Formula (15) thus becomes

$$A_{p_0} \cap Q_{\alpha, m_0} = \varnothing$$

Since $x_\gamma \in Q_{\alpha, m_0}$, we must have $x_\gamma \notin A_{p_0}$. However, this contradicts the choice of x_γ.

THEOREM 8 Under the given assumption, each set that is abundant everywhere can be decomposed into a denumerable number of disjoint sets, each of which is abundant everywhere.

Proof. We first show any set S that is abundant everywhere can be decomposed into two disjoint subsets of the same nature.

Let $\mathcal{N}$ be the family of all regions N having the property that there exist at least two disjoint subsets of S each of which is abundant everywhere in N. Since S is abundant everywhere, it follows from Theorem 7 that every region contains a region in $\mathcal{N}$. We apply Lemmas 3 and 4 of Chapter 1, Section II to obtain a nonempty, disjoint subfamily $\mathcal{M} = \{M_k : k \in K\}$ of $\mathcal{N}$. For each $k \in K$, let S_{1k} and S_{2k} be disjoint subsets of $S \cap M_k$, each of which is abundant everywhere in M_k. We show that the sets

$$T_1 = \bigcup_{k \in K} S_{1k} \quad \text{and} \quad U_1 = S - T_1$$

are abundant everywhere.

Suppose that A is any region. Then there exists a region $M_j \in \mathcal{M}$ such that $A \cap M_j$ contains a region B. Each of the sets S_{1j} and S_{2j} being abundant everywhere in M_j, these sets are abundant in B and hence also abundant in A. Therefore, both T_1 and U_1 are abundant in every region A.

Having thus established the decomposition of S into two disjoint subsets T_1 and U_1 which are abundant everywhere, we proceed by induction to define, for each natural number n, two disjoint subsets T_{n+1} and U_{n+1} of U_n

which are abundant everywhere. Placing

$$S_1 = T_1 \cup \left(S - \bigcup_{n=1}^{\infty} T_n\right)$$

and

$$S_n = T_n$$

for $n \geqslant 2$, we obtain the desired decomposition of S.

THEOREM 9 Under the given assumption, every abundant set S contains a set which is not a Baire set relative to S.[1]

Proof. Let S be abundant everywhere in a region A and let S_1, S_2 be two disjoint subsets of S both of which are abundant everywhere in a region $B \subset A$. We show that S_1 is not a Baire set relative to S.

Assume to the contrary that $S_1 = E \cap S$ for some Baire set E. Then we have $S_1 \subset E$, so E is abundant everywhere in B. Since E is a Baire set, $B-E$ must be a meager set. From the inclusion $S_2 \subset S-S_1$ we obtain

$$B \cap S_2 \subset (B \cap S)-S_1 = (B \cap S)-(E \cap S) = (B \cap S)-E \subset B-E$$

This implies that $B \cap S_2$ is a meager set, contradicting the fact that S_2 is abundant everywhere in B.

COROLLARY 10 Under the given assumption, a necessary and sufficient condition that a set S be meager is that every subset of S be a Baire set relative to S.

D. Quotient algebras

DEFINITION A quotient algebra $\mathscr{A}/\mathscr{I}$ is called separable[2] if there is a countable set of nonzero elements in $\mathscr{A}/\mathscr{I}$ with the property that for every nonzero element $[S]$ of $\mathscr{A}/\mathscr{I}$ there is at least one element $[D]$ of this countable set such that $[D] < [S]$.

In a straightforward manner one can readily establish the following theorem.

THEOREM 11 A category base is separable if and only if the quotient algebra of all Baire sets modulo the ideal of meager sets is a separable quotient algebra.

---

[1]Cf. Sierpiński (1956, Supplément).
[2]Cf. Horn and Tarski (1948) and Sikorski (1964, pp. 94–95).

REMARK   As seen from the remark following Theorem 5, the category base of all compact sets in $\mathbb{R}^n$ of positive Lebesgue measure is not separable. This fact, combined with Theorem 11 and the invariance of separability under isomorphisms, yields Theorem 17 of Chapter 1, Section III.

DEFINITION   A quotient algebra $\mathscr{A}/\mathscr{I}$ is called atomless if for every nonzero element $[S]$ of $\mathscr{A}/\mathscr{I}$ there is a nonzero element $[T]$ of $\mathscr{A}/\mathscr{I}$ such that $[T] < [S]$ and $[T] \neq [S]$.

The quotient algebra of all sets with the Baire property in $\mathbb{R}^n$ modulo sets of the first category and the quotient algebra of all Lebesgue measurable sets in $\mathbb{R}^n$ modulo sets of Lebesgue measure zero are both atomless quotient algebras. We note the following important result.[1]

THEOREM 12   Any two separable, atomless, complete quotient algebras are isomorphic.

[1]Cf. Sikorski (1964, p. 155).

4

Cluster Points and Topologies

I. CLUSTER POINTS

We assume throughout this part that $(X,\mathscr{C})$ is a given category base and $\mathscr{I}$ is a given ideal of subsets of X.

A. Definition

DEFINITION   A point $x \in X$ is called a cluster point for a set S if there is a neighborhood A of x such that for every neighborhood B of x contained in A, we have $S \cap B \notin \mathscr{I}$.

The notion of a cluster point is a generalization of the classical notion of a limit point. The latter notion is obtained from the former for $X = \mathbb{R}^n$ by taking $\mathscr{I}$ to be the ideal of all finite subsets of X and taking $\mathscr{C}$ to be the family of all open rectangles or, equivalently, taking $\mathscr{C}$ to be the family of all closed rectangles.[1]

For the ideal of all meager sets, a point x is a cluster point for a set S if and only if S is locally abundant at x.

[1]Cf. Chapter 1, Section I.D.

B. Derivation operation

DEFINITION The set of all points of X that are cluster points for a given set S is called the derived set of S and is denoted by $D(S)$. The function $D(S)$ thus defined for every set $S \subset X$ is called the derivation operation determined by $\mathscr{C}$ and $\mathscr{I}$.

The derivation operation has the following properties:

(1) If $S \in \mathscr{I}$, then $D(S) = \varnothing$.
(2) If $S \subset T$, then $D(S) \subset D(T)$.
(3) $D(S \cup T) = D(S) \cup D(T)$.

More generally, for any finite index set F,

$$D\left(\bigcup_{i \in F} S_i\right) = \bigcup_{i \in F} D(S_i)$$

(4) For any index set I,

$$\bigcup_{\alpha \in I} D(S_\alpha) \subset D\left(\bigcup_{\alpha \in I} S_\alpha\right)$$

(4a) If I is any index set and $\mathscr{H}$ is the family of all finite subsets of I, then

$$D\left(\bigcup_{\alpha \in I} S_\alpha\right) = \bigcup_{\alpha \in I} D(S_\alpha) \cup \left[\bigcap_{H \in \mathscr{H}} D\left(\bigcup_{\alpha \in I - H} S_\alpha\right)\right]$$

As a particular case, we have

(4b) For any sequence $\langle S_n \rangle_{n \in \mathbb{N}}$ of subsets of X,

$$D\left(\bigcup_{n=1}^{\infty} S_n\right) = \bigcup_{n=1}^{\infty} D(S_n) \cup \left[\bigcap_{n=1}^{\infty} D\left(\bigcup_{k=n}^{\infty} S_k\right)\right]$$

(5) For any index set I,

$$D\left(\bigcap_{\alpha \in I} S_\alpha\right) \subset \bigcap_{\alpha \in I} D(S_\alpha)$$

(6) $D(S) - D(T) \subset D(S - T)$.

The verification of these properties is straightforward. We limit our verification here to the property (4a).

From properties (4) and (2) we see that the right-hand term is a subset of the left-hand term. To establish the opposite inclusion, first consider an arbitrary finite set $H \subset I$.

Using property (3) we obtain

$$D\left(\bigcup_{\alpha\in I} S_\alpha\right) = \bigcup_{\alpha\in H} D(S_\alpha) \cup D\left(\bigcup_{\alpha\in I-H} S_\alpha\right)$$

Consequently,

$$D\left(\bigcup_{\alpha\in I} S_\alpha\right) \subset \bigcup_{\alpha\in I} D(S_\alpha) \cup D\left(\bigcup_{\alpha\in I-H} S_\alpha\right)$$

This being true for every finite subset H of I, we have

$$D\left(\bigcup_{\alpha\in I} S_\alpha\right) \subset \bigcup_{\alpha\in I} D(S_\alpha) \cup \left[\bigcap_{H\in\mathcal{X}} D\left(\bigcup_{\alpha\in I-H} S_\alpha\right)\right]$$

With respect to the derivation operation D we define the following concepts:[1]

DEFINITION   A set S is D-closed if $D(S) \subset S$. A set S is D-homogeneous if $S \subset D(S)$. A set S is D-invariant if $D(S) = S$.

THEOREM 1   The D-closed sets have the following properties:
   (i) $\emptyset$ and X are D-closed sets.
   (ii) Every set in $\mathcal{S}$ is a D-closed set.
   (iii) The union of any finite family of D-closed sets is a D-closed set.
   (iv) The intersection of any family of D-closed sets is a D-closed set.

DEFINITION   A set is called D-open if its complement is D-closed.

THEOREM 2   The D-open sets have the properties
   (i) $\emptyset$ and X are D-open sets.
   (ii) The intersection of any finite family of D-open sets is a D-open set.
   (iii) The union of any family of D-open sets is a D-open set.

C. Adherent ideals

DEFINITION   The given ideal $\mathcal{S}$ is called an adherent ideal if

(7) $S-D(S)\in\mathcal{S}$ for every set S.

According to Theorem 12 of Chapter 1, Section II, the ideal of all meager sets is an adherent ideal for any category base.

For adherent ideals there are the further properties:

(8) $S\in\mathcal{S}$ if and only if $D(S) = \emptyset$.
(9) $D(S) \subset D(D(S))$; i.e., $D(S)$ is a D-homogeneous set.

[1]Cf. Sierpiński (1926b).

Property (8) follows from properties (1) and (7) and the decomposition

(10) $$S = [S \cap D(S)] \cup [S - D(S)]$$

Property (9) is obtained by taking $T = D(S)$ in property (6).

Following Cantor, we call the sets $S \cap D(S)$ and $S - D(S)$, occurring in the decomposition (10), the D-coherence and D-adherence, respectively.[1] An adherent ideal is thus one that contains all adherences.

THEOREM 3 If $\mathcal{I}$ is an adherent ideal, then every set is representable as a disjoint union of a D-homogeneous set and a set in $\mathcal{I}$.

Proof. For any given set S we use the decomposition (10). By virtue of properties (1), (3), and (7), we have

$$S \cap D(S) \subset D(S) = D[S \cap D(S)] \cup D[S - D(S)] = D[S \cap D(S)]$$

Therefore, $S \cap D(S)$ is a D-homogeneous set.

THEOREM 4 If $\mathcal{I}$ is an adherent ideal, then every D-closed set is representable as a disjoint union of a D-invariant set and a set in $\mathcal{I}$.

Proof. We again use the decomposition (10). As a consequence of property (2) and the assumption that S is D-closed, we have

$$D[S \cap D(S)] \subset D(S) = S \cap D(S)$$

Hence, combining this with the result of the preceding proof, we obtain

$$D[S \cap D(S)] = S \cap D(S)$$

This theorem is a generalization of the classical Cantor-Bendixson Theorem according to which every closed set S in $\mathbb{R}^n$ is representable as the union of two disjoint sets

$$S = P \cup Q$$

where P is either empty or a perfect set and Q is a countable set.[2] Cantor further asserted that this decomposition is unique, but Vivanti was the first one to prove this assertion.[3] Vivanti's proof is based on the following fact:

THEOREM 5 If P is a D-invariant set and U is a D-closed subset of P, then the set $V = P - U$ is D-homogeneous.

Proof. Suppose that $x \in V$. Since $V \subset P \subset D(P)$, there is a region A containing x such that for every region $B \subset A$, if $x \in B$, then $P \cap B \notin \mathcal{I}$. Because $x \notin U$ and

[1] Cf. Cantor (1966, p. 265).
[2] See Corollary 17 of Section III.
[3] Vivanti (1899).

U is D-closed, $x \notin D(U)$. Accordingly, there is a region $A_0 \subset A$ such that $x \in A_0$ and $U \cap A_0 \in \mathcal{I}$. If B_0 is any subregion of A_0 containing x, then from the equality

$$P \cap B_0 = (V \cap B_0) \cup (U \cap B_0)$$

and the fact that $P \cap B_0 \notin \mathcal{I}$, we obtain $V \cap B_0 \notin \mathcal{I}$. Therefore, $x \in D(V)$.

THEOREM 6   If $\mathcal{I}$ is any ideal of subsets of X, then a D-closed set can be represented as a disjoint union of a D-invariant set and a set in $\mathcal{I}$ in at most one way.

Proof. Let S be any D-closed set. Assume that $S = P \cup R = Q \cup T$, where P, Q are D-invariant sets, R, T belong to $\mathcal{I}$, and $P \cap R = \varnothing$, $Q \cap T = \varnothing$. Since $P — (P \cap Q) \subset T$ and $T \in \mathcal{I}$ we can apply Theorem 5 to obtain

$$P — (P \cap Q) \subset D[P — (P \cap Q)] \subset D(T) = \varnothing$$

Hence, $P = P \cap Q$. A similar argument yields $Q = P \cap Q$. Therefore, $P = Q$ and $R = T$.

We now give a characterization for D-closed sets.

THEOREM 7   If $\mathcal{I}$ is an adherent ideal, then a necessary and sufficient condition that a set S be D-closed is that it be representable in the form

$$S = P \cup Q$$

where P is a D-invariant set and $Q \in \mathcal{I}$.

Proof. According to Theorem 4, every D-closed set has such a representation. On the other hand, if S is any set with such a representation, then

$$D(S) = D(P) \cup D(Q) = D(P) = P \subset P \cup Q = S$$

whence S is D-closed.

II. TOPOLOGIES

A. Definition and examples

DEFINITION   A pair $(X, \mathcal{G})$ where X is a set and $\mathcal{G}$ is a family of subsets of X is called a topology if the following conditions are satisfied:
(a) $\varnothing$ and X belong to $\mathcal{G}$.
(b) The intersection of any finite subfamily of $\mathcal{G}$ is a member of $\mathcal{G}$.
(c) The union of any subfamily of $\mathcal{G}$ is a member of $\mathcal{G}$.

REMARK   Condition (a) is actually a consequence of conditions (b) and (c) when we consider the empty subfamily of $\mathcal{G}$.

The family of all open sets in $\mathbb{R}^n$ is the principal example of a topology. As an intuitive guide we thus call the sets belonging to $\mathscr{G}$, for any topology $(X,\mathscr{G})$, open sets and call their complements closed sets. The family of all closed sets is denoted by $\mathscr{G}$. When we speak of open or closed sets in $\mathbb{R}^n$ we do so with reference to the usual topology of $\mathbb{R}^n$ as determined in Chapter 1, Section I.A, unless otherwise specified.[1]

As every topology is a category base, we have the associated concepts of singular, meager, abundant, comeager, and Baire sets. In accordance with our guide we call these nowhere dense sets, set of the first category, sets of the second category, residual sets, and sets with the Baire property, respectively, for any topology $(X,\mathscr{G})$. When referring to such sets in $\mathbb{R}^n$ we do so with respect to the usual topology in $\mathbb{R}^n$, unless stated otherwise.

Any trivial base, discrete base, cofinite base, and cocountable base, with $\varnothing$ adjoined, is a topology, called the indiscrete topology, discrete topology, cofinite topology, and cocountable topology, respectively.

The usual topology for $\mathbb{R}$ can be generalized to other ordered sets. Let X be an ordered set with ordering relation $<$, let $\mathscr{A}$ denote the family of all sets having one of the forms

$$\{x\in X: x < a\} \qquad \{x\in X: a < x < b\} \qquad \{x\in X: x > b\}$$

where $a,b\in X$ with $a < b$, and let $\mathscr{G}$ denote the family of all sets representable as the union of some subfamily of $\mathscr{A}$. Then $(X,\mathscr{G})$ is a topology, called the order topology for X.

Let $(X,\mathscr{G})$ be a given category base, let $\mathscr{I}$ be an ideal of subsets of X, and let D be the associated derivation operation. According to Theorem 2 of Section I, the D-open sets constitute a topology. We shall call this the D-topology. The nowhere dense sets, sets of the first category, sets of the second category, sets which have the Baire property, and so on, for this topology will be called, respectively, D-nowhere dense sets, sets of the first D-category, sets of the second D-category, sets that have the D-Baire property, and so on. In the case that $\mathscr{I}$ is the σ-ideal of all meager sets for the given category base, the D-topology is called the basic topology. It can be proven that every finite category base is equivalent to its basic topology.[2]

B. Topological bases

DEFINITION Let $(X,\mathscr{G})$ be a topology. A subfamily $\mathscr{B}$ of $\mathscr{G}$ is called a topological base for $(X,\mathscr{G})$ if every set in $\mathscr{G}$ is the union of some subfamily of $\mathscr{B}$.

[1]For a detailed discussion of the evolution of topological concepts, see Slagle (1988).
[2]Morgan (1983).

Equivalently, a subfamily $\mathscr{B}$ of $\mathscr{G}$ is a topological base for $(X,\mathscr{G})$ if for every set $G\in\mathscr{G}$ and every point $x\in G$, there exists a set $B\in\mathscr{B}$ such that $x\in B$ and $B\subset G$.

We note that if $\mathscr{B}$ is a topological base for the topology $(X,\mathscr{G})$, then $(X,\mathscr{B})$ is a category base that is equivalent to $(X,\mathscr{G})$, although $(X,\mathscr{B})$ need not be itself a topology.

EXAMPLE The topology of open sets in $\mathbb{R}^n$ has a countable topological base consisting of all open rectangles with rational vertices. However, its basic topology does not have a countable topological base. We note that both topologies do have a countable quasi-base; i.e., both are separable[1] topologies.

THEOREM 1 If $(X,\mathscr{G})$ is a topology that has a countable topological base, then every family $\mathscr{H}$ of open sets contains a countable subfamily $\mathscr{H}_0$ such that $\bigcup\mathscr{H}_0 = \bigcup\mathscr{H}$.[2]

Proof. Let $\mathscr{B}$ be a countable topological base for $(X,\mathscr{G})$. For each point $x\in\bigcup\mathscr{H}$ we choose an open set $G_x\in\mathscr{H}$ which contains x and then choose a set $B_x\in\mathscr{B}$ such that $x\in B_x$ and $B_x\subset G_x$. The family $\mathscr{B}_0 = \{B_x:x\in\bigcup\mathscr{H}\}$ is countable, since it is a subfamily of $\mathscr{B}$. Hence, there is a countable set $I\subset\bigcup\mathscr{H}$ such that $\mathscr{B}_0 = \{B_x:x\in I\}$. Then $\mathscr{H}_0 = \{G_x:x\in I\}$ is a countable subfamily of $\mathscr{H}$ with $\bigcup\mathscr{H}_0 = \bigcup\mathscr{H}$.

COROLLARY 2 If $(X,\mathscr{G})$ is a topology that has a countable topological base, then every family $\mathscr{E}$ of closed sets contains a countable subfamily $\mathscr{E}_0$ such that $\bigcap\mathscr{E}_0 = \bigcap\mathscr{E}$.

THEOREM 3 If $(X,\mathscr{G})$ is a topology that has a countable topological base, then every σ-ideal of subsets of X is an adherent ideal.

Proof. Let $\mathscr{I}$ be a given σ-ideal of subsets of X. Suppose that S is any subset of X. For each point $x\in S-D(S)$ we choose an open set G_x containing x such that $S\cap G_x\in\mathscr{I}$. Applying Theorem 1 to the family $\mathscr{H} = \{G_x:x\in S-D(S)\}$, we obtain a countable set $M\subset S-D(S)$ such that $\bigcup\mathscr{H} = \bigcup_{x\in M} G_x$. Then

$$S-D(S)\subset \bigcup_{x\in S-D(S)} (S\cap G_x) = \bigcup_{x\in M} (S\cap G_x)$$

The latter union being a countable union of sets belonging to the σ-ideal $\mathscr{I}$, we conclude that $S-D(S)\in\mathscr{I}$.

[1]Our definition of separability, as given in Chapter 3, is not the same as that given in textbooks on topology.
[2]Cf. Lindelöf (1903) and Young (1903a).

Of course, the σ-ideal of sets of the first category is always an adherent ideal for any topology.

III. TOPOLOGIES GENERATED BY A TOPOLOGY AND AN IDEAL

Throughout this part we assume a fixed topology $(X, \mathcal{G})$ and a fixed ideal $\mathcal{I}$ of subsets of X.[1]

A. General properties

The definition of cluster point is now expressible in the equivalent form: A point $x \in X$ is a cluster point for a set S if and only if $S \cap G \notin \mathcal{I}$ for every open set G containing x.

We have the following properties for arbitrary subsets S of X.

(1) Every closed set is a D-closed set.

For suppose that S is a closed set and $x \in D(S)$. Then for every open set G containing x, we have $S \cap G \neq \emptyset$. Because $X—S$ is an open set and $S \cap (X—S) = \emptyset$, we must have $x \notin X—S$. Thus, $D(S) \subset S$.

(2) Every open set is a D-open set.
(3) $D(S)$ is a closed set.

To establish (3), we associate with each point $x \in X—D(S)$ an open set G_x containing x such that $S \cap G_x \in \mathcal{I}$. For any point $y \in G_x$ we have $y \in X—D(S)$. Therefore,

$$X—D(S) = \bigcup_{x \in X—D(S)} G_x$$

The set $X—D(S)$ is thus an open set. Hence, $D(S)$ is a closed set.

(4) $D(D(S)) \subset D(S)$.

From (9) and (4) of Section I we obtain

(5) If $\mathcal{I}$ is any adherent ideal, then

$$D(D(S)) = D(S)$$

For adherent ideals the D-closed sets have a simple structure.

[1] The basic references for the following sections are Vaidyanathaswamy (1944; 1960, Chap. VIII); see also Burstin (1914, Secs. 1, 2), Alexandrow (1916), Hobson (1927, Secs. 138–141), Hashimoto (1952, 1954, 1976), Freud (1958), Martin (1961), Hayashi (1964, 1974), Kuratowski (1966, Secs. 4–12, 23), Mycielski (1969), and Samuels (1975).

THEOREM 1   If $\mathscr{I}$ is an adherent ideal, then a set S is D-closed if and only if it is representable in the form

$$S = F \cup I$$

where F is a closed set and $I \in \mathscr{I}$.

Proof. Since each closed set F is D-closed, as is also each set $I \in \mathscr{I}$, the set $F \cup I$ is D-closed.

   Conversely, if S is any D-closed set, then, upon setting $F = D(S)$ and $I = S{-}D(S)$, we have $S = F \cup I$, which is a representation of the desired form.

COROLLARY 2   If $\mathscr{I}$ is an adherent ideal, then a set S is D-open if and only if it is representable in the form

$$S = G{-}I$$

where G is an open set and $I \in \mathscr{I}$.

   If the topology $(X,\mathscr{G})$ has a countable topological base and $\mathscr{I}$ is a σ-ideal, then, in view of Theorem 3 of Section II, Corollary 2 implies the family of sets S of the forms $S = G{-}I$, with $G \in \mathscr{G}$ and $I \in \mathscr{I}$, is a topology. If $(X,\mathscr{G})$ does not have a countable topological base, then the family of all sets of the form $G{-}I$, where G is an open set and I belongs to a given σ-ideal $\mathscr{I}$, is not necessarily a topology, although it will always be a category base.

EXAMPLE   Let X denote the set of all countable ordinal numbers, let $(X,\mathscr{G})$ denote the order topology for X, and let $\mathscr{I}$ be the σ-ideal of all countable subsets of X. For each ordinal number $\alpha < \Omega$, define

$$G_\alpha = \{\xi \in X : \xi < \alpha + 1\}$$
$$I_\alpha = \{\xi \in X : \xi \text{ is odd and } \xi \leqslant \alpha\}$$

The sets G_α are open sets and the sets I_α are members of $\mathscr{I}$. The set

$$S = \bigcup_{\alpha < \Omega} (G_\alpha{-}I_\alpha)$$

which contains all limit ordinals but contains no odd ordinal numbers, is not representable in the form $G{-}I$, where G is open and $I \in \mathscr{I}$. This assertion is a consequence of the fact that if I is any member of $\mathscr{I}$ and α is any limit ordinal larger than the supremum of I, then every open set G containing α must contain some odd ordinal number that does not belong to I.

   We next establish some facts concerning adherent ideals for which X is D-homogeneous.

THEOREM 3 If $\mathcal{S}$ is an adherent ideal, then a necessary and sufficient condition that X be D-homogeneous is that no nonempty open set belong to $\mathcal{S}$.

Proof. If G is a nonempty open set that belongs to $\mathcal{S}$, then no point of G belongs to $D(X)$, so $X—D(X) \neq \varnothing$ and X is not D-homogeneous. On the other hand, if X is not D-homogeneous, then, by property (3), the set $G = X—D(X)$ is a nonempty open set that belongs to $\mathcal{S}$.

THEOREM 4 If $\mathcal{S}$ is an adherent ideal and X is D-homogeneous, then
  (i) Every nowhere dense set is D-nowhere dense.
  (ii) Every set in $\mathcal{S}$ is D-nowhere dense.

Proof. We shall prove (i) and (ii) together.
   Suppose that $S = P \cup J$, where P is nowhere dense and $J \in \mathcal{S}$. Let G^* be any nonempty D-open set. By Corollary 2, we have $G^* = G—I$, where G is a nonempty open set and $I \in \mathcal{S}$. The set P being nowhere dense, there is a nonempty open set $H \subset G$ such that $H \cap P = \varnothing$. According to Theorem 3 and Corollary 2, the set $H^* = H—(I \cup J)$ is a nonempty D-open set. We have $H^* \subset G^*$ and $H^* \cap (P \cup J) = \varnothing$. Thus, S is a D-nowhere dense set.

COROLLARY 5 If $\mathcal{S}$ is an adherent ideal and X is D-homogeneous, then every set of the first category is a set of the first D-category.

   This corollary immediately yields

THEOREM 6 Assume that $\mathcal{S}$ is an adherent ideal and X is D-homogeneous. If the topology $(X,\mathcal{G})$ is point-meager, then the D-topology is also point-meager.

THEOREM 7 Assume that $\mathcal{S}$ is an adherent ideal and X is D-homogeneous. If the topology $(X,\mathcal{G})$ satisfies CCC, then the D-topology also satisfies CCC.

Proof. Suppose that $\{G_\alpha^* : \alpha \in J\}$ is any family of nonempty, disjoint D-open sets. According to Corollary 2, for each index $\alpha \in J$, we have $G_\alpha^* = G_\alpha—I_\alpha$, where G_α is a nonempty open set and $I_\alpha \in \mathcal{S}$. For any different indices $\alpha, \beta \in J$,

$$(G_\alpha \cap G_\beta)—(I_\alpha \cup I_\beta) = (G_\alpha—I_\alpha) \cap (G_\beta—I_\beta) = \varnothing$$

This implies that $G_\alpha \cap G_\beta \subset I_\alpha \cup I_\beta$. By Theorem 3 we must have $G_\alpha \cap G_\beta = \varnothing$. The family $\{G_\alpha : \alpha \in J\}$ is thus a family of nonempty, disjoint open sets. Because $\mathcal{G}$ satisfies CCC, the index set J must be countable.

THEOREM 8 Assume that $\mathcal{S}$ is an adherent ideal and X is D-homogeneous. If the topology $(X,\mathcal{G})$ is separable, then the D-topology is separable.

Proof. Let $\mathcal{B}$ be a countable quasi-base for $(X, \mathcal{G})$. We know from property (2) that each set in $\mathcal{B}$ is a D-open set. We show $\mathcal{B}$ is also a quasi-base for the D-topology; i.e., every D-open set of the second D-category is everywhere of the second D-category in some set belonging to $\mathcal{B}$.

Suppose that E is any D-open set of the second D-category. By the Fundamental Theorem, E is everywhere of the second D-category in some D-open set G^*. According to Corollary 2, $G^* = G-I$ for some nonempty open set G and some set $I \in \mathcal{I}$. Since G^* is a subset of G and is of the second D-category, Corollary 5 implies that G is of the second category. Hence, G is everywhere of the second category in some set $B \in \mathcal{B}$. We show that E is everywhere of the second D-category in the D-open set B.

Let C^* be any nonempty D-open set contained in B. Then $C^* = C-J$, where C is a nonempty open set and $J \in \mathcal{I}$. Because $B \cap C$ is a nonempty open subset of B and G is everywhere of the second category in B, the set $G \cap (B \cap C)$ is nonempty, as is also the open set $G \cap C$. According to Theorem 3 and Corollary 2, the set $H^* = (G \cap C)-(I \cup J)$ is a nonempty D-open set contained in G^*. As E is everywhere of the second D-category in G^*, the set $E \cap H^*$ is of the second D-category. Therefore, $E \cap C^*$ is of the second D-category.

B.   The trivial ideal

Let $\mathcal{I}_0$ denote the trivial ideal whose only member is the empty set. The associated derivation operation $D_0(S)$ is also denoted by $\bar{S}$ and is called the closure of S.[1] The cluster points for this ideal are called accumulation points. Thus, a point $x \in X$ is an accumulation point for a set S if and only if every open set containing x has a nonempty intersection with S.

One can readily verify the following property, for arbitrary sets S, which implies $\mathcal{I}_0$ is an adherent ideal and X is D_0-homogeneous.

(6) $S \subset \bar{S}$.

Specializing properties (3) and (5), we have for every set S,

(7) $\bar{S}$ is a closed set.

(8) $\overline{(\bar{S})} = \bar{S}$.

We also have an extension of property (3) of Section I.

THEOREM 9   Let $\mathcal{M} = \{S_\alpha : \alpha \in I\}$ be a family of sets such that for every point $x \in X$, there is an open set G containing x which has a nonempty

[1]See Kuratowski (1966, Secs. 4–9).

intersection with at most a finite number of sets in $\mathcal{M}$. Then

(9) $$\overline{\bigcup_{\alpha \in I} S_\alpha} = \bigcup_{\alpha \in I} \bar{S}_\alpha$$

Proof. In view of property (4) of Section I we need only show that $\overline{\bigcup_{\alpha \in I} S_\alpha} \subset \bigcup_{\alpha \in I} \bar{S}_\alpha$.

Suppose that $x \in \overline{\bigcup_{\alpha \in I} S_\alpha}$. Let G be an open set containing x such that for some finite subset F of I we have $G \cap S_\alpha = \varnothing$ for all $\alpha \in I{-}F$. We assert that there is an index $\alpha \in F$ such that $x \in \bar{S}_\alpha$.

Assume to the contrary that $x \notin \bigcup_{\alpha \in F} \bar{S}_\alpha$. Then the set $H = G{-}\bigcup_{\alpha \in F} \bar{S}_\alpha$ is an open set containing x and $H \cap (\bigcup_{\alpha \in I} S_\alpha) = \varnothing$, contradicting our supposition $x \in \overline{\bigcup_{\alpha \in I} S_\alpha}$.

From property (6) we see that the D_0-closed sets are the same as the D_0-invariant sets. That is,

(10)   A set S is D_0-closed if and only if $S = \bar{S}$.

Using property (2) of Section I and property (1) above, we derive

(11)   If F is a closed set and $S \subset F$, then $\bar{S} \subset F$.

THEOREM 10   The D_0-closed sets coincide with the closed sets.

Proof. Use properties (1), (10), and (7).

COROLLARY 11   The D_0-topology is the same as the original topology $(X, \mathcal{G})$.

THEOREM 12   If $\mathcal{J}$ is any ideal of subsets of X with corresponding derivation operation $D(S)$, then for every set S
   (i) $D(S) \subset \bar{S}$.
   (ii) $\overline{D(S)} = D(S)$.

Proof. (i) is readily verified, while (ii) is a consequence of properties (3), (1), and (10).

C.   The ideal of finite sets

Let $\mathcal{J}_f$ denote the ideal of all finite subsets of X. The associated derivation operation $D_f(S)$ will henceforth also be denoted by S'. The cluster points for this ideal are called limit points. Thus, a point $x \in X$ is a limit point for a set S if and only if for every open set G containing x, the set $S \cap G$ is infinite. We note that every limit point for a set is an accumulation point for that set.

The points of a set S are divided into limit points and isolated points; a point $x \in S$ being an isolated point of S if there is an open set containing x which contains only finitely many points of S. In the case of the usual topology for $\mathbb{R}^n$, a point x belonging to a set S is an isolated point of S if and only if there is an open set containing x which contains no other point of S.

The ideal $\mathscr{I}_f$ is not in general an adherent ideal. For instance, if $(X, \mathscr{G})$ is the discrete topology for an infinite set X, then $D_f(S) = \varnothing$ for every set S. Hence, for any infinite set S we have $S—D_f(S) \notin \mathscr{I}_f$.

From property (1) of Section I, we obtain

(12) Every finite set is D_f-closed.

THEOREM 13 A necessary and sufficient condition in order that the D_f-closed sets coincide with the closed sets is that every finite set be a closed set.

Proof. Assume first of all that the condition holds. Let S be any D_f-closed set and suppose that $x \notin S$. Then $x \notin S'$. Consequently, there is an open set G containing x such that the set $F = S \cap G$ is finite. As $x \notin F$ the set $H = G—F$ is an open set containing x and we have $S \cap H = \varnothing$. This implies that $x \notin \bar{S}$. Hence, $\bar{S} \subset S$. Applying properties (6) and (10) and Theorem 10, we see that S is a closed set. On the other hand, property (1) states that every closed set is D_f-closed. Therefore, the D_f-closed sets coincide with the closed sets.

Conversely, assuming that the D_f-closed sets are the same as the closed sets, property (12) implies that every finite set is a closed set.

COROLLARY 14 A necessary and sufficient condition in order that the D_f-topology coincide with the original topology $(X, \mathscr{G})$ is that the complement of each finite set be an open set.

THEOREM 15 If $(X, \mathscr{G})$ is a topology for which every finite set is a closed set, then, for every set S,

$$\bar{S} = S \cup S'$$

Proof. From property (6) and Theorem 12(i) we obtain $S \cup S' \subset \bar{S}$. For the reverse inclusion, suppose that $x \in \bar{S}$ and $x \notin S$. We have to show $x \in S'$; i.e., for every open set G containing x, the set $S \cap G$ is infinite.

Assume to the contrary that there is an open set G containing x for which the set $F = S \cap G$ is finite. Then $X—(F—\{x\})$ is an open set. Hence, $H = G—(F—\{x\})$ is an open set containing x. As $x \in \bar{S}$ we have $S \cap H \neq \varnothing$. Since $x \in H$ and $x \notin S$, there is a point $y \neq x$ such that $y \in S \cap H$. Then $y \in S \cap (G—F)$. But $S \cap (G—F) = \varnothing$. Thus, our assumption $x \notin S'$ leads to a contradiction. We conclude that $\bar{S} \subset S \cup S'$.

DEFINITION A D_f-homogeneous set is called a dense-in-itself set; that is, S is dense-in-itself if $S \subset S'$. A nonempty D_f-invariant set is called a perfect set; i.e., S is perfect if $S \neq \emptyset$ and $S = S'$. A set that contains no nonempty dense-in-itself set is called a separated set.[1]

Properties (1), (3), and (4) of Section I imply that:

(13) Every nonempty dense-in-itself set is infinite.
(14) The union of any family of dense-in-themself sets is a dense-in-itself set.
(15) The union of a finite family of perfect sets is a perfect set.
(16) Every finite set is a separated set.
(17) Every subset of a separated set is a separated set.

The union of all dense-in-themself subsets of a set S is a dense-in-itself set and is the largest dense-in-itself subset of S. Every set S can thus be decomposed into a dense-in-itself set and a separated set.

(18) If X is dense-in-itself, then every open set is dense-in-itself, as is also every D_f-open set.

For suppose that G is an open set. By property (2), G is D_f-open and consequently $X—G$ is D_f-closed. Hence, $(X—G)' \subset X—G$. Using property (6) of Section I we get

$$G \subset X—(X—G)' \subset X'—(X—G)' \subset (X—(X—G))' = G'$$

(19) If S is a nonempty dense-in-itself set, then $\bar{S}$ is a perfect set.

For, from properties (6) and (2) of Section I, we obtain $S \subset S' \subset (\bar{S})'$. According to property (3), the set $(\bar{S})'$ is a closed set containing S. Hence, by property (11), we have $\bar{S} \subset (\bar{S})'$; that is, $\bar{S}$ is dense-in-itself. Finally, using properties (7) and (1), we get $(\bar{S})' \subset \bar{S}$. Thus, $\bar{S}$ is a perfect set.

We now extend the preceding definition.

DEFINITION Let $\mathscr{I}$ be any ideal containing the ideal $\mathscr{I}_f$ of all finite sets. A D-homogeneous set is called a D-dense-in-itself set. A nonempty D-invariant set is called a D-perfect set.

Then property (5) immediately yields

(20) If $\mathscr{I}$ is an adherent ideal containing all finite sets, then for every set S, either $D(S) = \emptyset$ or $D(S)$ is a D-perfect set.

Using property (1) and Theorems 4 and 6 of Section I, we obtain the following general result.

---

[1] Terminology of Cantor; see Cantor (1966, p. 228).

THEOREM 16 If $\mathcal{S}$ is an adherent ideal containing all finite sets, then every closed set S is uniquely representable as a disjoint union

$$S = P \cup Q$$

where P is either empty or a D-perfect set and Q is a set in $\mathcal{S}$.[1] The set P corresponding to the closed set S is called the D-kernel of S.

We note two consequences of this theorem for the usual topology for $\mathbb{R}^n$.

COROLLARY 17 Every uncountable closed set in $\mathbb{R}^n$ is uniquely representable as the disjoint union of a perfect set and a countable set.

COROLLARY 18 Every closed set in $\mathbb{R}^n$ of positive Lebesgue measure is uniquely representable as the disjoint union of a perfect set P that has positive measure in every open set containing a point of P and a set Q of Lebesgue measure zero.

D. The ideal of sets of the first category

Let $\mathcal{S}$ denote the σ-ideal of sets of the first category for the topology $(X, \mathcal{G})$. Then the D-topology is the basic topology. From Theorem 3 we obtain

THEOREM 19 $(X, \mathcal{G})$ is a Baire topology[2] if and only if X is locally of the second category at every point; i.e., $D(X) = X$.

Using this result, we derive the following consequences of Theorems 7 and 8.

THEOREM 20 If $(X, \mathcal{G})$ is a Baire topology satisfying CCC, then the basic topology satisfies CCC.

THEOREM 21 If $(X, \mathcal{G})$ is a separable Baire topology, then the basic topology is separable.

Furthermore, we have

THEOREM 22 Every Baire topology satisfying CCC is equivalent to its basic topology.

Proof. Let $(X, \mathcal{G})$ be a Baire topology satisfying CCC and let $(X, \mathcal{G}^*)$ denote the associated basic topology.

Suppose that S is a D-nowhere dense set. Let G be any nonempty open set. In view of property (2), there exists a nonempty D-open set $H^* \subset G$ such

---

[1]Cantor (1883; 1966, p. 193), Bendixson (1883), Fréchet (1906, p. 19; 1910, p. 8), and Freud (1958).
[2]An extensive study of Baire topologies is presented in Haworth and McCoy (1977).

that $S \cap H^* = \varnothing$. According to Corollary 2, $H^* = H—I$, where H is a nonempty open set and I is a set of the first category. From the equality

$$S \cap H = [(S \cap H) \cap I] \cup [S \cap (H—I)]$$

we see that $S \cap H$ is a set of the first category. Setting $N = H \cap G$, we thus see every nonempty open set G contains a nonempty open set N such that $S \cap N$ is a set of the first category.

Denoting by $\mathcal{N}$ the family of all nonempty open sets N such that $S \cap N$ is of the first category, we apply Lemmas 3 and 4 of Chapter 1, Section II to obtain a disjoint subfamily $\mathcal{M} = \{N_\alpha : \alpha \in K\}$ of $\mathcal{N}$ such that $X—\bigcup_{\alpha \in K} N_\alpha$ is a nowhere dense set. Because $\mathcal{G}$ satisfies CCC, the index set K is countable. From the equality

$$S = \left[\bigcup_{\alpha \in K} (S \cap N_\alpha) \right] \cup \left(S—\bigcup_{\alpha \in K} N_\alpha \right)$$

we thus see S is a set of the first category. Therefore, every D-nowhere dense set, as well as every set of the first D-category, is a set of the first category.

The foregoing result, in conjunction with Corollary 5, yields $\mathfrak{M}(\mathcal{G}) = \mathfrak{M}(\mathcal{G}^*)$. Utilizing Corollary 2 and Theorems 6 and 7 of Chapter 1, Section III, we obtain $\mathfrak{B}(\mathcal{G}) = \mathfrak{B}(\mathcal{G}^*)$.

We now obtain two characterizations of the sets having the Baire property with respect to the given topology $(X,\mathcal{G})$.[1]

THEOREM 23 A set S has the Baire property if and only if $D(S)—S$ is of the first category.

Proof. Suppose that S has the Baire property. If S is of the first category, then $D(S)—S = \varnothing$ is of the first category. Suppose, on the other hand, that S is of the second category.

Assume to the contrary that $D(S)—S$ is a set of the second category. Then $D(S)—S$ is everywhere of the second category in some nonempty open set G. The set $G \cap D(S)$ being of the second category, there exists a point $x \in G \cap D(S)$. Since G is an open set containing x, the set $G \cap S$ is of the second category. The set $X—S$, which contains $D(S)—S$, is everywhere of the second category in G. Hence, by Theorem 2 of Chapter 1, Section III, the set $G—(X—S) = G \cap S$ is of the first category. We thus have a contradiction. Therefore, $D(S)—S$ is a set of the first category.

Conversely, suppose that $D(S)—S$ is of the first category. We have the identity

$$S = [D(S)—(D(S)—S)] \cup [S—D(S)]$$

---

[1]Cf. Kuratowski (1933a, pp. 51–52; 1966, pp. 88–90).

where $D(S)$ is a closed set, while $D(S)$—S and S—$D(S)$ are sets of the first category. According to Theorem 6 of Chapter 1, Section III, S has the Baire property.

THEOREM 24   A set S has the Baire property if and only if $D(S) \cap D(X$—$S)$ is nowhere dense; i.e., every nonempty open set contains a point at which either S or X—S is locally of the first category.

Proof. Suppose that $D(S) \cap D(X$—$S)$ is not nowhere dense. Then there is a nonempty open set G having the property that for every nonempty open set $H \subset G$, we have $[D(S) \cap D(X$—$S)] \cap H \neq \varnothing$. Now, $D(S) \cap H \neq \varnothing$ implies that $S \cap H$ is of the second category and $D(X$—$S) \cap H \neq \varnothing$ implies that $(X$—$S) \cap H$ is also of the second category. Hence, for every nonempty open set $H \subset G$, both $S \cap H$ and $(X$—$S) \cap H$ are sets of the second category. Therefore, S does not have the Baire property.

Conversely, suppose that S does not have the Baire property. Let G be a nonempty open set such that for every nonempty open set $H \subset G$, both $S \cap H$ and $(X$—$S) \cap H$ are of the second category. From the equality

$$S \cap H = [S \cap D(S) \cap H] \cup [(S$—$D(S)) \cap H]$$

it follows that $D(S) \cap H$ is of the second category. This being true for every H, the set $D(S)$ is everywhere of the second category in G. The set $D(S)$ is a closed set and accordingly has the Baire property. Hence, by Theorem 2 of Chapter 1, Section III, G—$D(S)$ is a set of the first category. Similarly, G—$D(X$—$S)$ is of the first category. From the equality

$$G = [G \cap D(S) \cap D(X$—$S)] \cup [(G$—$D(S)) \cap D(X$—$S)]$$
$$\cup [D(S) \cap (G$—$D(X$—$S))] \cup [(G$—$D(S)) \cap (G$—$D(X$—$S))]$$

we see that $G \cap D(S) \cap D(X$—$S)$ is of the second category. Therefore, $D(S) \cap D(X$—$S)$ is not a nowhere dense set.

E.   Regular and nonregular points

We assume here that the topology $(X,\mathscr{G})$ has a countable topological base $\mathscr{B}$, $\mathscr{I}$ is a σ-ideal of subsets of X containing all countable sets, D is the corresponding derivation operation, and $(X,\mathscr{A})$ is a σ-field containing $\mathscr{G}$ and $\mathscr{I}$.

DEFINITION   A point $x \in X$ is called a regular point for a set S if there is an open set G containing x such that $S \cap G \in \mathscr{A}$. Otherwise, x is called a nonregular point for S.

According to Theorem 1 of Chapter 1, Section 1, in the case that $\mathscr{A} = \mathfrak{B}(\mathscr{G})$ the regular points for a set S are the points at which S has the Baire property locally.

If $(\mathbb{R}^n, \mathscr{G})$ is the usual topology for $\mathbb{R}^n$, $\mathscr{A}$ is the σ-field of Lebesgue measurable sets in $\mathbb{R}^n$, and $\mathscr{I}$ is the σ-ideal of all sets of Lebesgue measure zero, then the regular and nonregular points are called points of measurability and nonmeasurability, respectively.[1]

For each set S, we denote by $N(S)$ the set of all points of X which are nonregular points for S. That is, $x \in N(S)$ if and only if $S \cap G \notin \mathscr{A}$ for every open set G containing x.

The following properties hold for any set S.

(21) $N(S) = N(X—S)$.
(22) $N(S) \subset D(S)$.
(23) $S—N(S) \in \mathscr{A}$.
(24) $N(S \cap N(S)) = N(S)$.
(25) $S \in \mathscr{A}$ if and only if $N(S) = \varnothing$.
(26) $N(S)$ is a closed set.

Proof. (21) and (22) are easily verified.

(23). For each point $x \in S—N(S)$ there is an open set G_x containing x such that $S \cap G_x \in \mathscr{A}$. For any point $y \in S \cap G_x$ we have $y \in S—N(S)$. Therefore,

$$S—N(S) = \bigcup_{x \in S—N(S)} (S \cap G_x)$$

Applying Theorem 1 of Section II to the family $\mathscr{H} = \{G_x : x \in S—N(S)\}$, we obtain a countable set $M \subset S—N(S)$ with

$$\bigcup \mathscr{H} = \bigcup_{x \in M} G_x$$

Then

$$S—N(S) = \bigcup_{x \in M} (S \cap G_x)$$

The latter union being a countable union of sets belonging to $\mathscr{A}$, we conclude that $S—N(S) \in \mathscr{A}$.

(24). If G is any open set, then it follows from the equality

$$G \cap S = [G \cap S \cap N(S)] \cup [G \cap (S—N(S))]$$

and property (23) that $G \cap S \notin \mathscr{A}$ if and only if $G \cap (S \cap N(S)) \notin \mathscr{A}$. This means that $N(S) = N(S \cap N(S))$.

---

[1]These notions were investigated in Alexandrow (1916), Wilkosz (1920, Sec. 2), Zermelo (1927, Sec. 3), Blumberg (1935), Zaubek (1943), and Albanese (1974).

(25). If $S \in \mathscr{A}$, then it follows from the assumption $\mathscr{G} \subset \mathscr{A}$ that $N(S) = \varnothing$. On the other hand, if $N(S) = \varnothing$, then the decomposition

$$S = [S \cap N(S)] \cup [S-N(S)]$$

and property (23) yield $S \in \mathscr{A}$.

(26). Suppose that $x \in \overline{N(S)}$ and let G be any open set containing x. Then there exists a point $y \in G \cap N(S)$. Since G is an open set containing y and $y \in N(S)$, we have $G \cap S \notin \mathscr{A}$. Therefore, $x \in N(S)$, and consequently, $\overline{N(S)} \subset N(S)$. Using property (6), we obtain $N(S) = \overline{N(S)}$. Applying property (10) and Theorem 10, we conclude that $N(S)$ is a closed set.

THEOREM 25   Every set $S \notin \mathscr{A}$ can be represented as the disjoint union of a set that is nonregular at each of its own points and a set that belongs to $\mathscr{A}$.[1]

Proof. We use the decomposition

$$S = [S \cap N(S)] \cup [S-N(S)]$$

Property (24) yields $S \cap N(S) \subset N(S) = N(S \cap N(S))$, so $S \cap N(S)$ is nonregular at each of its own points, and property (23) yields $S-N(S) \in \mathscr{A}$.

As particular instances of this theorem we note:

(i)   Every set in $\mathbb{R}^n$ that does not have the Baire property is representable as the disjoint union of a set with the Baire property and a set that does not have the Baire property locally at any of its own points.

(ii) Every set in $\mathbb{R}^n$ that is not Lebesgue measurable is representable as the disjoint union of a Lebesgue measurable set and a set every point of which is a point of nonmeasurability.

IV.   TOPOLOGICAL PROPERTIES

We assume throughout this part a given topology $(X, \mathscr{G})$.

A.   General properties

We first establish an elementary fact that is useful for simplifying proofs.

THEOREM 1   A necessary and sufficient condition that a set H be an open set is that for each point $x \in H$ there exist an open set $G \subset H$ which also contains x.

---

[1]Cf. Alexandrow (1916), Wilkosz (1920, Sec. 2), Zermelo (1927, Sec. 3), Blumberg (1935, Theorem I), and Zaubek (1943).

Proof. Any open set obviously satisfies this condition.

Assume that H is a set satisfying the condition. For each point $x \in H$ choose an open set G_x such that $x \in G_x$ and $G_x \subset H$. Then $H = \bigcup_{x \in H} G_x$. Therefore, H is an open set.

The family $\mathscr{G}_\delta$ of all countable intersections of open sets and the family $\mathscr{F}_\sigma$ of all countable unions of closed sets have the following properties:

(1) The family $\mathscr{G}_\delta$ is closed under finite unions and countable intersections.
(2) The family $\mathscr{F}_\sigma$ is closed under finite intersections and countable unions.
(3) The complement of a $\mathscr{G}_\delta$-set is an $\mathscr{F}_\sigma$-set.
(4) The complement of an $\mathscr{F}_\sigma$-set is a $\mathscr{G}_\delta$-set.
(5) Each $\mathscr{G}_\delta$-set is representable as the intersection of a descending sequence of open sets.
(6) Each $\mathscr{F}_\sigma$-set is representable as the union of an ascending sequence of closed sets.

In the case of the usual topology for $\mathbb{R}^n$, we have

THEOREM 2 Every open set in $\mathbb{R}^n$ is an $\mathscr{F}_\sigma$-set.

Proof. Assume that G is a nonempty open set. For each point $x \in G$ let F_x be a closed rectangle with rational vertices such that $x \in F_x$ and $F_x \subset G$. Then the family $\{F_x : x \in G\}$ is countable and $G = \bigcup_{x \in G} F_x$.

COROLLARY 3 Every closed set in $\mathbb{R}^n$ is a $\mathscr{G}_\delta$-set.

Using properties (6), (7), (11) of Section III.A, it is seen that the closure $\bar{S}$ of a set is the smallest closed set containing S; i.e.,

$$\bar{S} = \bigcap \{F : F \text{ is a closed set and } S \subset F\}$$

The topological dual of the closure of S is called the interior of S.

DEFINITION The interior of a set S, denoted by Int(S), is the largest open set contained in S; i.e.,

$$\text{Int}(S) = \bigcup \{G : G \text{ is an open set and } G \subset S\}$$

These notions are related in the following manner:

THEOREM 4 For any set S, $\text{Int}(S) = X - \overline{(X - S)}$.

Proof. The set $\overline{X - S}$ is a closed set containing $X - S$ and its complement is an open set contained in S. Hence, $X - \overline{(X - S)} \subset \text{Int}(G)$.

Suppose that G is any open set contained in S. Then $X - G$ is a closed set containing $X - S$, so $\overline{X - S} \subset X - G$ and $G \subset X - \overline{(X - S)}$. Hence, $\text{Int}(S) \subset X - \overline{(X - S)}$.

We note some properties of the interior operation that are dual to properties of the closure operation. For any subsets S, T of X, we have

(7) $\text{Int}(X) = X$.
(8) $\text{Int}(S) \subset S$.
(9) $\text{Int}(S \cap T) = \text{Int}(S) \cap \text{Int}(T)$.
(10) $\text{Int}(\text{Int}(S)) = \text{Int}(S)$.

A composition of these two operations leads to a characterization of nowhere dense sets.

THEOREM 5 A set S is nowhere dense if and only if $\text{Int}(\bar{S}) = \varnothing$.

Proof. Assume that $x \in \text{Int}(\bar{S})$ and let G be an open set such that $x \in G$ and $G \subset \bar{S}$. Every nonempty open set contained in G contains a point of $\bar{S}$ and hence also contains a point of S. Therefore, S is not nowhere dense.

Assume that S is not a nowhere dense set. Then there exists a nonempty open set G every nonempty open subset of which has a nonempty intersection with S. This implies that every point of G belongs to $\bar{S}$, so $G \subset \text{Int}(\bar{S})$. Therefore, $\text{Int}(\bar{S}) \neq \varnothing$.

REMARK The characterization of nowhere dense sets given in this theorem is generally used as the definition of nowhere dense in topology texts. However, this characterization obscures the depth of the analogies between Baire category and Lebesgue measure concepts.

B. Relativization

For a set $P \subset X$ we define

$$\mathcal{G}_P = \mathcal{G} \cap P = \{G \cap P : G \in \mathcal{G}\}$$

$(P, \mathcal{G}_P)$ is a topology, called the relativization[1] of $(X, \mathcal{G})$ to P, or the relative topology for P. In the case that $P \in \mathcal{G}$ we have

$$\mathcal{G}_P = \{G \in \mathcal{G} : G \subset P\}$$

DEFINITION If $P \subset X$ and π is a property of sets, then we say a set $S \subset X$ has the property π in P, or relative to P, if $S \cap P$ has the property π with respect to the relativized topology $(P, \mathcal{G}_P)$.

For subsets S, P, Q of X with $S \subset P \subset Q$ we have:

(ρ1) S is open in P if and only if there is an open set G such that $S \cap P = G \cap P$.

---

[1]The relativization of topological concepts apparently originated in Baire (1899a).

(ρ2) If S is open in P and P is open in Q, then S is open in Q.

(ρ3) S is closed in P if and only if there is a closed set F such that $S \cap P = F \cap P$.

(ρ4) If S is closed in P and P is closed in Q, then S is closed in Q.

Similar statements hold for $\mathscr{G}_\delta$-sets and $\mathscr{F}_\sigma$-sets.

DEFINITION   A set S is dense in a set P if every open set containing a point of P contains a point of S. A set S is everywhere dense in a set P if every nonempty set in $\mathscr{G}_P$ contains at least one point of S.

EXAMPLE   For the usual topology for $\mathbb{R}$, the set $\mathbb{Q}$ of rational numbers is dense in the set of irrational numbers, but $\mathbb{Q}$ is not everywhere dense (in fact, it is nowhere dense) in the set of irrational numbers.

REMARK   If S is everywhere dense in P, then S is dense in P. If $S \subset P$ and S is dense in P, then S is everywhere dense in P.

REMARK   According to the preceding definition every set is dense in itself. However, a set need not be dense-in-itself as defined in Section III.C (i.e., S is dense-in-itself if every open set containing a point of S contains infinitely many points of S). Thus, the hyphenated and unhyphenated terms are not identical. Of course, every dense-in-itself set is dense in itself.

It is a simple matter to verify the following properties for subsets of X.

(ρ5) S is dense in P if $P \subset \bar{S}$.

(ρ6)   If R is dense in S and S is dense in T, then R is dense in T.

(ρ7) S is dense in $\bar{S}$.

(ρ8) S is dense in $\text{Int}(\bar{S})$.

(ρ9)   If R is dense in S, then R is dense in $\bar{S}$.

(ρ10)   If S is dense in an open set G, then $S \cap G$ is dense in G.

THEOREM 6   If the topology $(X, \mathscr{G})$ has a countable topological base, then every set P has a countable subset that is everywhere dense in P.

Proof. Let $\mathscr{B}$ be a countable topological base for $(X, \mathscr{G})$. We form a countable set D by choosing one point from each of the nonempty sets $B \cap P$ with $B \in \mathscr{B}$. Then D is everywhere dense in P.

We note that the sets which are dense in no nonempty open set are the same as the nowhere dense sets.

THEOREM 7   If F is a closed set whose complement is everywhere dense in a set P, then F is nowhere dense in P.

Proof. If H is any nonempty set in $\mathscr{G}_P$, then $H \cap (X - F)$ is a nonempty set in $\mathscr{G}_P$ which is contained in H and disjoint from F.

THEOREM 8 Suppose that K, L, M, N are subsets of X with $K \subset L \subset M \subset N$.

(i) If L is nowhere dense in M, then K is nowhere dense in N.

(ii) If L is of the first category in M, then K is of the first category in N.

Proof. (i). It suffices to show that L is nowhere dense in N.

Suppose that G^* is a nonempty set in $\mathscr{G}_N$. We have to determine a nonempty set $H^* \in \mathscr{G}_N$ satisfying $H^* \subset G^*$ and $H^* \cap L = \varnothing$. Now, if $G^* \cap M = \varnothing$, then it suffices to take $H^* = G^*$. We therefore assume that $G^* \cap M \neq \varnothing$. Because L is nowhere dense in M and $G^* \cap M$ is a nonempty set in $\mathscr{G}_M$, there is then a nonempty set $J^* \in \mathscr{G}_M$ such that $J^* \subset G^* \cap M$ and $J^* \cap L = \varnothing$. We have $G^* = G \cap N$ and $J^* = J \cap M$, where G and J are open sets in $\mathscr{G}$. It then suffices to take $H^* = (G \cap J) \cap N$.

(ii). We have $L = \bigcup_{n=1}^{\infty} L_n$ where each set L_n is nowhere dense in M and $K = \bigcup_{n=1}^{\infty} K_n$, where $K_n = K \cap L_n$ for each n. By (i) each set K_n is nowhere dense in N. Hence, K is of the first category in N.

THEOREM 9 Suppose that K, L, M, N are subsets of X with $K \subset L \subset M \subset N$, where L is dense in M and M is open in N. Then

(i) K is nowhere dense in N if and only if K is nowhere dense in L.

(ii) K is of the first category in N if and only if K is of the first category in L.[1]

Proof. (i). In view of the preceding theorem, we have only to show that K is nowhere dense in L whenever K is nowhere dense in N.

Let G^* be any nonempty set in $\mathscr{G}_L$. Then $G^* = G \cap L$ where $G \in \mathscr{G}_N$. The set $J = G \cap M$ is a nonempty set in $\mathscr{G}_N$. Since K is assumed to be nowhere dense in N, there is a nonempty set $H \in \mathscr{G}_N$ such that $H \subset J$ and $H \cap K = \varnothing$. The set $H^* = H \cap L$ is a subset of G^* that belongs to $\mathscr{G}_L$ and is disjoint from K. Since H is a nonempty subset of M that belongs to $\mathscr{G}_N$ and L is dense in M, the set H^* is nonempty. We thus see K is nowhere dense in L.

(ii) is an obvious consequence of (i).

We conclude this section with a decomposition theorem.[2]

THEOREM 10 Each set of the first category is representable as the disjoint union of a nowhere dense set and a set that is of the first category in itself.

Proof. For a set S of the first category we use the decomposition

(*) $$S = [S \cap \text{Int}(\bar{S})] \cup [S - \text{Int}(\bar{S})]$$

[1]Oxtoby (1980, p. 97).
[2]Oxtoby (personal communication, 1987); see also Oxtoby (1980, p. 97).

We show that the set S—Int($\bar{S}$) is nowhere dense. Suppose that G is a nonempty open set and for every nonempty open set $H \subset G$ we have $H \cap S \neq \varnothing$. Then every point of G is an accumulation point for S and thus is an element of $\bar{S}$. Since G is an open set contained in $\bar{S}$, we have $G \subset \text{Int}(\bar{S})$ and consequently $G \cap [S—\text{Int}(\bar{S})] = \varnothing$. Therefore, if G is any nonempty open set, then either $G \cap [S—\text{Int}(\bar{S})] = \varnothing$ or there exists a nonempty open set $H \subset G$ such that $H \cap S = \varnothing$, so that $H \cap [S—\text{Int}(\bar{S})] = \varnothing$. We conclude that S—Int($\bar{S}$) is nowhere dense.

To show that the set $T = S \cap \text{Int}(\bar{S})$ is of the first category in itself we first recall that the set S is dense in the set Int($\bar{S}$). Hence, according to property ($\rho 10$), T is dense in Int($\bar{S}$). Applying Theorem 9 with $K = T$, $L = T$, $M = \text{Int}(\bar{S})$, $N = X$, we conclude that T is of the first category in itself.

C. The Banach Category Theorem

The topological version of the Fundamental Theorem, which has become known as the Banach Category Theorem,[1] is virtually the only nonelementary theorem that is true for every topology. First established by Baire in a classical setting, Banach's generalization was the crucial factor necessary to extend most of the theory concerning Baire's category concepts to arbitrary topologies. In establishing it Banach had in a sense "broken the ice."[2] In this section we give several variant forms of the Banach Category Theorem. The first form, which is the topological specialization of the Fundamental Theorem, is essentially the form in which Baire stated the theorem for the so-called Baire space of all infinite sequences of natural numbers.[3]

THEOREM 11   Any set of the second category is everywhere of the second category in some nonempty open set.

The next form can also be obtained from the solution of a generalization of a game of Mazur.[4] Such a generalization led to the formulation of the axioms for a category base.

We recall that a set is locally of the first category at a point x if there is an open set containing x in which S is of the first category. A set S is locally of the

[1]Cf. Oxtoby (1957; 1980, Chap. 16).
[2]Marczewski (1948, p. 96).
[3]Cf. Baire (1899b, p. 948), Lebesgue (1905, pp. 185–186), and Hahn (1932, 19.5.42).
[4]Cf. Mycielski, Świerczkowski, and Zięba (1956), Oxtoby (1957, 1980), Volkmann (1959), and Ulam (1964a). A measure-theoretic version of Mazur's game is also given in Mycielski (1966). Generalizations of Mazur's game appear in Oxtoby (1957), Morgan (1974), Yates (1976), Kechris (1977), and Lisagor (1979, 1981).

second category at a point x if S is of the second category in every open set containing x.

THEOREM 12 Any set of the second category is locally of the second category at every point of some nonempty open set.

Proof. Let S be a set of the second category. According to Theorem 11, there is a nonempty open set G in which S is everywhere of the second category. Suppose that x is any point of G. If H is any open set containing x, then $K = G \cap H$ is a nonempty open subset of G. Hence, $S \cap K$ is a set of the second category, as is also the set $S \cap H$. Thus, S is locally of the second category at every point of G.

The following variant form is that enunciated by Banach. Although Banach formulated this theorem in a more restrictive setting, his method of proof readily extends to topologies.[1] His proof is basically the same as the proof given above for the Fundamental Theorem. As noted by Marczewski,[2] Banach's proof "combines a remarkable ingenuity with a quite astonishing simplicity; it is impossible to forget it!"

THEOREM 13 A set that is locally of the first category at each of its own points is a set of the first category.

Proof. We establish the contrapositive. Assume that S is a set of the second category. By Theorem 12, S is locally of the second category at every point of some nonempty open set G. Let x be a point of G. Then $S \cap G$ is of the second category and hence is nonempty. Therefore, S contains a point y at which S is not locally of the first category.

The remaining variations involve set-theoretical operations.

THEOREM 14 If $\mathscr{H}$ is a family of sets of the first category, each of which is open in the union $U = \bigcup \mathscr{H}$, then U is also of the first category.[3]

Proof. If x is any point of U, then there is a set $H \in \mathscr{H}$ containing x. The set H being open in U, there is an open set G containing x such that $H = G \cap U$. Therefore, U is locally of the first category at each of its points x. By Theorem 13, U is a set of the first category.

THEOREM 15 The union of any family of open sets of the first category is a set of the first category.[4]

[1]See Banach (1930), Hahn (1932, Sec. 19.5), and Banach (1967, pp. 345–347).
[2]Cf. Marczewski (1948, p. 96) and Ulam (1946, p. 601).
[3]Kuratowski (1933a, pp. 44–45; 1966, p. 82).
[4]Oxtoby (1980, p. 62). For a measure analogue, see Marczewski and Sikorski (1948) and Oxtoby (1980, Chap. 16).

Proof. Apply Theorem 14.

THEOREM 16 If $\mathscr{H}$ is any family of disjoint open sets and R is a set such that $H—R$ is of the first category for each $H \in \mathscr{H}$, then $(\bigcup \mathscr{H})—R$ is a set of the first category.[1]

Proof. Let $U = (\bigcup \mathscr{H})—R = \bigcup_{H \in \mathscr{H}} (H—R)$. For each $H \in \mathscr{H}$ we have $H—R = H \cap U$. Hence, each set $H—R$ is open in U. Applying Theorem 14 to the family $\mathscr{H}^* = \{H—R : H \in \mathscr{H}\}$, we see that U is a set of the first category.

We have established the following implications between the foregoing theorems:

$$11 \Rightarrow 12 \Rightarrow 13 \Rightarrow 14 \Rightarrow 15$$

We now establish the implication

$$15 \Rightarrow 11$$

and thus obtain the equivalence of these five theorems, without use of the Axiom of Choice.

Proof. Suppose that S is a set which is not everywhere of the second category in any nonempty open set. Then every nonempty open set G contains a nonempty open set H such that $H \cap S$ is of the first category. We show that S must be a set of the first category.

Let $\mathscr{H}$ denote the family of all nonempty open sets H in which S is of the first category and which have the additional property that for every nonempty open set $K \subset H$, the set $K \cap S$ is nonempty. Define $U = \bigcup (\mathscr{H} \cap S)$. For each $H \in \mathscr{H}$ we have $H \cap S = H \cap U$, so $H \cap S$ is open in U. Setting $K = H \cap S, L = U, M = \bigcup \mathscr{H}$, and $N = X$ in Theorem 9(ii), we see that $H \cap S$ is of the first category in U for every $H \in \mathscr{H}$. By Theorem 15, the set U is of the first category in itself. Hence, according to Theorem 8, U is a set of the first category.

Suppose that G is any nonempty open set. Let H be a nonempty open subset of G in which S is of the first category. If $H \notin \mathscr{H}$, then there is a nonempty open set $K \subset H$ which is disjoint from S and hence is also disjoint from $S—U$. If $H \in \mathscr{H}$, then H is disjoint from $X—\bigcup \mathscr{H}$ and hence is also disjoint from $S—U$. We thus see that $S—U$ is a nowhere dense set.

From the equality $S = U \cup (S—U)$ we now conclude that S is a set of the first category.

To establish the equivalence of Theorem 16 with the other five theorems it appears necessary to utilize the Axiom of Choice. As we have already

[1] Cf. Sikorski (1964, p. 75).

proved the implications

$$11 \Rightarrow 12 \Rightarrow 13 \Rightarrow 14 \Rightarrow 16$$

we have only to prove the implication

$$16 \Rightarrow 11$$

Proof. Suppose that S is a set which is not everywhere of the second category in any nonempty open set. Then every nonempty open set contains a nonempty open set in which S is of the first category. Apply Lemmas 3 and 4 of Chapter 1, Section II to obtain a disjoint family $\mathscr{H}$ of nonempty open sets such that $X-\bigcup\mathscr{H}$ is nowhere dense and $H \cap S$ is of the first category for each $H \in \mathscr{H}$. Applying Theorem 16 with $R = X-S$, we see that $S \cap (\bigcup\mathscr{H})$ is a set of the first category. From the inclusion

$$S \subset [S \cap (\bigcup\mathscr{H})] \cup (X-\bigcup\mathscr{H})$$

we conclude that S is a set of the first category.

D. Sets with the Baire property

For a topology the assumption of CCC in Theorem 5 of Chapter 1, Section II, and Theorems 7 and 14 of Chapter 1, Section III may be deleted.

THEOREM 17 Every nowhere dense set is contained in a closed, nowhere dense set. Every set of the first category is contained in an $\mathscr{F}_\sigma$-set of the first category.

Proof. Just as in the proof of Theorem 5 of Chapter 1, Section II, we first consider a nowhere dense set S, let $\mathscr{N}$ be the family of all nonempty open sets disjoint from S, and apply Lemmas 3 and 4 of Chapter 1, Section II to obtain a disjoint subfamily $\mathscr{M}$ of $\mathscr{N}$ such that $X-\bigcup\mathscr{M}$ is nowhere dense. The set $\bigcup\mathscr{M}$ is an open set and S is contained in the closed, nowhere dense set $X-\bigcup\mathscr{M}$. This fact then implies that every set of the first category is contained in $\mathscr{F}_\sigma$-set of the first category.

In addition to the two characterizations of the sets having the Baire property given in Section III.D, we have

THEOREM 18 The following statements are equivalent for a set S.

(i) S has the Baire property.
(ii) $S = (G-P) \cup R$, where G is an open set and P, R are sets of the first category.
(iii) $S = (F-Q) \cup T$, where F is a closed set and Q, T are sets of the first category.

(iv) S is the union of a $\mathcal{G}_\delta$-set and a set of the first category.

(v) S is the difference of an $\mathcal{F}_\sigma$-set and a set of the first category.

Proof. The proof is completely analogous to the proof of Theorem 7 of Chapter 1, Section III, with the following modifications: We have

$$S = \left[\left(\bigcup_{\alpha \in I} M_\alpha \right) - \bigcup_{\alpha \in I} (M_\alpha - S) \right] \cup \left(S - \bigcup_{\alpha \in I} M_\alpha \right)$$

where the set $S - \bigcup_{\alpha \in I} M_\alpha$ has been established to be of the first category. The set $\bigcup_{\alpha \in I} M_\alpha$ will now be an open set. Applying Theorem 16, we see that $\bigcup_{\alpha \in I} (M_\alpha - S)$ is a set of the first category.

The implication (i) $\Rightarrow$ (ii) being thereby established, we obtain the implication (ii) $\Rightarrow$ (iv) as before, utilizing Theorem 17 in lieu of Theorem 5 of Chapter 1, Section II. The remaining implications are established as before.

For the next two theorems we denote by D the derivation operation corresponding to the σ-ideal $\mathcal{J}$ of sets of the first category. That is, $D(S)$ is the set of all points at which S is locally of the second category.

THEOREM 19   Every set has a hull that is an $\mathcal{F}_\sigma$-set.

Proof. Let S be an arbitrary set. We know from Theorem 12 of Chapter 1, Section II that $S - D(S)$ is a set of the first category. According to Theorem 17, there is an $\mathcal{F}_\sigma$-set K of the first category containing $S - D(S)$. The set $E = K \cup D(S)$ will therefore be an $\mathcal{F}_\sigma$-set containing S.

Suppose now that F is any set with the Baire property containing S. Then $D(S) \subset D(F)$ and consequently

$$E - F = (K - F) \cup [D(S) - F] \subset K \cup [D(S) - S]$$

The set K is of the first category and, by Theorem 23 of Section III, the set $D(S) - S$ is also of the first category. Therefore, $E - F$ is a set of the first category.

We next obtain a characterization of the sets having the Baire property with respect to the topology $(X, \mathcal{G})$ in terms of closed sets for the basic topology.[1]

THEOREM 20   A necessary and sufficient condition that a set have the Baire property is that it be representable as the difference of two D-closed sets.

Proof. It follows from Theorem 1 of Section III and Theorem 6 of Chapter 1,

[1]Cf. Hashimoto (1976).

Section III that if a set is representable as the difference of two D-closed sets, then it has the Baire property.

Conversely, suppose that S is a set which has the Baire property. Then $X—S$ also has the Baire property. From Theorem 18 we have

$$X—S = (G—P) \cup R$$

where G is an open set and P, R are sets of the first category. This implies that

$$S = [(X—G) \cup P]—R$$

The set $X—G$ being closed, it follows from Theorem 1 of Section III that the set $(X—G) \cup P$ is D-closed, as is also the set R. Therefore, S is the difference of two D-closed sets.

THEOREM 21 If S and T have the Baire property and are locally of the first category at the same points, then $S \approx T$.[1]

Proof. Suppose that T is of the second category in an open set G. Then T is everywhere of the second category in some nonempty open set $H \subset G$. Choose a point $x \in H$. As T is locally of the second category at x, so also is the set S. This implies that $S \cap G$ is of the second category. We thus see S is of the second category in every open set in which T is of the second category.

By Theorem 3 of Chapter 1, Section III, $T—S$ is of the first category. Similarly, it is seen that $S—T$ is of the first category. Therefore, $S \approx T$.

REMARK This result is not valid for all category bases. For example, let $\mathscr{C}$ be the family of all compact sets of positive Lebesgue measure in $\mathbb{R}$ and let $S = [0,1]$, $T = [1,2]$. Both S and T are locally abundant at every point, but $S \not\approx T$.

E. Functions with the Baire property

The Baire functions for a topology are called functions that have the Baire property. Thus, a function $f: X \to \mathbb{R}$ has the Baire property if and only if the set $f^{-1}(I)$ has the Baire property for every open interval $I \subset \mathbb{R}$. A function $f: X \to \mathbb{R}$ is called a continuous function if and only if $f^{-1}(I)$ is an open set for every open interval I. We note that in both of these defining conditions, the intervals I may be confined to open intervals with rational endpoints. Obviously, every continuous function has the Baire property.

The foregoing notions may be relativized. For a function $f: X \to \mathbb{R}$ and a set $P \subset X$, we denote by f_P the restriction of f to the set P. We say that f has the Baire property on P if for every open interval I the set

---

[1]Cf. Nikodym (1925, Lemme I).

$f_P^{-1}(I) = P \cap f^{-1}(I)$ has the Baire property in P. We say that f is continuous on P if $f_P^{-1}(I)$ is open in P for every open interval I; i.e., if the restriction $f_P: P \to \mathbb{R}$ is continuous with respect to the relative topology for P.

According to the following theorem,[1] the functions with the Baire property are "almost continuous functions."

THEOREM 22 A necessary and sufficient condition that a function $f: X \to \mathbb{R}$ have the Baire property is the existence of a set M of the first category such that f is continuous on $X - M$.

Proof. Suppose that $f: X \to \mathbb{R}$ has the Baire property. Let $\langle I_n \rangle_{n \in \mathbb{N}}$ be an enumeration of all open intervals with rational endpoints. According to Theorem 18, for each $n \in \mathbb{N}$ there exists an open set G_n and sets P_n, R_n of the first category such that

$$f^{-1}(I_n) = (G_n - P_n) \cup R_n$$

The set $M = \bigcup_{n=1}^{\infty} (P_n \cup R_n)$ is a meager set. For any open interval I_n with rational endpoints the set

$$f_{X-M}^{-1}(I_n) = f^{-1}(I_n) - M = [(G_n - P_n) \cup R_n] - M = G_n - M$$

is open in $X - M$. Therefore, f is continuous on $X - M$.

Conversely, suppose that M is a set of the first category and $f: X \to \mathbb{R}$ is a function continuous on $X - M$. If I is any open interval with rational endpoints, then

$$f^{-1}(I) = [f^{-1}(I) \cap M] \cup [f^{-1}(I) - M] = [f^{-1}(I) \cap M] \cup f_{X-M}^{-1}(I)$$

From the assumed continuity of f, the set $f^{-1}(I)$ is open in $X - M$. Hence, there exists an open set G such that $f_{X-M}^{-1}(I) = G - M$. This set has the Baire property in X, as does also the first category set $f^{-1}(I) \cap M$. Hence, the set $f^{-1}(I)$ has the Baire property. We thus see that the function f has the Baire property.

F. The absolute Baire property

DEFINITION   A set is a scattered set if it is nowhere dense in every perfect set.[2]

In this definition the term "perfect" can be replaced by "dense-in-itself."

[1]Cf. Sierpiński (1924a, 1928a), Nikodym (1929), Kuratowski (1930; 1966, p. 400), Kunugui (1937), Inagaki (1954), Hansell (1971, Theorem 9), Pol (1976), Emeryk, Frankieweicz, and Kulpa (1979), and Frankiewicz (1982).
[2]Denjoy (1915); see also Semadeni (1959).

THEOREM 23   A set is scattered if and only if it is nowhere dense in every dense-in-itself set.

Proof. A set that is nowhere dense in every dense-in-itself set is obviously scattered.

Assume that S is a scattered set and Q is a dense-in-itself set. Without loss of generality, we can assume that Q is nonempty. According to property (19) in Section III, the set $P = \bar{Q}$ is a perfect set. Suppose that G is any open set with $G \cap Q \neq \emptyset$. Then $G \cap P \neq \emptyset$. Since S is nowhere dense in P, there is an open set H such that $H \cap P \neq \emptyset$, $H \cap P \subset G \cap P$, and $(H \cap P) \cap S = \emptyset$. This implies that $H \cap Q \neq \emptyset$, $H \cap Q \subset G \cap Q$, and $(H \cap Q) \cap S = \emptyset$. Therefore, S is nowhere dense in Q.

Concerning the set-theoretical structure of the family of scattered sets, it is clear that

THEOREM 24   The family of scattered sets forms an ideal.

The relationship between the scattered sets and the separated sets is clarified in the following two theorems.[1]

THEOREM 25   Every scattered set is a separated set.

Proof. Assume that S is a set which is not a separated set. Then S contains a nonempty dense-in-itself set D and $P = \bar{D}$ is a perfect set. For any open set G, $G \cap P \neq \emptyset$ implies that $G \cap D \neq \emptyset$ and consequently, $G \cap S \neq \emptyset$. This means that S is not scattered set.

THEOREM 26   For a topology $(X, \mathcal{G})$ for which every finite set is a closed set, the separated sets coincide with the scattered sets.

Proof. Assume that $(X, \mathcal{G})$ is such a topology and S is a set that is not scattered. Let P be a perfect set for which $S \cap P$ is not nowhere dense in P. Then there is a nonempty set G^* in $\mathcal{G}_p$ such that $S \cap H^* \neq \emptyset$ for every nonempty set $H^* \subset G^*$ belonging to $\mathcal{G}_p$. We show that $S \cap G^*$ is dense-in-itself.

Let x be any point in $S \cap G^*$ and let J be any open set containing x. Then $J \cap G^*$ is a nonempty set in $\mathcal{G}_p$. The set P being dense-in-itself, it follows from property (18) in Section III that $J \cap G^*$ is dense-in-itself and, consequently, is an infinite set. For any finite set F, the set $H^* = (J \cap G^*) - F$ is a nonempty subset of G in $\mathcal{G}_p$. Accordingly, $S \cap H^* \neq \emptyset$. This implies that $J \cap (S \cap G^*)$ is infinite. Therefore, $S \cap G^*$ is a nonempty dense-in-itself subset of S. We conclude that S is not a separated set.

[1]Cf. Fréchet (1927, 1929).

EXAMPLE In the case of the indiscrete topology for the set N, every nonempty finite set is a separated set that is not a scattered set.

As an immediate consequence of Theorem 23 we obtain

THEOREM 27 If X is dense-in-itself, then every scattered set is nowhere dense.

DEFINITION A set is always of the first category if it is of the first category in every perfect set.[1]

THEOREM 28 The sets always of the first category form a σ-ideal containing all scattered sets.

THEOREM 29 If X is dense-in-itself, then each set always of the first category is a set of the first category.

DEFINITION A set has the absolute Baire property if it has the Baire property in every perfect set.

THEOREM 30 The family of sets that have the absolute Baire property forms a σ-field which contains the open sets and the sets always of the first category.

THEOREM 31 The family of sets that have the absolute Baire property is closed under operation $\mathscr{A}$.

THEOREM 32 If X is dense-in-itself, then every set having the absolute Baire property has the Baire property.

We note that the sets always of the first category and the sets having the absolute Baire property have as measure analogues the absolute null sets and the absolute measurable sets.[2]

DEFINITION A function $f: X \to \mathbb{R}$ has the absolute Baire property it is has the Baire property on every perfect set.

THEOREM 33 Every continuous function has the absolute Baire property.

THEOREM 34 If X is dense-in-itself, then every function having the absolute Baire property has the Baire property.

More generally, one can define extended real-valued functions having the absolute Baire property and verify that most of the results of Sections IV.A to IV.C of Chapter 1 are valid.

---

[1]Cf. Luzin (1914, 1921).
[2]Cf. Szpilrajn (1937), Morgan (1976, 1978b, 1979, 1984), and the references cited therein.

5

Perfect Bases

We assume in this chapter that the underlying space $X = \mathbb{R}^n$ unless otherwise specified.

I. PERFECT SETS AND BASES

A. Definition and examples

DEFINITION A category base $(X,\mathscr{C})$ consisting of perfect sets is called a perfect base if the following condition is satisfied:

    (+) For every region A and every point $x \in A$ there is a descending sequence $\langle A_n \rangle_{n \in \mathbb{N}}$ of regions such that $x \in A_n$, $A_n \subset A$, and $\operatorname{diam}(A_n) \leqslant 1/n$ for each $n \in \mathbb{N}$.

    An equivalent characterization is given in the following theorem, whose proof is elementary.

THEOREM 1 A necessary and sufficient condition that a category base consisting of perfect sets by a perfect base is that for each region A, each positive integer n, and each pair x_1, x_2 of different points in A, there exist

144

disjoint regions A_1, A_2 such that $x_i \in A_i$, $A_i \subset A$, and $\operatorname{diam}(A_i) \leqslant 1/n$ for $i = 1, 2$.

In subsequent sections of this chapter we assume that a given perfect base $(X,\mathscr{C})$ is under consideration.

EXAMPLE $\mathscr{C}$ is the family of all closed rectangles in $\mathbb{R}^n$.

EXAMPLE $\mathscr{C}$ is the family of all compact perfect sets in $\mathbb{R}^n$ that have positive Lebesgue measure in every open set containing one of their points.

EXAMPLE   (Assume CH.) $\mathscr{C}$ is the family of all perfect sets in $\mathbb{R}^n$ that have positive Hausdorff measure in every open set containing one of their points, for a given function $h \in \mathscr{H}_c$.

EXAMPLE   (Assume CH.) $\mathscr{C}$ is the family of all perfect sets in $\mathbb{R}^n$ that have positive Hausdorff dimension in every open set containing one of their points.

The latter three examples are obtainable in a natural manner from the related examples given in Chapter 1, Section I.C. Assume that $(X,\mathscr{C})$ is a point-meager, Baire base consisting of closed sets, which contains all closed rectangles. We know from Theorem 16 of Chapter 4, Section III that every abundant closed set A is uniquely representable in the form $A = A^* \cup M$, where M is a meager set and A^* is a perfect set, called the kernel of A and characterized by the property that for any point $x \in X$

(*) $x \in A^*$ if and only if, for every open set G containing x, the set $A \cap G$ is abundant.

In the case that A is a closed rectangle we have $A^* = A$. Let $\mathscr{C}^*$ denote the family of all compact perfect subsets A^* of $\mathbb{R}^n$ having the property (*). Assume, in addition to the assumption above, that $\mathscr{C}^* \subset \mathscr{C}$. Then we have

THEOREM 2 $(X,\mathscr{C}^*)$ is a perfect base that is equivalent to $(X,\mathscr{C})$.

Proof. Because every closed rectangle belongs to $\mathscr{C}^*$ we have $X = \bigcup \mathscr{C}^*$. Applying Theorem 2 of Chapter 1, Section I, and Theorem 1 of Chapter 1, Section III, we see that $(X,\mathscr{C}^*)$ is a category base equivalent to $(X,\mathscr{C})$. As every set in $\mathscr{C}^*$ is a perfect set, we have only to verify the condition $(+)$ holds. For this we take E_n to be closed square having the given point x as center with diameter of magnitude $1/n$ and let A_n be the kernel of the set $A \cap E_n$.

In addition to the foregoing examples of perfect bases, one can construct various hybrids.

EXAMPLE $\mathscr{C}$ is the family of all product sets $A \times B$ in $\mathbb{R}^2$, where A is a compact perfect subset of $\mathbb{R}$ having positive Lebesgue measure in every open interval containing one of its points and B is a bounded closed subinterval of

$\mathbb{R}$. We note that if S is a meager set, then almost[1] every vertical section of S is a set of the first category, and if S is a Baire set, then almost every vertical section of S is a set that has the Baire property.[2]

EXAMPLE $\mathscr{C}$ consists of all compact perfect subsets of the set of negative real numbers that have positive Lebesgue measure in every open interval containing one of their points and of all bounded, closed subintervals of the set of nonnegative real numbers.

A further example of a perfect base, that of all perfect sets in $\mathbb{R}^n$, is discussed in Section I.E.

REMARK The notion of a perfect base and the resultant theory can be extended to include additional examples of interest:[3] for instance, the family $\mathscr{C}$ of all sets in $\mathbb{R}^n$ representable in the form $A—M$, where A is a closed line segment and M is a countable set (assuming CH). For this category base, the meager sets are the subsets of $\mathbb{R}^n$ that are of the first category on every line segment and the Baire sets are the sets that have the Baire property on every line segment. Analogously, we have the category base of all sets in $\mathbb{R}^n$ representable in the form $A—M$, where A is a compact, perfect subset of some line segment L that has positive linear Lebesgue measure in every open subinterval of L containing a point of A and M is a countable set (assuming CH). We note that there exists a subset of $\mathbb{R}^2$ which has linear Lebesgue measure zero on every line but is nonmeasurable with respect to planar Lebesgue measure.[4]

B. Basic properties

THEOREM 3 Every perfect base is a point-meager, Baire base.

Proof. We know from property (13) of Chapter 4, Section III that each region is an infinite set. It then follows from Theorem 1 that each set consisting of a single point is singular. Therefore, the base is point-meager.

The fact that a perfect base is a Baire base is a consequence of the classical theorem[5] that a descending sequence of nonempty compact sets has a nonempty intersection.

---

[1]That is, with the exception of a set of Lebesgue measure zero.
[2]See Morgan (1982, Sec. 5).
[3]See Morgan (1986).
[4]Sierpiński (1920c).

[5]Although commonly called the Cantor Intersection Theorem, Cantor established this result only for a sequence of successive derived sets of a bounded set. The theorem for an arbitrary sequence of compact sets was first established by Baire. See Cantor (1966, p. 225), Baire (1899a, p. 48), Young and Young (1906, p. 26), Grattan-Guinness (1971, p. 127), and Morgan (1978a).

COROLLARY 4   Every closed rectangle is an abundant set.

THEOREM 5   If F is a closed set that has a nonempty intersection with every subregion of a region A, then $A \subset F$.

Proof. Suppose that $x \in A$ and let $\langle A_n \rangle_{n \in \mathbb{N}}$ be a descending sequence of regions satisfying the condition $(+)$. Then $\langle A_n \cap F \rangle_{n \in \mathbb{N}}$ is a descending sequence of nonempty, bounded closed sets. Hence, there is a point belonging to all the sets $A_n \cap F$. But the only point belonging to every set A_n is the point x. Therefore, $x \in F$.

COROLLARY 6   Every closed set that is not singular contains a region.

COROLLARY 7   If A and B are regions whose symmetric difference is meager, then $A = B$.

Proof. If C is any subregion of A, then it follows from the equality

$$C = (C \cap B) \cup (C-B)$$

and the fact that $C-B$, as a subset of $A-B$, is meager that $C \cap B \neq \varnothing$. This means that $A \subset B$. Similarly, it is seen that $B \subset A$. Thus, $A = B$.

C. Perfect sets

This section is devoted primarily to establishing properties involving perfect sets in $\mathbb{R}^n$ which are utilized to prove subsequent theorems about perfect bases. We begin with the notion of a condensation point.[1]

DEFINITION   A point $x \in X$ is called a condensation point for a set S if every open rectangle containing x contains uncountably many points of S.

Obviously, a condensation point for a set is a limit point for that set.

THEOREM 8   In every uncountable set there are uncountably many points that are condensation points for that set.

Proof. Apply Theorem 3 of Chapter 4, Section II to the σ-ideal $\mathscr{I}$ of all countable subsets of X.

THEOREM 9   The set of all points of a set S that are condensation points for S is dense-in-itself.

Proof. This is a consequence of the proof of Theorem 3 of Chapter 4, Section I.

COROLLARY 10   Every uncountable set contains an uncountable dense-in-itself set.

---

[1]Cf. Cantor (1966, pp. 264–275), Lindelöf (1903), and Young (1903a).

COROLLARY 11   Every separated set is countable.

According to the Cantor-Bendixson Theorem, every uncountable closed set F is representable in the form $F = P \cup Q$, where P is a perfect set (called the perfect kernel of F), the set Q is countable, and $P \cap Q = \varnothing$.

THEOREM 12   The set of all condensation points for an uncountable closed set F coincides with the perfect kernel of F.

Proof. Let F^* denote the set of all condensation points for F. We know from Theorem 8 that F^* is nonempty. Because F is closed, each condensation point for F, being a limit point for F, must belong to F, so $F^* \subset F$. Hence, by Theorem 9, F^* is dense-in-itself. Every limit point for F^*, being a condensation point for F, belongs to F^*, whence F^* is a closed set. Therefore, F^* is a perfect subset of F.

We have the decomposition $F = F^* \cup (F - F^*)$. From Theorem 3 of Chapter 4, Section II we know that $F - F^*$ is a countable set. We conclude from the uniqueness of the Cantor-Bendixson decomposition, established in Theorem 6 of Chapter 4, Section I that F^* coincides with the perfect kernel of F.

THEOREM 13   Every point of a perfect set is a condensation point for that set.

Proof. Let x be a point of a perfect set P and let E be any open rectangle containing x. Assume to the contrary that $P \cap E$ is a countable set. As x is a limit point for P, the set $P \cap E$ is denumerable. Let

(*) $x_1, x_2, \ldots, x_n, \ldots$ $(n \in \mathbb{N})$

be an enumeration of all points of $P \cap E$. Define E_1 to be an open rectangle such that $P \cap E_1 \neq \varnothing$, $x_1 \notin \bar{E}_1$, $\bar{E}_1 \subset E$, and $\operatorname{diam}(\bar{E}_1) \leqslant 1$. Continuing inductively, for each $n \in \mathbb{N}$ we define E_{n+1} to be an open rectangle such that $P \cap E_{n+1} \neq \varnothing$, $x_{n+1} \notin \bar{E}_{n+1}$, $\bar{E}_{n+1} \subset E_n$, and $\operatorname{diam}(\bar{E}_{n+1}) \leqslant 1/(n+1)$. We thereby determine a descending sequence $\langle \bar{E}_n \rangle_{n \in \mathbb{N}}$ of closed rectangles with $\lim_{n \to \infty} \operatorname{diam}(\bar{E}_n) = 0$. Let y denote the point belonging to all these rectangles.

Every open rectangle containing y contains a rectangle E_n that contains infinitely many points of P. Hence, y is a limit point for P. But y is a point of E that does not occur among the points in the enumeration (*), so $y \notin P$. This yields the contradiction that P is not a closed set.

Suppose that for each element of $\sigma \in \mathbb{B}^F$ is associated a nonempty closed set F_σ such that the following conditions hold for every $n \in \mathbb{N}$:

(i) For each $\mu \in \mathbb{B}^\infty$, $F_{\mu|n+1} \subset F_{\mu|n}$.
(ii) If $\sigma, \tau \in \mathbb{B}^n$ and $\sigma \neq \tau$, then F_σ and F_τ are disjoint.
(iii) For each $\sigma \in \mathbb{B}^n$, $\operatorname{diam}(F_\sigma) \leqslant 1/n$.

Then the family $\{F_\sigma : \sigma \in \mathbb{B}^F\}$ is called a dyadic schema. A dyadic schema determines a set P defined by

$$P = \bigcap_{n=1}^{\infty} \bigcup_{\sigma \in \mathbb{B}^n} F_\sigma$$

THEOREM 14 The set P determined by a dyadic schema is a perfect set.

Proof. It is a simple matter to verify that P is a nonempty closed set. It remains to show that P is dense-in-itself.

Suppose that x is any point of P and G is any open set containing x. Using condition (iii), choose $n \in \mathbb{N}$ and $\sigma \in \mathbb{B}^n$ so that $x \in F_\sigma$ and $F_\sigma \subset G$. For each sequence $v \in \mathbb{B}_\sigma^\infty$, let x_v denote the unique point belonging to the set $\bigcap_{n=1}^{\infty} F_{v|n}$. Then $\{x_v : v \in \mathbb{B}_\sigma^\infty\}$ is an infinite subset of $G \cap P$. Hence, P is dense-in-itself.

THEOREM 15 Every uncountable $\mathscr{G}_\delta$-set contains a perfect set.[1]

Proof. Let H be an uncountable $\mathscr{G}_\delta$-set. Then $H = \bigcap_{n=1}^{\infty} G_n$, where each set G_n is open. We know from Corollary 10 that H contains an infinite dense-in-itself set D.

Choose two points $d_0, d_1 \in D$ and two open rectangles E_0, E_1 containing d_0, d_1 whose closures F_0, F_1 are disjoint subsets of G_1 each with diameter $\leqslant 1$. Continuing inductively, suppose that for each $\sigma \in \mathbb{B}^n$ we have already determined the point $d_\sigma \in D$, the open rectangle E_σ containing d_σ, and the closure F_σ of E_σ with $F_\sigma \subset G_n$ and $\operatorname{diam}(F_\sigma) \leqslant 1/n$. The set $D \cap E_\sigma$ being nonempty and dense-in-itself, we can choose two points $d_{\sigma 0}, d_{\sigma 1}$ in this set and two open rectangles $E_{\sigma 0}, E_{\sigma 1}$ containing these points whose closure $F_{\sigma 0}, F_{\sigma 1}$ are disjoint subsets of $G_{n+1} \cap E_\sigma$ each with diameter $\leqslant 1/(n+1)$.

We thus obtain a dyadic schema $\{F_\sigma : \sigma \in \mathbb{B}^F\}$ that determines a perfect set contained in H.

COROLLARY 16 Every perfect set contains a perfect set determined by a dyadic schema.

THEOREM 17 Every perfect set has the power of the continuum.[2]

Proof. It suffices to show that any perfect set has a subset which has the power of the continuum. This follows from Corollary 16 and the fact that there is a one-to-one correspondence between the set of points belonging to the perfect set determined by a dyadic schema and the set $\mathbb{B}^\infty$, which has the power of the continuum.

[1]Young (1903b).
[2]Cf. Cantor (1966, pp. 252–257).

COROLLARY 18 Every uncountable closed set has the power of the continuum.

Proof. Use the Cantor-Bendixson Theorem.

We shall denote by $\mathscr{P}$ the family of all perfect sets. As pointed out in Chapter 1, Section I.A, the family of all closed sets has the power of the continuum. Similarly, we have

THEOREM 19 The family $\mathscr{P}$ of all perfect sets has the power of the continuum.

THEOREM 20 Every perfect set contains continuum many disjoint perfect sets.[1]

Proof. Let Q be a given perfect set. By Corollary 16, Q contains a perfect set P determined by a dyadic schema $\{F_\sigma : \sigma \in \mathbb{B}^F\}$.

Suppose that $\mu = \langle \mu_k \rangle_{k \in \mathbb{N}}$ is a fixed sequence in $\mathbb{B}^\infty$. For each $n \in \mathbb{N}$ and each $\tau = \langle \nu_1, \ldots, \nu_n \rangle \in \mathbb{B}^n$, define $E_\tau = F_{\mu_1 \nu_1 \mu_2 \nu_2 \cdots \mu_n \nu_n}$. Then $\{E_\tau : \tau \in \mathbb{B}^F\}$ is a dyadic schema that determines the perfect set $P_\mu = \bigcap_{n=1}^{\infty} \bigcup_{\tau \in \mathbb{B}^n} E_\tau$ which is a subset of Q. For any two different sequences $\mu, \mu' \in \mathbb{B}^\infty$ the sets $P_\mu, P_{\mu'}$ are disjoint. As $\mathbb{B}^\infty$ has the power of the continuum, the family $\{P_\mu : \mu \in \mathbb{B}^\infty\}$ consists of continuum many disjoint perfect subsets of Q.

COROLLARY 21 If a set of power less than the power of the continuum is removed from a given perfect set, then the set remaining contains a perfect set.

We now utilize Theorem 20 to establish a general set-theoretical result.[2]

PROPOSITION 22 The union of a countable family of arbitrary sets each of which has power less than the power of the continuum, itself has power less than the power of the continuum.

Proof. We treat only the case of a denumerable family $\{S_n : n \in \mathbb{N}\}$ of sets which have power less than that of the continuum.

Assume to the contrary that there exists a one-to-one function ϕ mapping $\bigcup_{n=1}^{\infty} S_n$ onto $\mathbb{R}$. Let P be a given compact, perfect set. As $\phi(S_1)$ has power less than that of the continuum, it follows from Theorem 20 that there is a perfect set $P_1 \subset P$ that is disjoint from $\phi(S_1)$. Continuing inductively, for each $n \in \mathbb{N}$ we use Theorem 20 to obtain a perfect set $P_{n+1} \subset P_n$ which is disjoint from $\phi(S_{n+1})$. The intersection of the descending sequence $\langle P_n \rangle_{n \in \mathbb{N}}$ of compact sets thus determined contains a point that belongs to none of the sets $\phi(S_n)$ and hence does not belong to $\bigcup_{n=1}^{\infty} \phi(S_n) = \phi(\bigcup_{n=1}^{\infty} S_n)$. Thus, ϕ cannot be onto $\mathbb{R}$.

[1]Cf. Mahlo (1913, Satz 3).
[2]Cf. Luzin and Sierpiński (1917b).

This proposition may be reformulated in an equivalent form.

PROPOSITION 23 The union of fewer than continuum many countable sets has power less than the power of the continuum.

We have the following further consequence of Theorem 20.

THEOREM 24 If P is a perfect set and $\mathscr{R}$ is a family of perfect sets that has power less than the power of the continuum, then there exists a perfect set $Q \subset P$ which is nowhere dense relative to each set in $\mathscr{R}$.

Proof. Let $\mathscr{R} = \{P_\alpha : \alpha \in I\}$. We know from Theorem 6 of Chapter 4, Section IV that for each $\alpha \in I$, there is a countable set $D_\alpha \subset P_\alpha$ which is everywhere dense in P_α. By virtue of Proposition 23 and Theorem 20, there is a perfect set $Q \subset P$ that contains no point of the set $D = \bigcup_{\alpha \in I} D_\alpha$. For each $\alpha \in I$, the complement of Q contains D_α and hence is everywhere dense in P_α. It thus follows from Theorem 7 of Chapter 4, Section IV that Q is nowhere dense in each set P_α.

COROLLARY 25 Every perfect set P contains a perfect set that is nowhere dense in P.

THEOREM 26 The following properties hold for a perfect set P:

(i) Every finite set is nowhere dense in P.
(ii) Every countable set is of the first category in P.
(iii) If M is a set of the first category in P, then $P-M$ contains a perfect set.
(iv) P is of the second category in itself.
(v) If S has the Baire property in P, then either $S \cap P$ or $(X-S) \cap P$ contains a perfect set.

Proof. The proofs of (i) and (ii) are straightforward. Property (iv) is an immediate consequence of (iii). It remains to prove (iii) and (v).

Suppose $M = \bigcup_{n=1}^\infty S_n$, where each set S_n is nowhere dense in P. Let G be an open rectangle with $G \cap P \neq \emptyset$. Then there exists an open rectangle H such that $H \cap P \neq \emptyset$, $H \cap P \subset G \cap P$, and $(H \cap P) \cap S_1 = \emptyset$. Define G_0, G_1 to be open rectangles whose closures E_0, E_1 are disjoint subsets of H, each with diameter $\leqslant 1$, with $G_0 \cap P \neq \emptyset$, $G_1 \cap P \neq \emptyset$. Assume that $\sigma \in \mathbb{B}^n$ and we have already defined the open rectangle G_σ and its closure E_σ with $G_\sigma \cap P \neq \emptyset$. Then there exists an open rectangle H_σ such that $H_\sigma \cap P \neq \emptyset$, $H_\sigma \cap P \subset G_\sigma \cap P$, and $(H_\sigma \cap P) \cap S_{n+1} = \emptyset$. We then define $G_{\sigma 0}, G_{\sigma 1}$ to be open rectangles whose closures $E_{\sigma 0}, E_{\sigma 1}$ are disjoint subsets of H_σ, each with diameter $\leqslant 1/(n+1)$, with $G_{\sigma 0} \cap P \neq \emptyset$, $G_{\sigma 1} \cap P \neq \emptyset$. In this manner we determine closed sets E_σ for all $\sigma \in \mathbb{B}^F$. Setting $F_\sigma = E_\sigma \cap P$ for each σ we obtain a dyadic schema $\{F_\sigma : \sigma \in \mathbb{B}^F\}$ of nonempty closed subsets of P which determines a perfect subset of P disjoint from M. Thus, property (iii) is established.

Turning to property (v), assume first that there is an open rectangle G with $G \cap P \neq \emptyset$ such that $S \cap (G \cap P)$ is of the first category in P or, equivalently stated, $S \cap G$ is of the first category in P. Taking $M = S \cap G$ in the proof of property (iii), we obtain a perfect set $R \subset P—M$. But the construction also yields $R \subset G$. Hence, $R \subset (X—S) \cap P$.

On the other hand, if there is no such an open rectangle G, then S is everywhere of the second category in P. Since S has the Baire property in P, the set $M = P—S$ is of the first category in P. Applying property (iii) we see that the set $P—M = S \cap P$ contains a perfect set.

DEFINITION A set that contains no perfect set is called a totally imperfect set.

The question of the existence of uncountable, totally imperfect sets remained unanswered for more than twenty years, until Bernstein proved that $\mathbb{R}$ can be decomposed into two totally imperfect sets.[1]

DEFINITION A set that contains no perfect set and whose complement also contains no perfect set is called a Bernstein set.

Equivalently, a set S is a Bernstein set if both S and $X—S$ contain at least one point of each perfect set. In view of the fact that every perfect set contains continuum many disjoint perfect sets, any set containing at least one point of each perfect set must actually contain continuum many points of each perfect set.

The following theorem is a generalization of Bernstein's decomposition theorem.

THEOREM 27 For each cardinal number $\mathfrak{m}$ satisfying $2 \leqslant \mathfrak{m} \leqslant 2^{\aleph_0}$, the space X can be decomposed into $\mathfrak{m}$ disjoint Bernstein sets.[2]

Proof. Let

(1) $x_1, x_2, \ldots, x_a, \ldots \quad (\alpha < \Lambda)$

(2) $P_1, P_2, \ldots, P_a, \ldots \quad (\alpha < \Lambda)$

[1]Cf. Scheeffer (1884, p. 287) and Bernstein (1908). We note that an earlier constructed set was actually the first uncountable totally imperfect set, but this property was not established until after Bernstein's proof. See Hardy (1903, 1905) and Luzin (1921). See Brown and Cox (1982) for a survey of results concerning totally imperfect sets.
[2]Cf. Bernstein (1908), Mahlo (1913, Aufgabe 9), Burstin (1914, Sec. 4, Zweiter Satz, Dritter Satz; 1916b);Luzin and Sierpiński (1917b), and Kondô (1937, Sec. 2).

be transfinite enumerations of all points of X and all perfect subsets of X, respectively.

Define y_{11} to be the first point of the enumeration (1) that belongs to P_1. Define y_{21} and y_{22} to be the first two points in (1) belonging to $P_2-\{y_{11}\}$. Continuing by transfinite induction, assume that $1 < \alpha < \Lambda$ and the points $y_{\gamma\beta}$ have already been defined for all ordinal numbers γ, β with $\beta \leqslant \gamma < \alpha$. The set of points $y_{\gamma\beta}$ already determined having power $<2^{\aleph_0}$, while P_α has power $2^{\aleph_0}$, there are $2^{\aleph_0}$ points y in (1) that belong to the set

$$P_\alpha-\{y_{\gamma\beta}:\beta \leqslant \gamma < \alpha\}$$

Denoting by $\mathfrak{n}$ the cardinal number of the order type α, we select from the points y belonging to this set the first $\mathfrak{n}$ points, according to the enumeration (1), and arrange the $\mathfrak{n}$ points selected in the form of a sequence $y_{\alpha1}, y_{\alpha2}, \ldots, y_{\alpha\alpha}$. In this manner we determine points $y_{\alpha\beta}$ for all ordinal numbers α, β with $\beta \leqslant \alpha < \Lambda$. Since each point of X belongs to $2^{\aleph_0}$ different perfect sets, it is clear that each point of X must occur among the defined points $y_{\alpha\beta}$.

For each $\beta < \Lambda$, define

$$S_\beta = \{y_{\alpha\beta}:\beta \leqslant \alpha < \Lambda\}$$

In view of the fact that each perfect set contains continuum many disjoint perfect sets, each of the sets S_β contains at least one point of each perfect set.

In the case that $\mathfrak{m} = 2^{\aleph_0}$, the sets S_β provide the desired decomposition of X into $\mathfrak{m}$ disjoint Bernstein sets. For $2 \leqslant \mathfrak{m} < 2^{\aleph_0}$, the desired decomposition is obtained by choosing $\mathfrak{m}$ of the sets S_β and adjoining all other sets S_β to one of the $\mathfrak{m}$ chosen sets.

Sierpiński proved:[1] If $2^{\aleph_0} = \aleph_1$, then the unit interval $[0,1]$ can be represented as the union of $2^{2^{\aleph_0}}$ sets, each of Lebesgue outer measure 1 and of the second category in every interval, such that the intersection of any two different sets has power $<2^{\aleph_0}$ (cf. Theorem 5 of Chapter 6, Section III). The sets are necessarily nonmeasurable and do not have the Baire property.

Without assuming CH, Sierpiński proved[2] that there exists such a representation of the unit interval into more than $2^{\aleph_0}$ sets. By a modification of his argument, one can obtain in general

THEOREM 28   The space X is representable as the union of more than $2^{\aleph_0}$ Bernstein sets, the intersection of any two different sets of which has power $<2^{\aleph_0}$.

[1]Sierpiński (1929b).
[2]Sierpiński (1937b) and Harazišvili (1979, Teorema 1).

We omit the proof of this theorem, as we shall further extend this result below.[1]

Bernstein sets are important theoretically in resolving many questions. Consider, for instance, the question: Does there exist a sequence $\langle P_n \rangle_{n \in \mathbb{N}}$ of perfect sets with the property that any set having the Baire property relative to each of these sets must have the Baire property relative to every perfect set, i.e., have the absolute Baire property? Using a Bernstein set, one obtains[2]

THEOREM 29   For every sequence $\langle P_n \rangle_{n \in \mathbb{N}}$ of perfect sets there exists a set S that has the Baire property relative to each set P_n, but does not have the absolute Baire property.

Proof. Let P be any perfect set. According to Theorem 24, there exists a perfect set $Q \subset P$ that is nowhere dense relative to each set P_n. If S is a Bernstein set, then $S \cap Q$ has the Baire property relative to each set P_n. However, in view of Theorem 26(v), S does not have the Baire property relative to Q.

D.   The linear continuum

In this section we establish a fundamental relationship between perfect subsets of $\mathbb{R}$ and sets having the same order type as the linear continuum. This relationship is utilized to eliminate numerically defined sets of real numbers that occurred in earlier proofs of some theorems, resulting in an extension of the domain of validity of those theorems to additional classifications of point sets.[3]

We first recall some facts about ordered sets.

DEFINITION   A subset D of an ordered set L is said to be ordinally dense in L if for all elements $x, y \in L$ with $x < y$ there is an element $d \in D$ satisfying $x < d < y$. In the case that $D = L$ we merely say that L is ordinally dense.

DEFINITION   Two ordered sets L and L' are called isomorphic if there is a one-to-one mapping ϕ of L onto L' such that if $x, y \in L$ and $x < y$, then $\phi(x) < \phi(y)$.

PROPOSITION 30   Any two ordinally dense, denumerable sets with no smallest elements and no largest elements are isomorphic.[4]

[1]See Theorem 24 of Section II.
[2]Cf. Szpilrajn (1935, Sec. 1.3).
[3]Compare the proofs of Theorems 17 and 20 of Section III with those given in Ruziewicz (1935), Sierpiński (1935c), Hong and Tong (1983), and Ruziewicz and Sierpiński (1933).
[4]Cf. Cantor (1966, pp. 304–306) and Sierpiński (1958, pp. 209–211).

DEFINITION   A pair (A,B) of nonempty, disjoint subsets of an ordered set L is called a Dedekind cut of L if $A \cup B = L$ and every element of A is smaller than every element of B.

The set of real numbers $\mathbb{R}$, considered as an ordered set L, satisfies the following conditions defining the linear continuum:

(λ_1) L has no smallest element and no largest element.
(λ_2) There is a denumerable subset D of L that is ordinally dense in L.
(λ_3) If (A,B) is any Dedekind cut of L, then either A has a largest element or B has a smallest element.

These properties uniquely characterize the linear continuum as an ordered set.

PROPOSITION 31   Any two ordered sets L and L' satisfying the conditions (λ_1)–(λ_3) are isomorphic. Moreover, if D and D' are ordinally dense subsets of L and L', respectively, then any isomorphism between D and D' can be extended to an isomorphism between L and L'.[1]

DEFINITION   A point $x \in \mathbb{R}$ is called a right-hand limit point for a set $S \subset \mathbb{R}$ if x is a limit point for the set $\{y \in S : y > x\}$, but is not a limit point for the set $\{y \in S : y < x\}$. The notion of a left-hand limit point is defined in a dual manner. The right-hand and left-hand limit points for a set are collectively called unilateral limit points for the set. A limit point for a set that is not a unilateral limit point for the set is called a bilateral limit point.

One can establish a one-to-one correspondence associating with each unilateral limit point for a set S a rational number not belonging to S. Consequently, the set of all unilateral limit points for a set is a countable set.

We now establish the relationship between perfect subsets of $\mathbb{R}$ and the linear continuum.[2]

THEOREM 32   Every perfect set $P \subset \mathbb{R}$ is representable as the union of two disjoint sets L and M, where L is isomorphic to $\mathbb{R}$ and M is a countable set.

Proof. Delete from P the smallest and largest elements of P, as well as all right-hand limit points for P, if such exist. The set M of points removed is countable. We show that the set $L = P—M$ satisfies conditions (λ_1)–(λ_3) and is accordingly isomorphic to $\mathbb{R}$.

Suppose that $x \in L$. Since x is not the smallest element of P (if such exists) there is a point $y \in P$ such that $y < x$. By virtue of Theorem 13, the set

---

[1] Cf. Cantor (1966, pp. 310–311), Sierpiński (1958, pp. 216–218), and Morgan (1976, p. 434).
[2] Cf. Bettazzi (1888).

$\{y \in L : y < x\}$ is nonempty. Hence, x is not a smallest element of L. Similarly, it is seen that L has no largest element. Thus, condition (λ_1) is satisfied.

Let D be the set consisting of all rational numbers belonging to open intervals contained in P and all left-hand limit points for L. The set of rational numbers in D and the set of all left-hand limit points for L are both countable sets. The first of these two sets can only be empty or infinite and, if it is empty, then P is a nowhere dense perfect set whose complement contains denumerably many disjoint open intervals with left-hand endpoints that belong to L, so the second set is infinite. Thus, D is a denumerable set.

Suppose that $x, y \in L$ and $x < y$. If there is an open interval $I \subset P$ lying between x and y, then there is a point $d \in D \cap I$ such that $x < d < y$. Suppose that there is no such interval. Then $P \cap [x,y]$ is a nowhere dense set which, because y is not a right-hand limit point for P, must be infinite. This implies that there is a left-hand limit point d for L such that $x < d < y$. We thus see that condition (λ_2) is satisfied.

Let (A,B) be a Dedekind cut of L. Setting $a = \sup A$ and $b = \inf B$, we have $a, b \in P$. If $a = b$, then since no bilateral limit points for P were removed from P, we must have $a \in A$ or $a \in B$. Assume therefore that $a < b$. Then no point of P will lie in the interval (a,b), so a is a left-hand limit point for P which is clearly neither a smallest element of P nor a largest element for P. This means that $a \in L$, whence $a \in A$ or $a \in B$. We conclude that the condition (λ_3) is also satisfied.

E. Marczewski's classification

THEOREM 33 $(X, \mathscr{P})$ is a perfect base.

Proof. Axiom 1 is obviously satisfied; i.e., $X = \bigcup \mathscr{P}$. To verify Axiom 2, assume that $A \in \mathscr{P}$ and $\mathscr{D}$ is any nonempty, disjoint subfamily of $\mathscr{P}$ which has power less than the power of $\mathscr{P}$. According to Theorem 19, $\mathscr{D}$ has power less than the power of the continuum.

Suppose that $A \cap (\bigcup \mathscr{D})$ contains a set in $\mathscr{P}$. Then it follows from Theorem 17 and Proposition 23 that there is a set $D \in \mathscr{D}$ such that $A \cap D$ is uncountable. Applying the Cantor-Bendixson Theorem, we see that $A \cap D$ contains a set in $\mathscr{P}$.

Suppose that $A \cap (\bigcup \mathscr{D})$ contains no set in $\mathscr{P}$. Then $A \cap (\bigcup \mathscr{D})$ has power less than the power of the continuum. By virtue of Corollary 21, there is a set in $\mathscr{P}$ that is contained in A and is disjoint from every set in $\mathscr{D}$.

Assume now that $A \in \mathscr{P}$ and $x \in A$. For each $n \in \mathbb{N}$, let E_n be the closed square having the point x as center with diameter of magnitude $1/n$ and let A_n be the set of all condensation points for the set $A \cap E_n$. We know from Theorem 13 that x is a condensation point for A and hence also for $A \cap E_n$, so

$x \in A_n$. From Theorem 12 we see that A_n is a perfect set for each $n \in \mathbb{N}$. We thus determine a descending sequence $\langle A_n \rangle_{n \in \mathbb{N}}$ of sets satisfying the condition (+) in the definition of a perfect base.

The singular, meager, abundant, and Baire sets for the category base $(X,\mathscr{P})$ constitute Marczewski's classification of sets.[1] The Baire sets for this category base are called Marczewski sets, while the singular, meager, and abundant sets are called Marczewski singular, Marczewski meager, and Marczewski abundant sets. The theorems of this section pertain only to this category base.

REMARK It can be shown[2] that for every perfect set P there is a function $h \in \mathscr{H}_c$ such that P has positive measure for the Hausdorff measure μ^h. Hence, $(X,\mathscr{P})$ is equivalent to the last example of Chapter 1, Section II.A.

THEOREM 34 Every set whose power is less than the power of the continuum is singular.

Proof. Apply Theorem 20.

THEOREM 35 If $\mathscr{D}$ is a family of disjoint perfect sets, then there exists a singular set containing a point from each set in $\mathscr{D}$.[3]

Proof. In view of Theorem 34 we may assume that $\mathscr{D}$ has the power of the continuum; say $\mathscr{D} = \{D_\alpha : \alpha < \Lambda\}$. Let $\mathscr{E}$ denote the family of all perfect sets whose intersection with each set D_α has power less than the power of the continuum. If $\mathscr{E} = \varnothing$, then we merely choose one point from each set D_α to form a singular set. Assume therefore that $\mathscr{E} \neq \varnothing$. As seen from Theorem 20, $\mathscr{E}$ must then have the power of the continuum, say, $\mathscr{E} = \{E_\alpha : \alpha < \Lambda\}$. By virtue of Proposition 23, for each $\alpha < \Lambda$ we can choose a point $x_\alpha \in D_\alpha - (\bigcup_{\beta < \alpha} E_\beta)$ to form a set $S = \{x_\alpha : \alpha < \Lambda\}$ having the power of the continuum. It can be seen without difficulty that S is a singular set.

Combining Theorems 20 and 35 yields

THEOREM 36 There exists a singular set having the power of the continuum.[4]

THEOREM 37 The following are equivalent for a set S:[5]
  (i) S is a singular set.
  (ii) S is a meager set.

[1] Szpilrajn (Marczewski) (1935).
[2] Cf. Rogers (1970, Chap. 2, Sec. 3.4).
[3] Cf. Walsh (1984, Theorem 2.2).
[4] Miller (1984, Theorem 5.10) and Walsh (1984).
[5] Szpilrajn (1935, Secs. 3.1, 3.2, 5.1).

(iii) The union of S and any given totally imperfect set is a totally imperfect set.

(iv) Every perfect set contains a perfect set P such that the set $S \cap P$ is of the first category relative to P.

Proof. (i) $\Rightarrow$ (ii) is trivial, while (ii) $\Rightarrow$ (i) is established by construction of a dyadic schema in a given perfect set. Thus, (i) $\Leftrightarrow$ (ii). We complete the proof of the theorem by establishing the implications (i) $\Rightarrow$ (iii) $\Rightarrow$ (iv) $\Rightarrow$ (i).

(i) $\Rightarrow$ (iii). Assume that S is a singular set and T is a set such that $S \cup T$ is not totally imperfect. Let Q be a perfect subset of $S \cup T$. Then there is a perfect set $R \subset Q$ that is disjoint from S. This implies that $R \subset T$, so T is not totally imperfect.

(iii) $\Rightarrow$ (iv). We prove the contrapositive. Assuming that $\neg$(iv), there exists a perfect set Q such that for every perfect set $P \subset Q$ the set $S \cap P$ is nonempty. This implies that $T = Q - S$ is a totally imperfect set. We have $Q \subset S \cup T$, so $S \cup T$ is not totally imperfect. Thus, $\neg$(iii) holds.

(iv) $\Rightarrow$ (i). Suppose that Q is any given perfect set. Then there is a perfect set $P \subset Q$ such that $S \cap P$ is of the first category relative to P. Applying Theorem 26(iii), there is a perfect set $R \subset P - (S \cap P)$. Hence, $R \subset Q$ and $R \cap S = \varnothing$. We conclude that S is a singular set.

NOTE. For additional equivalent forms see Corollary 40 and Theorem 13 of Section II.

COROLLARY 38   Every set that is always of the first category is a singular set.

THEOREM 39   The following are equivalent for a set S:[1]

(i) S is a Baire set.

(ii) Every perfect set contains a perfect set P such that either $P \subset S$ or $P \cap S = \varnothing$.

(iii) Every perfect set contains a perfect set P such that the set $S \cap P$ is open in P.

(iv) Every perfect set contains a perfect set P such that the set $S \cap P$ has the Baire property relative to P.

Proof. (i) $\Rightarrow$ (ii). Assume that S is a Baire set and let R be any perfect set. Then there is a perfect set $Q \subset R$ such that either $S \cap Q$ or $(X - S) \cap Q$ is a meager set. We can replace the word "meager" here by "singular," since every meager set is singular. Consequently, there is a perfect set $P \subset Q$ such that either $P \cap S = \varnothing$ or $P \cap (X - S) = \varnothing$; i.e., $P \cap S = \varnothing$ or $P \subset S$.

The implications (ii) $\Rightarrow$ (iii) $\Rightarrow$ (iv) are obvious.

(iv) $\Rightarrow$ (i). Let Q be a given perfect set. Let P be a perfect subset of Q such that $S \cap P$ has the Baire property in P. According to Theorem 26(v), either

[1]Szpilrajn (1935, Secs. 2.3, 5.1).

$S \cap P$ or $(X—S) \cap P$ contains a perfect set R, so that either $R \cap (X—S) = \varnothing$ or $R \cap S = \varnothing$.

COROLLARY 40   A set is meager if and only if it is a totally imperfect Baire set.

COROLLARY 41   Every set that has the absolute Baire property is a Baire set.

In contrast with the classifications of Baire and Lebesgue, there does not exist an uncountable rare set for Marczewski's classification.

THEOREM 42   Every abundant set contains an uncountable singular set.

Proof. Let S be an abundant set. Then there is a perfect set P every perfect subset of which has a nonempty intersection with S. We know from Theorem 11 of Chapter 1, Section III that $(P, \mathscr{P}_P)$ is a category base. Let $\mathscr{N}$ denote the family of all perfect subsets of P that are nowhere dense in P and let $Y = \bigcup \mathscr{N}$. According to Theorem 25, every perfect subset of P contains a set in $\mathscr{N}$. Apply Lemmas 3 and 4 of Chapter 1, Section II to the category base $(P, \mathscr{P}_P)$ to obtain a disjoint subfamily $\mathscr{M}$ of $\mathscr{N}$ such that $Y—\bigcup \mathscr{M}$ is $\mathscr{N}$-singular and having the property that for every perfect set $N \subset P$ there exists a set $M \in \mathscr{M}$ such that $N \cap M$ contains a perfect set. Lemma 3 of Chapter 1, Section II and Theorem 11 of Chapter 1, Section III further reveal that $P—\bigcup \mathscr{M}$ is singular with respect to $\mathscr{P}$.

The family $\mathscr{M}$ must be uncountable since, otherwise, it follows from Theorem 26 that $P—\bigcup \mathscr{M}$ is an uncountable $\mathscr{G}_\delta$-set which, according to singular. Form a set T consisting of one and only one point from each of the sets $S \cap M$. Obviously, T is an uncountable subset of $S \cap P$. Using the Cantor-Bendixson Theorem and the property given in the preceding paragraph, it is readily seen that T is a singular set.

COROLLARY 43   If every singular subset of a set S is countable, then S is a singular set.

II. BAIRE SETS

In the remaining sections of this chapter, a given perfect base $(X, \mathscr{C})$ is presupposed.

A. Meager sets

We first show that Marczewski's classification is the only one determined by a perfect base for which there are no singular perfect sets.

THEOREM 1   There exists a singular perfect set for a perfect base $(X,\mathscr{C})$ if and only if $(X,\mathscr{C})$ is not equivalent to $(X,\mathscr{P})$.

Proof. If there are no singular perfect sets, then, by Corollary 6 of Section I, every perfect set contains a $\mathscr{C}$-region. As every $\mathscr{C}$-region trivially contains a perfect set, we conclude from Theorem 1 of Chapter 1, Section III that $(X,\mathscr{C})$ and $(X,\mathscr{P})$ are equivalent.

Suppose on the other hand that $(X,\mathscr{C})$ and $(X,\mathscr{P})$ are equivalent. Then each perfect set, being $\mathscr{P}$-abundant, is $\mathscr{C}$-abundant. Hence, there are no perfect sets that are $\mathscr{C}$-singular.

COROLLARY 2   If $\mathscr{C}$ satisfies CCC, then there exists a singular perfect set.

THEOREM 3   If $(X,\mathscr{C})$ is not equivalent to $(X,\mathscr{P})$, then there exists a $\mathscr{C}$-singular set that is not a Marczewski set.[1]

Proof. By Theorem 1 there exists a $\mathscr{C}$-singular perfect set Q. If S is a Bernstein set, then $S \cap Q$ is a $\mathscr{C}$-singular set. However, $S \cap Q$ is not a Marczewski set, since there is no perfect set $P \subset Q$ such that $P \subset S \cap Q$ or $P \cap (S \cap Q) = \emptyset$, so Theorem 39(ii) of Section I does not hold.

B.   Baire sets

THEOREM 4   Every closed set is a Baire set.

Proof. If a closed set F is abundant everywhere in a region A, then, according to Theorem 5 in Section I, we have $A \subset F$ and consequently $A - F = \emptyset$. We then conclude from Theorem 2 of Chapter 1, Section III that every closed set F is a Baire set.

COROLLARY 5   Every open set is a Baire set.

THEOREM 6   Assume CH. If $\mathscr{C}$ satisfies CCC, then every subset of a set S is a restricted Baire set relative to S if and only if S is a meager set.[2]

Proof. Apply Corollary 9 of Chapter 2, Section II.

The following general result is of especial importance.

THEOREM 7   Let $\langle E_{k,n} \rangle_{k,n \in \mathbf{N}}$ be a double sequence of Baire sets with the property that for each $k \in \mathbf{N}$, the sequence $\langle E_{k,n} \rangle_{n \in \mathbf{N}}$ is a descending sequence whose intersection is a meager set. Then for every abundant Baire set S there exists an increasing sequence $\langle n_k \rangle_{k \in \mathbf{N}}$ of natural numbers such that the set $S - \bigcup_{k=1}^{\infty} E_{k,n_k}$ contains a perfect set.

[1]Cf. Szpilrajn (1935, Secs. 5.1, 5.2).
[2]Cf. Eilenberg (1932, Corollaire 7) and Sierpiński (1934c, pp. 133–134).

Proof. Let A be a region in which S is abundant everywhere. Since S is a Baire set we have

$$A - S = \bigcup_{i=1}^{\infty} T_i$$

where each T_i is a singular set. Proceeding inductively, we will determine simultaneously an increasing sequence $\langle n_k \rangle_{k \in \mathbb{N}}$ of natural numbers and a dyadic schema $\langle A_\sigma : \sigma \in \mathbf{B}^F \rangle$ consisting of subregions of A.

We first show there is a subregion B of A and a natural number n_1 for which $B \cap E_{1,n_1}$ is a meager set. Assume that the contrary holds. Each of the sets $E_{1,n}$ is then abundant everywhere in A. These being Baire sets, the sets $A - E_{1,n}$ are meager for each n. From the equality

$$A = \left[A \cap \left(\bigcap_{n=1}^{\infty} E_{1,n} \right) \right] \cup \left[\bigcup_{n=1}^{\infty} (A - E_{1,n}) \right]$$

it follows that $\bigcap_{n=1}^{\infty} E_{1,n}$ is an abundant set, contradicting our hypothesis.

The existence of B and n_1 being thus established, we determine two disjoint subregions A_0, A_1 of B each of which is disjoint from T_1 and has diameter $\leqslant 1$. We have

$$A_0 \cap E_{1,n_1} = \bigcup_{i=1}^{\infty} T_{0,i} \qquad A_1 \cap E_{1,n_1} = \bigcup_{i=1}^{\infty} T_{1,i}$$

where the sets $T_{0,i}, T_{1,i}$ are singular for each i.

Suppose now that k is a given natural number greater than 1 and we have already determined the natural number n_k, the regions A_σ for $\sigma \in \mathbf{B}^k$, and the singular sets $T_{\sigma,i}$ for all $\sigma \in \mathbf{B}^k$ and all $i \in \mathbb{N}$. Take $\sigma \in \mathbf{B}^k$. There is a subregion B_σ of A_σ and a natural number $n_{k+1} > n_k$ for which $B_\sigma \cap E_{k+1,n_{k+1}}$ is a meager set. We then determine two disjoint subregions $A_{\sigma 0}, A_{\sigma 1}$ of B_σ, each of which is disjoint from all previously defined sets T with index $i \leqslant k + 1$ and has diameter $\leqslant 1/(k + 1)$, and place

$$A_{\sigma 0} \cap E_{k+1,n_{k+1}} = \bigcup_{i=1}^{\infty} T_{\sigma 0,i} \qquad A_{\sigma 1} \cap E_{k+1,n_{k+1}} = \bigcup_{i=1}^{\infty} T_{\sigma 1,i}$$

where the sets $T_{\sigma 0,i}, T_{\sigma 1,i}$ are singular for each $i \in \mathbb{N}$. In this manner we obtain the desired sequence $\langle n_k \rangle_{k \in \mathbb{N}}$ and the dyadic schema $\langle A_\sigma : \sigma \in \mathbf{B}^F \rangle$.

The perfect set

$$P = \bigcap_{k=1}^{\infty} \bigcup_{\sigma \in \mathbf{B}^k} A_\sigma$$

determined by this dyadic schema is contained in A and is disjoint from all the sets T. This implies that $P \subset A \cap S$ and $P \cap E_{k,n_k} = \varnothing$ for every $k \in \mathbb{N}$, so $P \subset S - \bigcup_{k=1}^{\infty} E_{k,n_k}$.

NOTE   For the category base $\mathscr{C}$ of all compact sets in $\mathbb{R}^n$ of positive Lebesgue measure (or equivalently, all compact perfect sets in $\mathbb{R}^n$ of positive Lebesgue measure in every open set containing one of their points) one can obtain the stronger conclusion that the perfect set P has positive measure. This can be established as follows. Let S be a measurable set of positive measure that is everywhere of positive measure in a set $A \in \mathscr{C}$. Then the set $T_0 = A - S$ has measure zero. Let $\alpha = \mu(A)$. The basic argument of the second paragraph of the preceding proof can be adapted to show that for each set $A^* \in \mathscr{C}_A$ there is a set $B^* \in \mathscr{C}_A$ and a natural number n_{B^*} such that $B^* \subset A^*$ and $B^* \cap E_{1,n_{B^*}}$ has measure zero. Moreover, we may assume that B^* is disjoint from T_0. Now, apply Lemmas 3 and 4 of Chapter 1, Section II to obtain a countable, disjoint family $\mathscr{M}$ consisting of sets B^* such that $A - \bigcup \mathscr{M}$ has measure zero. Choose a finite subfamily $\mathscr{B}$ of $\mathscr{M}$ with $\mu(\bigcup \mathscr{B}) > \alpha - \alpha/4$. Taking $B_1 = \bigcup \mathscr{B}$ and $n_1 = \sup\{n_{B^*} : B^* \in \mathscr{B}\}$, we have $B_1 \in \mathscr{C}_A$, $B_1 \cap T_0 = \varnothing$, $\mu(B_1) > \alpha - \alpha/4$, and the set $T_1 = B_1 \cap E_{1,n_1}$ has measure zero. Continue inductively to determine, at the $(k + 1)$st step, a set $B_{k+1} \in \mathscr{C}_{B_k}$ and $n_{k+1} > n_k$ such that $B_{k+1} \cap (\bigcup_{i=0}^{k} T_i) = \varnothing$, $\mu(B_{k+1}) > \alpha - (\alpha/4 + \alpha/8 + \cdots + \alpha/2^{k+2})$, and the set $T_{k+1} = B_{k+1} \cap E_{k+1,n_{k+1}}$ has measure zero. Then the set $C = \bigcap_{k=1}^{\infty} B_k$ will be a compact set of positive measure contained in $(S \cap A) - \bigcup_{k=1}^{\infty} E_{k,n_k}$. Finally, apply Corollary 18 of Chapter 4, Section III to obtain a perfect set $P \in \mathscr{C}_A$ which is a subset of C.

As one consequence of Theorem 7, we have

THEOREM 8   Every abundant Baire set contains a perfect set.

COROLLARY 9   A Baire set either contains a perfect set or its complement contains a perfect set.

NOTE   In the case of Lebesgue measure, every measurable set of positive measure actually contains a perfect set of positive measure. This is not the case for non-σ-finite Hausdorff measures. However, every abundant Hausdorff measurable set does contain a perfect set of positive measure.

THEOREM 10   If $\mathscr{C}$ satisfies CCC, then every abundant Baire set contains a meager perfect set.

Proof.   Use Theorem 8, Theorem 20 of Section 1, Theorem 4, and Theorem 5 of Chapter 1, Section III.

Corollary 9 leads to the existence of non-Baire sets.[1]

THEOREM 11   A Bernstein set is not a Baire set for any perfect base.

In fact, any subset of a Bernstein set that is a Baire set is a meager set. As the complement of a Bernstein set is also a Bernstein set, it follows from Theorem 4 of Chapter 2, Section I that a Bernstein set is abundant everywhere and its complement is also abundant everywhere. In the Lebesgue measure case, a set having this property is called a saturated nonmeasurable set.[2]

According to Theorem 11, decompositions of sets into Bernstein sets are, in reality, decompositions into non-Baire sets.

THEOREM 12   Every abundant set contains a set that is not a Baire set.[3]

Proof. Let R be an abundant set and let $X = S \cup T$ be a decomposition of X into two disjoint Bernstein sets as in the proof of Theorem 27 of Section I. At least one of the sets $R \cap S$ and $R \cap T$ must be abundant, but neither of these sets contains any perfect set. Hence, in view of Theorem 8, at least one of the sets $R \cap S$ and $R \cap T$ is not a Baire set.

REMARK   According to Theorem 6 of Chapter 2, Section IV, the space $\mathbb{R}^n$ is representable as the union of a set of Lebesgue measure zero and a set of the first category. In conjunction with Theorem 12, this implies that there is a set of Lebesgue measure zero which does not have the Baire property and there is a set of the first category which is not Lebesgue measurable.

Theorem 12 readily leads to

THEOREM 13   A necessary and sufficient condition that a set S be a meager set is that every subset of S be a Baire set.[4]

This result, in turn, implies

COROLLARY 14   Two perfect bases $(X,\mathscr{C})$ and $(X,\mathscr{D})$ are equivalent if and only if $\mathfrak{B}(\mathscr{C}) = \mathfrak{B}(\mathscr{D})$.

[1] Bernstein (1908). Concerning sets that are not Lebesgue measurable or do not have the Baire property, see Vitali (1905a), Lebesgue (1907b), Van Vleck (1908), Lennes (1913), Sierpiński (1938b, 1947, 1949a, 1949c), Kuratowski and Sierpiński (1941), Erdős (1950), Gomes (1952), Králik (1952), Swingle (1958, 1959), Semadeni (1964), Prikry (1973, 1976), Rosenthal (1975), Pawelska (1979), Grzegorek (1980), Krawczyk and Pelc (1980), Taylor (1980), and Pelc (1981).
[2] Halperin (1951).
[3] Rademacher (1916, Satz VI), Sierpiński (1920a), Wilkosz (1920, Sec. 3), Wolff (1923), Králik (1952), and Pu (1972).
[4] Szpilrajn (1930).

C. Rare sets

THEOREM 15 If $\mathscr{C}$ satisfies CCC, then any uncountable rare set is not a Baire set.

Proof. According to Theorem 8, any uncountable rare set that is a Baire set contains a perfect set. This perfect set contains continuum many disjoint perfect sets each of which is an abundant Baire set. This implies that $\mathscr{C}$ does not satisfy CCC.

Theorem 15, in conjunction with Theorem 1 of Chapter 2, Section III, yields the following two consequences:

(1) If $2^{\aleph_0} = \aleph_1$, then there exists an uncountable subset of $\mathbb{R}^n$ no uncountable subset of which has the Baire property.
(2) If $2^{\aleph_0} = \aleph_1$, then there exists an uncountable subset of $\mathbb{R}^n$ no uncountable subset of which is Lebesgue measurable.

On the other hand, there does not exist any uncountable subset of $\mathbb{R}^n$ having the property that every uncountable subset simultaneously lacks the Baire property and is nonmeasurable. This is because of the complementary nature of the category bases involved.

We note two additional consequences of Theorem 15.

THEOREM 16 Assume CH. If $\mathscr{C}$ satisfies CCC, then every abundant set contains a rare set that is not a Baire set.

THEOREM 17 If $\mathscr{C}$ satisfies CCC, then a set S is a rare set if and only if every Baire set contained in S is countable.

THEOREM 18 If $\mathscr{C}$ satisfies CCC, then every rare set is a Marczewski singular set.[1]

Proof. Let S be a rare set and let R be any perfect set. Since every perfect set is a Baire set and $\mathscr{C}$ satisfies CCC, it follows from Theorem 20 of Section I that there is a $\mathscr{C}$-singular perfect set $Q \subset R$. The set $S \cap Q$ being countable, we apply Theorem 20 of Section I again to obtain a perfect set $P \subset Q$ that is disjoint from S. We thus see that S is a Marczewski singular set.

COROLLARY 19 Assuming CH, there exists a Marczewski singular set that does not have the Baire property and which is not Lebesgue measurable.[2]

Proof. Take the union of a Mahlo-Luzin set and a Sierpiński set.

---

[1]Szpilrajn (1935, Sec. 5.3).
[2]Szpilrajn (1935, Sec. 5.3).

COROLLARY 20 Assuming CH, there exists a Marczewski singular set that does not have the absolute Baire property.

Proof. A Mahlo-Luzin set is such a set.

D. Invariant sets

Throughout this section, Φ will denote a group of one-to-one mappings of X onto itself with the group operation that of composition of mappings.

Using a modification of the proof of Theorem 27 of Section I we first obtain a stronger theorem that generalizes results of Van Vleck and Sierpiński on decompositions of $\mathbb{R}$ into disjoint invariant sets which are not Lebesgue measurable and do not have the Baire property.[1]

THEOREM 21 If Φ is countable, then X can be decomposed into $\mathfrak{m}$ disjoint Φ-invariant Bernstein sets, for each cardinal number $\mathfrak{m}$ satisfying $2 \leqslant \mathfrak{m} \leqslant 2^{\aleph_0}$.

Proof. Let

$$(1) \qquad\qquad x_1, x_2, \ldots, x_\alpha, \ldots \qquad (\alpha < \Lambda)$$

$$(2) \qquad\qquad P_1, P_2, \ldots, P_\alpha, \ldots \qquad (\alpha < \Lambda)$$

be transfinite enumerations of all points of X and all perfect subsets of X, respectively.

Define $y_{1,1}$ to be the first point in the enumeration (1) that belongs to P_1. Assume that $1 < \alpha < \Lambda$ and the points $y_{\gamma,\beta}$ have already been defined for all ordinal numbers γ, β with $\beta \leqslant \gamma < \alpha$. The set

$$R_{\alpha,1} = \{\phi(y_{\gamma,\beta}) : \beta \leqslant \gamma < \alpha \text{ and } \phi \in \Phi\}$$

having power $< 2^{\aleph_0}$, there are points y belonging to the set $P_\alpha - R_{\alpha,1}$. We define $y_{\alpha,1}$ to be the first such point y according to the enumeration (1). For $1 < \beta \leqslant \alpha$ we set

$$R_{\alpha,\beta} = R_{\alpha,1} \cup \{\phi(y_{\alpha,\xi}) : \xi < \beta \text{ and } \phi \in \Phi\}$$

and define $y_{\alpha,\beta}$ to be the first point in (1) belonging to $P_\alpha - R_{\alpha,\beta}$.

Having thus determined by transfinite induction points $y_{\alpha,\beta}$ for all ordinal numbers α, β with $\beta \leqslant \alpha < \Lambda$, we define for each $\beta < \Lambda$

$$S_\beta = \{\phi(y_{\alpha,\beta}) : \beta \leqslant \alpha < \Lambda \text{ and } \phi \in \Phi\}$$

The sets S_β are disjoint Φ-invariant Bernstein sets that yield the desired

---

[1]Van Vleck (1908) and Sierpiński (1917b).

decomposition of X in the manner described in the proof of Theorem 27 of Section I.

As will be seen from Theorem 27 of Chapter 6, Section II, the hypothesis that Φ be countable is essential in this theorem.

We next establish a general theorem valid for any group Φ of power $\leqslant 2^{\aleph_0}$, utilizing a proposition due to Banach.[1] In this proposition E denotes an arbitrary abstract set of power $\mathfrak{n} > \aleph_0$ and Θ denotes the smallest ordinal number of power $\mathfrak{n}$. The symbol Ψ denotes a family of power $\leqslant \mathfrak{n}$ consisting of one-to-one mappings whose domains and ranges are subsets of E.

NOTATION   For each mapping $\psi \in \Psi$ and each integer n, we denote by ψ^n the n-fold composition of the mapping ψ if n is positive, the identity mapping if n is zero, and the $|n|$-fold composition of the inverse mapping ψ^{-1} if n is negative.

PROPOSITION 22   The set E can be decomposed into a transfinite sequence $\langle T_\alpha \rangle_{\alpha < \Theta}$ of nonempty, disjoint sets satisfying the conditions
   (i) For each $\alpha < \Theta$, the set $\bigcup_{\beta \leqslant \alpha} T_\beta$ has power $< \mathfrak{n}$.
   (ii) If ψ is any mapping in Ψ with domain D, then there exists an ordinal number $\beta < \Theta$ such that

$$\psi(D \cap T_\alpha) \subset T_\alpha$$

   for all ordinal numbers $\alpha > \beta$.

Proof. The case in which Ψ is empty being easily dispensed with, we assume that Ψ is nonempty. Let

(1) $$x_1, x_2, \ldots, x_\alpha, \ldots \qquad (\alpha < \Theta)$$

(2) $$\psi_1, \psi_2, \ldots, \psi_\alpha, \ldots \qquad (\alpha < \Theta)$$

be arrangements of E and Ψ, respectively, into transfinite sequence of type Θ. [In the case that Ψ has power $< \mathfrak{n}$, one element of Ψ is repeated $\mathfrak{n}$ times to obtain the sequence (2).]

We first determine by transfinite induction a subsequence

(3) $$y_1, y_2, \ldots, y_\alpha, \ldots \qquad (\alpha < \Theta)$$

of (1). Define $y_1 = x_1$. Assume that $1 < \alpha < \Theta$ and that the elements y_β have been defined for all $\beta < \alpha$. For each $k \in \mathbb{N}$, the set of all $(k + 1)$-tuples consisting of ordinal numbers $< \alpha$ has power $< \mathfrak{n}$, and the set of all k-tuples of integers is denumerable. Consequently, the set of all elements x represen-

---

[1]Banach (1932a).

table in the form

(4) $$x = \psi_{\alpha_1}^{n_1} \psi_{\alpha_2}^{n_2} \cdots \psi_{\alpha_k}^{n_k}(y_\xi)$$

where $k \in \mathbb{N}$, ξ, $\alpha_1, \alpha_2, \ldots, \alpha_k$ are ordinal numbers $< \alpha$, and $n_1, n_2, \ldots, n_k$ are integers, has power $< n$. Accordingly, we are justified to define y_α to be the first element of the sequence (1) that is different from all the elements x representable in the form (4). From the equality $y_\alpha = \psi_\alpha^0(y_\alpha)$ it is clear that the elements of the transfinite sequence (3), thus determined, will be different from one another.

Place $R_1 = \varnothing$ and, for $1 < \alpha < \Theta$, define R_α to be the set of all elements x representable in the form (4). Then

(5) $$R_\beta \subset R_\alpha$$

whenever $\beta < \alpha < \Theta$,

(6) $$R_\alpha = \bigcup_{\beta < \alpha} R_\beta$$

for all limit ordinals α, and

(7) $$E = \bigcup_{\alpha < \Theta} R_\alpha$$

The sets T_α are now defined for $\alpha < \Theta$ by

(8) $$T_\alpha = R_{\alpha+1} - R_\alpha$$

These sets are nonempty since $y_\alpha = \psi_\alpha^0(y_\alpha)$ is an element of T_α for each $\alpha < \Theta$ and it readily follows from (5) that the sets T_α are disjoint. We also have

$$E = \bigcup_{\alpha < \Theta} T_\alpha$$

For, suppose that x is any element of E. By (7), there is a smallest ordinal number γ such that $x \in R_\gamma$. According to (6), γ cannot be a limit ordinal. Because $R_1 = \varnothing$ we must thus have $\gamma = \alpha + 1$ for some $\alpha < \Theta$. Hence, $x \in T_\alpha$ and the inclusion $E \subset \bigcup_{\alpha < \Theta} T_\alpha$ is established. The reverse inclusion being obvious, we have $E = \bigcup_{\alpha < \Theta} T_\alpha$.

In view of (5) and (8), the set $\bigcup_{\beta \leq \alpha} T_\beta$ is a subset of $R_{\alpha+1}$ for each $\alpha < \Theta$ and consequently has power $< n$. Condition (i) being thereby satisfied, we have only to verify condition (ii).

Assume that ψ is any mapping in Ψ with domain D. Then $\psi = \psi_\beta$ for some mapping ψ_β in the sequence (2). We show that $\psi(D \cap T_\alpha) \subset T_\alpha$ for all ordinal numbers $\alpha > \beta$.

Suppose that $x \in \psi(D \cap T_\alpha)$. Then

$$x = \psi_\beta(y)$$

for some element $y \in D \cap T_\alpha$. From (8) we see that

$$y = \psi_{\alpha_1}^{n_1} \psi_{\alpha_2}^{n_2} \cdots \psi_{\alpha_k}^{n_k}(y_\xi)$$

where $k \in \mathbb{N}$, $\xi, \alpha_1, \alpha_2, \ldots, \alpha_k$ are ordinal numbers $< \alpha + 1$, and $n_1, n_2, \ldots, n_k$ are integers. Hence,

$$x = \psi_\beta \psi_{\alpha_1}^{n_1} \psi_{\alpha_2}^{n_2} \cdots \psi_{\alpha_k}^{n_k}(y_\xi)$$

As $\beta < \alpha$, this yields $x \in R_{\alpha+1}$. Now, if x were also an element of R_α, then we would have

$$x = \psi_{\gamma_1}^{m_1} \psi_{\gamma_2}^{m_2} \cdots \psi_{\gamma_j}^{m_j}(y_\eta)$$

where $j \in \mathbb{N}$, $\eta, \gamma_1, \gamma_2, \ldots, \gamma_j$ are ordinal numbers $< \alpha$, and $m_1, m_2, \ldots, m_j$ are integers. This implies that

$$y = \psi_\beta^{-1} \psi_{\gamma_1}^{m_1} \psi_{\gamma_2}^{m_2} \cdots \psi_{\gamma_k}^{m_k}(y_\eta)$$

Because $\beta < \alpha$, this means that $y \in R_\alpha$, contradicting the fact that $y \in T_\alpha$. Therefore, we must have $x \notin R_\alpha$. We conclude that $\psi(D \cap T_\alpha) \subset T_\alpha$.

We utilize this proposition to prove[1]

THEOREM 23 If Φ has power $\leq 2^{\aleph_0}$, then for each cardinal number $\mathfrak{m}$, with $2 \leq \mathfrak{m} \leq 2^{\aleph_0}$, the space X can be decomposed into $\mathfrak{m}$ disjoint Bernstein sets S such that $S \triangle \phi(S)$ has power $< 2^{\aleph_0}$ for every mapping $\phi \in \Phi$.

Proof. We apply Proposition 22 with $E = X$, $\mathfrak{n} = 2^{\aleph_0}$, $\Theta = \Lambda$, and $\Psi = \Phi$. Let S be a nonempty set that is a union of sets T_α given in Proposition 22, say,

$$S = \bigcup_{\alpha \in I} T_\alpha$$

where I is a nonempty set of ordinal numbers $< \Lambda$. We first show $S \triangle \phi(S)$ has power $< 2^{\aleph_0}$ for each $\phi \in \Phi$.

Assume then that $\phi \in \Phi$ is arbitrary and let $\beta < \Lambda$ be an ordinal number such that $\phi(T_\alpha) = \phi(D \cap T_\alpha) \subset T_\alpha$ for all $\alpha > \beta$. Setting

$$L = \{\alpha \in I : \alpha \leq \beta\} \qquad M = \{\alpha \in I : \alpha > \beta\}$$

we have

$$\bigcup_{\alpha \in M} \phi(T_\alpha) \subset \bigcup_{\alpha \in M} T_\alpha \subset S$$

Since

$$\phi(S) = \left[\bigcup_{\alpha \in L} \phi(T_\alpha) \right] \cup \left[\bigcup_{\alpha \in M} \phi(T_\alpha) \right] \subset \left[\bigcup_{\alpha \in L} \phi(T_\alpha) \right] \cup S$$

---

[1]Cf. Banach (1932a, Remarque 1) and Sierpiński (1932b, Théorème II; 1936f, Théorème I; 1949b).

and $\bigcup_{\alpha \in L} \phi(T_\alpha)$ has power $< 2^{\aleph_0}$, the set $\phi(S)$—S has power $< 2^{\aleph_0}$. This being true for every $\phi \in \Phi$, the set $\phi^{-1}(S)$—S has power $< 2^{\aleph_0}$, as does the set $\phi[\phi^{-1}(S)$—$S] = S$—$\phi(S)$. Therefore, $S \triangle \phi(S)$ has power $< 2^{\aleph_0}$ for every $\phi \in \Phi$.

To complete the proof, we have only to show that X can be decomposed into m disjoint Bernstein sets, each of which is a union of certain sets T_α, whenever $2 \leqslant m \leqslant 2^{\aleph_0}$. We proceed in a manner analogous to that employed to prove Theorem 27 of Section I.

Let

(1) $T_1, T_2, \ldots, T_\alpha, \ldots$ $(\alpha < \Lambda)$

be the transfinite sequence of sets given in Proposition 22 and let

(2) $P_1, P_2, \ldots, P_\alpha, \ldots$ $(\alpha < \Lambda)$

be a transfinite enumeration of all perfect subsets of X.

Define $T_{1,1}$ to be the first set of the sequence (1) that has a nonempty intersection with P_1. Since each set T_α has power $< 2^{\aleph_0}$, there are continuum many sets in (1) which have a nonempty intersection with the set P_2—$T_{1,1}$. We define $T_{2,1}$ and $T_{2,2}$ to be the first two such sets in the enumeration (1). Continuing by transfinite induction, assume that $1 < \alpha < \Lambda$ and that the sets $T_{\gamma,\beta}$ have already been defined for all pairs of ordinal numbers γ, β with $\beta \leqslant \gamma < \alpha$. For each such pair of ordinal numbers there is an ordinal number $\mu < \Lambda$ such that $T_{\gamma,\beta} = T_\mu$, where T_μ occurs in (1). The set of all such numbers μ having power $< 2^{\aleph_0}$, there is an ordinal number $v < \Lambda$ that is larger than all these numbers μ. From the inclusion

$$\bigcup_{\beta \leqslant \gamma < \alpha} T_{\gamma,\beta} \subset \bigcup_{\mu \leqslant v} T_\mu$$

and condition (i) of Proposition 22, it follows that the set $\bigcup_{\beta \leqslant \gamma < \alpha} T_{\gamma,\beta}$ has power $< 2^{\aleph_0}$. Consequently, there are $2^{\aleph_0}$ sets T_ξ in (1) that have a nonempty intersection with the set P_α—$\bigcup_{\beta \leqslant \gamma < \alpha} T_{\gamma,\beta}$. Denoting by p the cardinal number of the order type α, we select from these sets T_ξ the first p sets, according to the enumeration (1), and arrange them in the form of a sequence $T_{\alpha,1}, T_{\alpha,2}, \ldots, T_{\alpha,\alpha}$.

For each ordinal number $\beta < \Lambda$ we now define

$$S_\beta = \bigcup_{\beta \leqslant \alpha < \Lambda} T_{\alpha,\beta}$$

The sets S_β are disjoint Bernstein sets furnishing the desired decomposition of X when $m = 2^{\aleph_0}$. For $2 \leqslant m < 2^{\aleph_0}$, the desired decomposition is obtained by selecting m of the sets S_β and adjoining all other sets S_β to one of the selected sets.

REMARK Sierpiński has called a linear set S a "Banach set" if both S and its complement have power $2^{\aleph_0}$ and S differs from each of its translates by at most a set of power $< 2^{\aleph_0}$. He has proved[1] that $\mathbb{R}$ can be represented as the union of more than $2^{\aleph_0}$ Banach sets, the intersection of any two different sets of which has power $< 2^{\aleph_0}$.

The following theorem generalizes Theorem 28 of Section I.

THEOREM 24 If Φ is countable, then X is representable as the union of more than $2^{\aleph_0}$ Φ-invariant Bernstein sets, the intersection of any two different sets of which has power $< 2^{\aleph_0}$.

Proof. We denote by Λ^+ the smallest ordinal number with power $> 2^{\aleph_0}$.

We start with a decomposition of X into $2^{\aleph_0}$ disjoint, Φ-invariant Bernstein sets

$$R_1, R_2, \ldots, R_\gamma, \ldots \qquad (\gamma < \Lambda)$$

the existence of which is guaranteed by Theorem 21. We proceed by transfinite induction to define a family $\mathscr{S} = \{S_\alpha : \alpha < \Lambda^+\}$ consisting of Φ-invariant Bernstein sets having the property that $R_\gamma \cap S_\alpha$ is a countable Φ-invariant set for all ordinal numbers $\gamma < \Lambda$, $\alpha < \Lambda^+$.

Let

$$P_1, P_2, \ldots, P_\gamma, \ldots \qquad (\gamma < \Lambda)$$

be an enumeration of all perfect subsets of X. Choose a point $x_{\gamma,1}$ in $R_\gamma \cap P_\gamma$ for each $\gamma < \Lambda$. Then

$$S_1 = \{\phi(x_{\gamma,1}) : \gamma < \Lambda \text{ and } \phi \in \Phi\}$$

is a Φ-invariant set for which $R_\gamma \cap S_1$ is a countable Φ-invariant set for each $\gamma < \Lambda$. Obviously, S_1 has at least one point in common with each perfect set. Since for each $\gamma < \Lambda$, R_γ is a Bernstein set and P_γ has $2^{\aleph_0}$ disjoint perfect subsets, the set $R_\gamma \cap P_\gamma$ has power $2^{\aleph_0}$, while $R_\gamma \cap S_1$ is countable. This implies that the complement of S_1 also contains at least one point of each perfect set. Hence, S_1 is a Bernstein set.

Assume that $1 < \alpha < \Lambda^+$ and the sets S_β have already been defined for all ordinal numbers $\beta < \alpha$. Let

$$T_1, T_2, \ldots, T_\xi, \ldots \qquad (\xi < \Lambda)$$

be an arrangement of all thus-far-defined sets S_β in a transfinite sequence of type Λ. [In the case that $\alpha < \Lambda$ one set S_β is repeated $2^{\aleph_0}$ times to obtain such an arrangement.] Suppose that γ is an ordinal number $< \Lambda$. By inductive

[1]Sierpiński (1936f, Théorème 2).

hypothesis, the set $R_\gamma \cap T_\xi$ is a countable Φ-invariant set for all $\xi < \Lambda$. The set $R_\gamma \cap (\bigcup_{\xi \leqslant \gamma} T_\xi)$, as a union of fewer than $2^{\aleph_0}$ countable sets, has power $< 2^{\aleph_0}$. As $R_\gamma \cap P_\gamma$ has power $2^{\aleph_0}$, we can choose a point $x_{\gamma,\alpha}$ in the set

$$(R_\gamma \cap P_\gamma) - \left(\bigcup_{\xi \leqslant \gamma} T_\xi \right)$$

for each $\gamma < \Lambda$. We define

$$S_\alpha = \{\phi(x_{\gamma,\alpha}): \gamma < \Lambda \text{ and } \phi \in \Phi\}$$

Then S_α is a Φ-invariant set for which $R_\gamma \cap S_\alpha$ is a countable Φ-invariant set for each $\gamma < \Lambda$. Clearly, S_α is a Bernstein set.

The family $\mathscr{S} = \{S_\alpha : \alpha < \Lambda^+\}$ thereby determined consists of Φ-invariant Bernstein sets. Moreover, if $1 \leqslant \beta < \alpha < \Lambda^+$, then $S_\alpha \cap S_\beta$ must have power $< 2^{\aleph_0}$. For, in the construction of S_α, there is an ordinal number $\xi_0 < \Lambda$ such that $S_\beta = T_{\xi_0}$ and, accordingly,

$$S_\alpha \cap S_\beta \subset \{\phi(x_{\gamma,\alpha}): \gamma < \xi_0 \text{ and } \phi \in \Phi\}$$

To complete the proof we have only to adjoin the Φ-invariant set $X - \bigcup_{\alpha < \Lambda^+} S_\alpha$ to one of the sets S_α to obtain the desired representation of X.

In connection with the proof of this theorem we note that if α and β are any two different ordinal numbers $< \Lambda$, then the set $S_\alpha - S_\beta$ has at least one point in common with each perfect set. Hence, the defined family of sets yielding the representation of X has the property

(δ) If S and T are any two different sets in the family, then $S - T$ is a Bernstein set.

We shall now establish the existence of a family consisting of $2^{2^{\aleph_0}}$ sets that has the property (δ) (but does not necessarily have the property that the intersection of any two different sets has power $< 2^{\aleph_0}$).[1] The proof is based on the following result of Knaster.[2]

PROPOSITION 25 If M is any infinite set with power $\mathfrak{m}$, then there exists a family consisting of $2^{\mathfrak{m}}$ subsets of M none of which is a subset of any other.

Proof. By virtue of the equality $\mathfrak{m} + \mathfrak{m} = \mathfrak{m}$, the set M is decomposable into disjoint sets M_1 and M_2, each of which has power $\mathfrak{m}$. Let ϕ_1 be a one-to-one mapping of M onto M_1 and let ϕ_2 be a one-to-one mapping of M onto M_2. For each set $L \subset M$ define

$$\psi(L) = \phi_1(L) \cup \phi_2(M - L)$$

[1]Cf. Sierpiński (1937c).
[2]See Kuratowski (1926, p. 205) and Tarski (1930, pp. 230–231).

We show that the family of sets $\{\psi(L): L \subset M\}$ has the desired properties. To this end it suffices to prove that if L and K are any two different subsets of M, then $\psi(L)-\psi(K) \neq \varnothing$. [The sets K and L being arbitrary, we must then also have $\psi(K)-\psi(L) \neq \varnothing$.]

Assume that $x \in L-K$. Then $y = \phi_1(x) \in \psi(L)$, but we cannot have $y \in \psi(K)$. For suppose that we did have $y \in \psi(K)$. Then either $y \in \phi_1(K)$ or $y \in \phi_2(M-K)$. Now, if $y \in \phi_1(K)$, then $y = \phi_1(z)$ for some element $z \in K$ and consequently, $\phi_1(x) = \phi_1(z)$ for $x \neq z$, contradicting the fact that ϕ_1 is one-to-one. On the other hand, if $y \in \phi_2(M-K)$, then $y \in \phi_1(M) \cap \phi_2(M)$, contradicting the disjointness of M_1 and M_2. We thus see that if $x \in L-K$, then $\phi_1(x) \in \psi(L)-\psi(K)$.

Similarly, we see that if $x \in K-L$, then $\phi_2(x) \in \psi(L)-\psi(K)$. Thus, if $L \neq K$, then $\psi(L)-\psi(K) \neq \varnothing$ for any subsets L, K of M.

THEOREM 26 If Φ is countable, then there exists a family of $2^{2^{\aleph_0}}$ Φ-invariant Bernstein sets with the property (δ) above.[1]

Proof. According to Theorem 21, X can be decomposed into a family $\mathcal{M}$ consisting of $2^{\aleph_0}$ disjoint, Φ-invariant Bernstein sets. Applying Proposition 25 to $\mathcal{M}$, we obtain a system $\mathfrak{S}$ consisting of $2^{2^{\aleph_0}}$ subfamilies of $\mathcal{M}$ none of which is contained in any other. Let

$$\mathcal{U} = \{\bigcup \mathcal{S} : \mathcal{S} \in \mathfrak{S}\}$$

The sets in $\mathcal{M}$ being Φ-invariant, each set in $\mathcal{U}$ is likewise Φ-invariant. That the family $\mathcal{U}$ has power $2^{2^{\aleph_0}}$ is a consequence of the fact that the sets in $\mathcal{M}$ are disjoint.

Suppose that S and T are any two different sets in $\mathcal{U}$. Then there exist families $\mathcal{S}, \mathcal{T} \in \mathfrak{S}$ such that $S = \bigcup \mathcal{S}$ and $T = \bigcup \mathcal{T}$. From $S \neq T$ we obtain $\mathcal{S} \neq \mathcal{T}$. According to the definition of $\mathfrak{S}$, we have $\mathcal{S}-\mathcal{T} \neq \varnothing$ and $\mathcal{T}-\mathcal{S} \neq \varnothing$. As $\mathcal{S}$ and $\mathcal{T}$ are subfamilies of $\mathcal{M}$, there exist Bernstein sets $M, N \in \mathcal{M}$ such that $M \in \mathcal{S}-\mathcal{T}$ and $N \in \mathcal{T}-\mathcal{S}$. The sets in $\mathcal{M}$ being disjoint, we have $M \subset S-T$ and $N \subset T-S$. The set $S-T$ contains the Bernstein set M and its complement contains the Bernstein set N. Therefore, $S-T$ is a Bernstein set. It is also clear from this argument that every set $S \in \mathcal{U}$ is a Bernstein set.

E. Quotient algebras

THEOREM 27 If $\mathscr{C}$ satisfies CCC, then the quotient algebra of all subsets of X modulo the ideal of meager sets is not complete.[2]

[1] Sierpiński (1937c).
[2] Sikorski (1949; 1964, pp. 78–79).

Proof. Let X be decomposed into a family $\mathscr{S}$ consisting of $2^{\aleph_0}$ disjoint Bernstein sets. By an argument similar to that used to prove Theorem 17 of Chapter 2, Section II, it is seen that $\mathscr{S}$ has no least upper bound with respect to the relation $\prec$ of inclusion modulo meager sets.

III. BAIRE FUNCTIONS

A. General properties

The general statements below are formulated in terms of functions $f: X \to \mathbb{R}$ with domain $X = \mathbb{R}^n$, except in Sections III.D and III.E, where we take $X = \mathbb{R}$.

THEOREM 1   If M is a meager set and f is a function that is continuous on $X-M$, then f is a Baire function.

Proof. The argument is a simple transposition of the second part of the proof of Theorem 22 of Chapter 4, Section IV.

COROLLARY 2   Every continuous function is a Baire function.

COROLLARY 3   If $f: \mathbb{R} \to \mathbb{R}$ is a monotone function, then f is a Baire function.

Proof. The set of points of discontinuity for a monotone function is a countable set.

THEOREM 4   There exists a function that is not a Baire function.

Proof. According to Theorem 3 of Chapter 1, Section IV, and Theorem 11 of Section II, the characteristic function of a Bernstein set is not a Baire function.

B. Continuity on perfect sets

According to Theorem 22 of Chapter 4, Section IV, we have

(τ_1) If f is a function having the Baire property, then there exists a set M of the first category such that f is continuous on the set $X-M$.

The Lebesgue measure analogue of this statement reads:

(μ_1) If f is a Lebesgue measurable function, then there exists a set M of Lebesgue measure zero such that f is continuous on the set $X-M$.

This statement is not valid, as can be seen by taking f to be the characteristic function of a nowhere dense, perfect set of positive Lebesgue measure. However, the following result is valid.

(μ'_1) If f is a Lebesgue measurable function, then for every positive number ε, there exists a closed set K such that $\mu(\mathbb{R}^n - K) < \varepsilon$ and f is continuous on K.[1]

This statement can also be expressed in the equivalent form:

(μ''_1) If f is a Lebesgue measurable function, then every measurable set E of positive measure contains a measurable set F of positive measure on which f is continuous.

We note that the sets E and F in this statement may be taken to be compact sets of positive measure.

We shall now establish a result which, although weaker, is valid for every perfect base.

THEOREM 5 If f is a Baire function, then every abundant Baire set contains a perfect set on which f is continuous.

Proof. Let f be a Baire function and suppose that S is a given abundant Baire set. In the event that there is a point $y \in \mathbb{R}$ for which $f_S^{-1}(y)$ is abundant, then $f_S^{-1}(y)$, being an abundant Baire set, contains a perfect set P and f_P is continuous on P. Having disposed of this situation, we assume henceforth that $f_S^{-1}(y)$ is a meager set for every $y \in \mathbb{R}$.

Suppose that H is an open subset of $\mathbb{R}$ such that $f_S^{-1}(H)$ is abundant in a given region A, and let H^* denote the set of all points $y \in H$ having the property that for every open interval I containing y, the set $f_S^{-1}(I)$ is abundant in A. We first show that H^* is nonempty.

Assume to the contrary that $H^* = \varnothing$. Let $\mathscr{I}$ denote the family of all open intervals with rational endpoints. For each $y \in H$ there will then be an open interval $I_y \in \mathscr{I}$ such that $y \in I_y$ and $f_S^{-1}(I_y)$ is meager in A. We have

$$A \cap f_S^{-1}(H) \subset \bigcup \{A \cap f_S^{-1}(I_y) : y \in H\}$$

The family $\{A \cap f_S^{-1}(I_y) : y \in H\}$ being countable, its union is a meager set. This implies that $A \cap f_S^{-1}(H)$ is meager, contradicting the supposition $f_S^{-1}(H)$ is abundant in A. We must therefore have $H^* \neq \varnothing$.

---

[1] Although usually attributed to Luzin, he was not the first one to prove this result. See Borel (1903), Lebesgue (1903), Vitali (1905b), and Luzin (1911, p. 280 Teorema 4 and p. 284); (1912). See also Sierpiński (1916b, 1922), Hahn (1921, pp. 568–570), Cohen (1927), Viola (1934), Scorza-Dragoni (1936), Hahn and Rosenthal (1948, p. 148), Kvačko (1958), Trjitzinsky (1960, Sec. 17), Blakney (1969), Vrkoč (1971), Feldman (1981), and Brown and Prikry (1987).

Suppose that I is any open interval for which $f_S^{-1}(I)$ is abundant in a given region A and suppose that $y \in I^*$. Taking $H = I - \{y\}$ above, it follows that there is a point $z \in I^*$ such that $z \neq y$. Consequently, there exist two disjoint open intervals I_0, I_1 contained in I such that both $f_S^{-1}(I_0)$ and $f_S^{-1}(I_1)$ are abundant in A. Moreover, the lengths of I_0 and I_1 can be made arbitrarily small.

With these preliminaries out of the way, we turn to the main part of the proof.

Because S is abundant and is representable as the countable union of sets $f_S^{-1}(I)$ for $I \in \mathscr{I}$, there exists an open interval I such that $f_S^{-1}(I)$ is an abundant set. Let A be a region in which $f_S^{-1}(I)$ is abundant everywhere. We proceed inductively to define a certain dyadic schema $\langle A_\sigma : \sigma \in \mathbb{B}^F \rangle$ of subregions of A.

The set S being a Baire set, we have

$$A - S = \bigcup_{i=1}^{\infty} T_i$$

where each set T_i is singular. As established above, there exist two disjoint open intervals I_0, I_1 contained in I, each with length $\leqslant 1$, for which both $f_S^{-1}(I_0)$ and $f_S^{-1}(I_1)$ are abundant in A. Let A_0 be a subregion of A having the properties: $f_S^{-1}(I_0)$ is abundant everywhere in A_0, $A_0 \cap T_1 = \varnothing$, and $\operatorname{diam}(A_0) \leqslant 1$. The set $f_S^{-1}(I_1)$ is abundant everywhere in a subregion B of A. We note that $A_0 \cap B$ can contain no region C. For, otherwise, both $f_S^{-1}(I_0)$ and $X - f_S^{-1}(I_0)$, which contains $f_S^{-1}(I_1)$, would be abundant everywhere in C, contradicting the fact that $f_S^{-1}(I_0)$ is a Baire set. Since $A_0 \cap B$ thus contains no region, there exists a region $A_1 \subset B$ that is disjoint from A_0 and that may be assumed to have the properties: $f_S^{-1}(I_1)$ is abundant everywhere in A_1, $A_1 \cap T_1 = \varnothing$, and $\operatorname{diam}(A_1) \leqslant 1$. We have

$$A_0 - f_S^{-1}(I_0) = \bigcup_{i=1}^{\infty} T_{0,0,i} \qquad A_0 \cap f_S^{-1}(I_1) = \bigcup_{i=1}^{\infty} T_{0,1,i}$$

$$A_1 \cap f_S^{-1}(I_0) = \bigcup_{i=1}^{\infty} T_{1,0,i} \qquad A_1 - f_S^{-1}(I_1) = \bigcup_{i=1}^{\infty} T_{1,1,i}$$

where all the sets T are singular.

Assume now that $n \in \mathbb{N}$ and for all $\sigma, \tau \in \mathbb{B}^n$, $i \in \mathbb{N}$, we have already defined the intervals I_σ, regions A_σ, and singular sets $T_{\sigma,\tau,i}$. Given $\sigma \in \mathbb{B}^n$, we define $I_{\sigma 0}, I_{\sigma 1}$ to be disjoint open intervals contained in I_σ, each with length $\leqslant 1/(n + 1)$, with both $f_S^{-1}(I_{\sigma 0})$ and $f_S^{-1}(I_{\sigma 1})$ abundant in A_σ. We define $A_{\sigma 0}, A_{\sigma 1}$ to be disjoint subregions of A_σ, each of diameter $\leqslant 1/(n + 1)$ and disjoint from all previously defined sets T with index $i \leqslant n + 1$, such that $f_S^{-1}(I_{\sigma 0})$ is abundant everywhere in $A_{\sigma 0}$ and $f_S^{-1}(I_{\sigma 1})$ is abundant everywhere in $A_{\sigma 1}$. Finally, for all σ', $\tau' \in \mathbb{B}^{n+1}$ with $\sigma' \neq \tau'$, set

$$A_{\sigma'} - f_S^{-1}(I_{\sigma'}) = \bigcup_{i=1}^{\infty} T_{\sigma',\sigma',i} \qquad A_{\sigma'} \cap f_S^{-1}(I_{\tau'}) = \bigcup_{i=1}^{\infty} T_{\sigma',\tau',i}$$

where all the sets T are singular.

Having thus determined the dyadic schema $\langle A_{\sigma} : \sigma \in \mathbb{B}^F \rangle$, we see that

$$P = \bigcap_{n=1}^{\infty} \bigcup_{\sigma \in \mathbb{B}^n} A_{\sigma}$$

is a perfect set contained in S. We proceed to show the function f is continuous on P.

First, we establish for every $n \in \mathbb{N}$ and every $\tau \in \mathbb{B}^n$ that $f_P^{-1}(I_{\tau})$ is open in P; i.e., there exists a set G open in X such that

$$f_P^{-1}(I_{\tau}) = P \cap G$$

For this we take G to be any open set containing A_{τ} which is disjoint from all the setsp $A_{\tau'}$ with $\tau' \in \mathbb{B}^n$ and $\tau' \neq \tau$.

Assuming that $x \in P$, there exists a sequence $\mu = \langle \mu_1, \mu_2, \ldots \rangle \in \mathbb{B}^{\infty}$ such that $x \in A_{\mu|k}$ for all $k \in \mathbb{N}$.

Suppose that $x \in f_P^{-1}(I_{\tau})$. Placing $\sigma = \mu|n$, we have $x \in A_{\sigma} \cap f_S^{-1}(I_{\tau})$. If we suppose that $\sigma \neq \tau$, then there is an index i such that $x \in T_{\sigma,\tau,i}$. However, for $k \in \mathbb{N}$ and $k > \max\{i,n\}$, we have $x \in A_{\mu|k}$ and $A_{\mu|k} \cap T_{\sigma,\tau,i} = \emptyset$. From this contradiction we see that $\sigma = \tau$. This means that $x \in A_{\tau}$, so that $x \in P \cap G$.

Conversely, suppose that $x \in G$. Then $\tau = \mu|n$, so $x \in A_{\tau}$. If we suppose that $x \notin f_S^{-1}(I_{\tau})$, then there is an index i such that $x \in T_{\tau,\tau,i}$. However, for $k \in \mathbb{N}$ and $k > \max\{i,n\}$, we have $x \in A_{\mu|k}$ and $A_{\mu|k} \cap T_{\tau,\tau,i} = \emptyset$. From this contradiction we conclude that $x \in f_S^{-1}(I_{\tau})$, so $x \in P \cap f_S^{-1}(I_{\tau}) = f_P^{-1}(I_{\tau})$.

Having thus established the special case that $f_P^{-1}(I_{\tau})$ is open in P for each $\tau \in \mathbb{B}^F$, we consider now the general case that H is any open interval. We assert that

$$f_P^{-1}(H) = \bigcup \{f_P^{-1}(I_{\tau}) : \tau \in \mathbb{B}^F \text{ and } I_{\tau} \subset H\}$$

For suppose that $x \in f_P^{-1}(H) = P \cap f^{-1}(H)$. Then $f(x) \in H$ and there is a sequence $\mu = \langle \mu_1, \mu_2, \ldots \rangle \in \mathbb{B}^{\infty}$ such that $f(x) \in I_{\mu|k}$ for all $k \in \mathbb{N}$. Because $\lim_{k \to \infty} \text{diam}(I_{\mu|k}) = 0$, there exists an $n \in \mathbb{N}$ for which $I_{\mu|n} \subset H$. This implies that x is an element of the right-hand side of the asserted equality. The converse being obviously true, the equality is established. This equality, in conjunction with the special case treated above, yields the desired conclusion that $f_P^{-1}(H)$ is open in P. We conclude that f is continuous on P.

We note that the characteristic function of a Sierpiński set is a nonmeasurable function satisfying the condition that in every Lebesgue measurable set of positive measure there is a perfect set on which the function is continuous. Analogously, the characteristic function of a Mahlo-Luzin set is a function that does not have Baire property, but does satisfy the condition that in every second category set with the Baire property there is a perfect set on which the function is continuous.

For the category base $(X,\mathscr{S})$ yielding Marczewski's classification, the converse of Theorem 5 is also true.

THEOREM 6 For the category base $(X,\mathscr{S})$, a necessary and sufficient condition that a function f be a Baire function is that every perfect set have a perfect subset on which f is continuous.[1]

Proof. The necessity of the condition is a consequence of Theorem 5.

To establish sufficiency, suppose that G is an open subset of $\mathbb{R}$. Let Q be a perfect subset of X. By hypothesis, there exists a perfect set $P \subset Q$ on which f is continuous. Consequently, the set $f_P^{-1}(G) = P \cap f^{-1}(G)$ is open in P. According to Theorem 39 of Section I, $f^{-1}(G)$ is a Baire set. Thus, f is a Baire function.

We note the following related results:

(1) Every function is continuous on an everywhere dense set.[2]
(2) There exists a function that is discontinuous on every set having the power of the continuum.[3]
(3) If $2^{\aleph_0} = \aleph_1$, then there exists an upper semicontinuous function of a real variable and an uncountable set E such that f is discontinuous on every uncountable subset of E.[4]

C. Boundedness

We first establish one consequence of Theorem 5.

THEOREM 7 If f is a Baire function, then there exists a perfect set on which f is bounded.

Proof. Let P be a perfect set on which f is continuous. The set P being uncountable, there exists a closed rectangle R for which the set $P \cap R$ is

[1]Szpilrajn (1935); see also Brown (1985) and Brown and Prikry (1987).
[2]Blumberg (1922, Theorem III); see also Sierpiński (1932e), Bradford and Goffman (1960), Brown (1971, 1977, 1983, 1985), Levy (1973, 1974), White (1974a, 1975a, 1975b, 1979), Weiss (1975, 1977), Alas (1976), Lukeš and Zajíček (1976), and Brown and Prikry (1987).
[3]Sierpiński and Zygmund (1923).
[4]Sierpiński (1936a, 1936b, 1936e, 1937a).

uncountable. According to the Cantor-Bendixson Theorem, the latter set contains a perfect set Q. We show that f is bounded on Q.

Assume to the contrary that f is not bounded on Q. Then there exists, for every natural number n, a point $x_n \in Q$ such that $|f(x_n)| > n$. By the Bolzano-Weierstrass Theorem, the bounded, infinite set $\{x_n : n \in \mathbb{N}\}$ has a limit point x_0, which is necessarily an element of Q. Let $\langle x_{n_k} \rangle_{k \in \mathbb{N}}$ be a subsequence of the sequence $\langle x_n \rangle_{n \in \mathbb{N}}$ that converges to x_0. The sequence $\langle f(x_{n_k}) \rangle_{k \in \mathbb{N}}$ does not converge to $f(x_0)$. This contradicts the fact that f is continuous on Q.

DEFINITION   A function g is said to majorize a function f if $f(x) \leqslant g(x)$ for every $x \in \mathbb{R}^n$.

Sierpiński established the existence of a (nonmeasurable) function that cannot be majorized by any Lebesgue measurable function.[1] More generally, we have for any perfect base

THEOREM 8   There exists a function that cannot be majorized by any Baire function.

This theorem is a consequence of Theorem 7 and the following result.

THEOREM 9   There exists a function that is not bounded above on any perfect set.

Proof. Let

(1) $$P_1, P_2, \ldots, P_\alpha, \ldots \qquad (\alpha < \Lambda)$$

be a well-ordering of all perfect subsets of $\mathbb{R}^n$ in which each perfect set is listed continuum many times, let

(2) $$r_1, r_2, \ldots, r_\alpha, \ldots \qquad (\alpha < \Lambda)$$

be a well-ordering of all points of $\mathbb{R}^n$, and let

(3) $$y_1, y_2, \ldots, y_\alpha, \ldots \qquad (\alpha < \Lambda)$$

be a well-ordering of $\mathbb{R}$.

Define x_1 to be the first element of (2) that belongs to P_1 and set $f(x_1) = y_1$. Continuing by transfinite induction, suppose that for $\alpha < \Lambda$ we have already defined x_β and $f(x_\beta)$ for all ordinal numbers $\beta < \alpha$. Define x_α to be the first element of (2) that belongs to $P_\alpha - \{x_\beta : \beta < \alpha\}$. If the set

$$S_\alpha = \{y \in \mathbb{R} : y > f(x_\beta) \text{ for all } x_\beta \in P_\alpha \text{ with } \beta < \alpha\}$$

is nonempty, then we define $f(x_\alpha)$ to be the first element of (3) that belongs to S_α; otherwise, $f(x_\alpha)$ is defined to be the first element of (3) that is different from

[1] Sierpiński (1918, p. 145).

all the elements $f(x_\beta)$ with $\beta < \alpha$. It is readily seen that the function f determined in this manner is defined for all points of $\mathbb{R}^n$.

Suppose that the function f is bounded above on some perfect set P by a number m. Since P occurs continuum many times in the transfinite sequence (1), there exists an increasing transfinite sequence $\langle \alpha_\xi \rangle_{\xi < \Lambda}$ of ordinal numbers such that $P = P\alpha_\xi$ for all $\xi < \Lambda$. By virtue of the definition of f, the transfinite sequence $\langle f(x\alpha_\xi) \rangle_{\xi < \Lambda}$ must be an increasing sequence. But this is impossible, since no uncountable set of real numbers is well-ordered numerically. We thus conclude that f is not bounded above on any perfect set.

NOTE Utilizing Proposition 22 of Section I, it can be shown that any function must be bounded on some set having the power of the continuum.

Sierpiński further established a special case of the following general theorem.[1]

THEOREM 10 If $(\mathbb{R}^n, \mathscr{C})$ and $(\mathbb{R}^n, \mathscr{D})$ are complementary bases, then there exists a $\mathscr{C}$-Baire function that cannot be majorized by any $\mathscr{D}$-Baire function.

Proof. Let $\mathbb{R}^n = M \cup N$, where M is $\mathscr{C}$-meager, N is $\mathscr{D}$-meager, and $M \cap N = \varnothing$. Let f be a function that is not bounded above on any perfect set and define a function g by

$$g(x) = \begin{cases} f(x), & x \in M \\ 0, & x \in N \end{cases}$$

As g is constant on the set N whose complement is $\mathscr{C}$-meager, g is a $\mathscr{C}$-Baire function.

Suppose that g is majorized by a $\mathscr{D}$-Baire function h. By Theorem 5, there is a perfect set $P \subset M$ on which h is continuous. Let Q be a compact perfect subset of P. Then h is bounded above on Q, as is also f. We thus have a contradiction.

D. Images and inverses

We assume throughout this section that $X = \mathbb{R}$.

THEOREM 11 There exists a Baire function whose range is not a Baire set.[2]

Proof. By Theorem 1 of Section II or Theorem 36 of Section I, there exists a meager set E having the power of the continuum. Let ϕ be a one-to-one

---

[1]Cf. Sierpiński (1918, pp. 143–145).
[2]Cf. Ballew (1974).

function mapping E onto a Bernstein set T. Define $f: X \to \mathbb{R}$ by

$$f(x) = \begin{cases} \phi(x), & x \in E \\ 0, & x \notin E \end{cases}$$

We know by virtue of Theorem 15 of Chapter 1, Section IV, that f is a Baire function. The range of f is the Bernstein set $T \cup \{0\}$, which is not a Baire set.

In the case of one-to-one Baire functions, Sierpiński has given the following characterization of the range.[1]

THEOREM 12 A necessary and sufficient condition that a linear set be the range of a one-to-one Baire function is that the set contain a perfect set.

Proof. Necessity: Suppose that f is a one-to-one Baire function with range T. Then f is continuous on a compact, perfect set P. The set $f(P)$, being an uncountable compact set, contains a perfect set $Q \subset T$.

Sufficiently: Suppose that T is a linear set which contains a perfect set Q. Let $\phi: X \to Q$ be an isomorphism mapping X onto a set $L \subset Q$ as in Theorem 32 in Section I. We know from Corollary 3 that ϕ is a Baire function. Let E be a meager subset of X having the power of the continuum. The set $S = (T - L) \cup \phi(E)$ also has the power of the continuum and, accordingly, there is a one-to-one function ψ mapping E onto S. Define $f: X \to T$ by

$$f(x) = \begin{cases} \psi(x), & x \in E \\ \phi(x), & x \notin E \end{cases}$$

Then f is a one-to-one function whose range is the set T. Since f is equal essentially everywhere to the Baire function ϕ, the function f is a Baire function.

Assuming CH, Sierpiński has shown[2] that there exists a sequence of functions $\langle f_n \rangle_{n \in \mathbb{N}}$ such that for any uncountable set S, each term of the sequence, with the possible exception of a finite number of terms, maps S onto $\mathbb{R}$. One cannot in general obtain such a sequence consisting of Baire functions if CH is assumed. This is a consequence of the following theorem.[3]

THEOREM 13 Assume CH. If $\mathscr{C}$ satisfies CCC, then for any family $\mathscr{F}$ of power of the continuum consisting of sets having the power of the continuum and for any family $\mathscr{M}$ of power of the continuum consisting of Baire

---

[1]Sierpiński (1933c).
[2]Cf. Braun and Sierpiński [1932, Proposition (R)] and Sierpiński (1933b; 1933d; 1956, pp. 12–14, 189–191).
[3]Sierpiński (1932c; 1956, Proposition C_{71}).

functions, there exists a set S of power of the continuum such that for every function $f \in \mathcal{M}$, the set $f(S)$ contains no set of the family $\mathcal{T}$.

Proof. The families $\mathcal{T}$ and $\mathcal{M}$ having power $2^{\aleph_0}$, it follows from CH and the equality $\aleph_1^2 = \aleph_1$ that we can arrange the sets in $\mathcal{T}$ in a transfinite sequence

$$T_1, T_2, \ldots, T_\alpha, \ldots \qquad (\alpha < \Omega)$$

and the functions in $\mathcal{M}$ in a transfinite sequence

$$f_1, f_2, \ldots, f_\alpha, \ldots \qquad (\alpha < \Omega)$$

such that for every set $T \in \mathcal{T}$ and every function $f \in \mathcal{M}$, there exists an ordinal number $\alpha < \Omega$ for which $T = T_\alpha$ and $f = f_\alpha$. We shall define by transfinite induction a transfinite sequence

$$E_1, E_2, \ldots, E_\alpha, \ldots \qquad (\alpha < \Omega)$$

of meager sets and a transfinite sequence of points

$$p_1, p_2, \ldots, p_\alpha, \ldots \qquad (\alpha < \Omega)$$

If $T_1 - f_1(X) \neq \varnothing$, then we define $E_1 = \varnothing$. Suppose that $T_1 - f_1(X) = \varnothing$, i.e., that $T_1 \subset f_1(X)$. Then for each element $t \in T_1$ the set $f_1^{-1}(t)$ is nonempty. Applying Theorem 5 of Chapter 1, Section III, we define E_1 to be a meager set belonging to the family $\{f_1^{-1}(t) : t \in T_1\}$. After determining E_1, we then define p_1 to be a point that does not belong to E_1.

Assume that $1 < \alpha < \Omega$ and the sets E_β and points p_β have already been defined for all ordinal numbers $\beta < \alpha$. If $T_\alpha - f_\alpha(X) \neq \varnothing$, then we define $E_\alpha = \varnothing$. Suppose, on the other hand, that $T_\alpha \subset f_\alpha(X)$. Then there are uncountably many sets $f_\alpha^{-1}(t)$ with $t \in T_\alpha$ which are nonempty meager sets. These sets being disjoint, we define E_α to be one such set that contains none of the points p_β with $\beta < \alpha$. Having determined E_α, we then define p_α to be a point that does not belong to the set $(\bigcup_{\beta < \alpha} E_\alpha) \cup \{p_\beta : \beta < \alpha\}$.

The set $S = \{p_\alpha : \alpha < \Omega\}$ thus obtained is uncountable, and since CH is assumed, has the power of the continuum.

Suppose that we have $T_\gamma \subset f_\xi(S)$ for $\gamma, \xi < \Omega$. Choose $\alpha < \Omega$ so that $T_\gamma = T_\alpha$ and $f_\xi = f_\alpha$. Then $T_\alpha \subset f_\alpha(S)$ and, because $f_\alpha(S) \subset f_\alpha(X)$, we have $T_\alpha \subset f_\alpha(X)$. According to our definition, $E_\alpha = f_\alpha^{-1}(t_0)$ for some element $t_0 \in T_\alpha$. Now, $t_0 \in T_\alpha$ and $T_\alpha \subset f_\alpha(S)$ implies that there is a point $p_\beta \in S$ such that $f_\alpha(p_\beta) = t_0$, so that $p_\beta \in f_\alpha^{-1}(t_0) = E_\alpha$. But no point p_β belongs to E_α! Our supposition $T_\gamma \subset f_\xi(S)$ for $\gamma, \xi < \Omega$ has led to a contradiction. We conclude that there is no function $f_\xi \in \mathcal{M}$ for which $f_\xi(S)$ contains a set $T_\gamma \in \mathcal{T}$.

The inverse of a Baire function need not be a Baire function.

THEOREM 14   There exists a one-to-one Baire function whose inverse function is not a Baire function.[1]

Proof. We first obtain a totally imperfect meager set S having the power of the continuum. If $\mathscr{C}$ is equivalent to $\mathscr{P}$, then we obtain S via Theorems 20 and 35 of Section I. If $\mathscr{C}$ is not equivalent to $\mathscr{P}$, then we obtain a singular perfect set M from Theorem 1 of Section II, decompose X into two disjoint Bernstein sets U, V by means of Theorem 27 of Section I, and take S to be one of the sets $M \cap U$, $M \cap V$. (It follows from Theorem 20 of Section I that both of the sets $M \cap U$, $M \cap V$ have the power of the continuum.)

Let ϕ be an isomorphism mapping $\mathbb{R}$ onto $\mathbb{R}$—$[0,1)$ as in Proposition 31 of Section I and let ψ be a one-to-one function mapping S onto the set $\phi(S) \cup [0,1)$. Define $f : \mathbb{R} \to \mathbb{R}$ by

$$f(x) = \begin{cases} \phi(x), & x \notin S \\ \psi(x), & x \in S \end{cases}$$

It follows from Corollary 3 and Theorem 15 of Chapter 1, Section IV, that f is a Baire function.

Now, if f^{-1} is a Baire function, then f^{-1} is continuous on some perfect set $Q \subset [0,1)$. Then $f^{-1}(Q)$ is an uncountable compact set that contains a perfect set P. But $P \subset f^{-1}(Q) \subset S$ contradicts the fact that S has no perfect subsets. We conclude that f^{-1} is not a Baire function.

E.   Composition

We turn to consider composition of functions, restricting our attention to functions of a real variable. As is seen from the following theorem, Theorem 4, and Theorem 17 below, there is essentially only one perfect base for which the composition of any two Baire functions will always be a Baire function.

THEOREM 15   For the category base $(X,\mathscr{P})$, the family of Baire functions is closed under composition.[2]

Proof. Let $h = g \circ f$, where f and g are Baire functions. We shall use the characterization given in Theorem 6.

Suppose that P is any given perfect set. Then there exists a compact perfect set $P_1 \subset P$ on which f is continuous. The set $f(P_1)$, as a continuous image of a compact set, is a closed set. We distinguish two cases.

Assume that $f(P_1)$ is not a perfect set. Then $f(P_1)$ contains an isolated point y_0. Consequently, there exists an open set G such that $G \cap P_1 \neq \varnothing$ and

[1]Cf. Szpilrajn (1935) and Sierpiński (1939c).
[2]Sierpiński (1935c).

$f(x) = y_0$ for every $x \in G \cap P_1$. Being an uncountable $\mathscr{G}_\delta$-set, $G \cap P_1$ contains a perfect set P_2. For every $x \in P_2$ we have $h(x) = g[f(x)] = g(y_0)$. As a constant function on P_2 the function h is continuous on the perfect set $P_2 \subset P$.

Assume on the other hand that $f(P_1)$ is a perfect set. Then there exists a perfect set $Q \subset f(P_1)$ on which g is continuous. It follows from the continuity of f on P_1 and the foregoing inclusion that $f_{P_1}^{-1}(Q) = P_1 \cap f^{-1}(Q)$ is an uncountable closed set which, according to the Cantor-Bendixson Theorem, contains a perfect set P_2. Since $P_2 \subset P_1$ and $f(P_2) \subset Q$, the function $h = g \circ f$ is continuous on the perfect set $P_2 \subset P$.

We note next an elementary fact.

THEOREM 16   If f is a Baire function and g is a continuous function, then the composition $g \circ f$ is a Baire function.

In this theorem the order of composition given is essential. Sierpiński has shown there exists a continuous function f and a Lebesgue measurable function g for which the composition $g \circ f$ is not a Lebesgue measurable function.[1] This is accomplished by defining an increasing continuous function f from $\mathbb{R}$ onto $\mathbb{R}$ which maps a certain everywhere dense $\mathscr{F}_\sigma$-set D whose complement has measure zero onto an everywhere dense $\mathscr{F}_\sigma$-set E of measure zero, choosing a nonmeasurable set S, and taking g to be the characteristic function of the set $E \cap f(S)$, which has measure zero. Then $g \circ f$ is the characteristic function of the nonmeasurable set $D \cap S$.

As pointed out by Ruziewicz, there are functions that are not representable as a Lebesgue measurable function or, more generally, as a Baire function of a continuous function; e.g., in view of Theorem 5, the characteristic function of a Bernstein set or any other function discontinuous on every perfect set cannot be so represented. On the other hand, he has proved that every function can be expressed as a Lebesgue measurable function of a Lebesgue measurable function.

THEOREM 17   There exists a fixed increasing Baire function ϕ which is continuous on the set of irrational numbers such that every function f is representable in the form $f = h \circ \phi$ for some Baire function h if and only if $(X, \mathscr{C})$ is not equivalent to $(X, \mathscr{P})$.[2]

---

[1]Sierpiński (1916a). See Sierpiński (1935b) for a survey of results concerning composition of functions.
[2]Ruziewicz (1935), Sierpiński [1935c, pp. 12–13 (category version)], and Hong and Tong (1983).

Proof. If $(X,\mathscr{C})$ is equivalent to $(X,\mathscr{P})$, then the composition of any two Baire functions is a Baire function, so any non-Baire function is not representable as such a composition.

Assume that $(X,\mathscr{C})$ is not equivalent to $(X,\mathscr{P})$. According to Theorem 1 of Section II, there then exists a singular, perfect set P. Let L and D be the sets given in the proof of Theorem 32 of Section I and let ϕ be an isomorphism mapping X onto L such that $\phi(\mathbb{Q}) = D$. For any given function f, define the function h for each $y \in \mathbb{R}$ by

$$h(y) = \begin{cases} f[\phi^{-1}(y)] & \text{if } y \in L \\ 0 & \text{if } y \notin L \end{cases}$$

For any $x \in X$ we have $\phi(x) \in L$ and $h[\phi(x)] = f(x)$, whence $f = h \circ \phi$.

We know as a result of Theorem 1 that ϕ and h are Baire functions. The set of all unilateral limit elements of L is a subset of D and ϕ is continuous for all $x \notin \phi^{-1}(D)$. Hence, ϕ is continuous on the set of irrational numbers.

The functions ϕ having the property that every function is representable as a Lebesgue measurable function of ϕ can be completely characterized.

THEOREM 18   In order that every real-valued function f defined on a set S have the form $f = h \circ \phi$, where h is a Baire function and ϕ is a fixed real-valued function defined on S, it is necessary and sufficient that ϕ satisfy the following conditions:

  (i) ϕ is one-to-one.
  (ii) Every subset of $\phi(S)$ is a restricted Baire set relative to $\phi(S)$.[1]

Proof. Sufficiency: Assume that ϕ satisfies the given conditions and let f be any real-valued function defined on S. Since ϕ is one-to-one, placing $h_0(y) = f[\phi^{-1}(y)]$ for each $y \in \phi(S)$, we obtain a real-valued function on $\phi(S)$. For each $x \in S$ we have $f(x) = h_0[\phi(x)]$. Applying Theorem 11 of Chapter 2, Section II, we obtain a Baire function h such that $h(x) = h_0(x)$ for each $x \in \phi(S)$. For every $x \in S$ we have $f(x) = h[\phi(x)]$.

Necessity: If $x_1, x_2 \in S$ and $x_1 \neq x_2$, then we cannot have $\phi(x_1) = \phi(x_2)$, lest every real-valued function f defined on S satisfy $f(x_1) = f(x_2)$, which is impossible. Therefore, ϕ is one-to-one.

Suppose that h_0 is any real-valued function defined on $\phi(S)$. Define a function f on S by $f(x) = h_0[\phi(x)]$ for each $x \in S$. By hypothesis, there exists a Baire function h such that $f(x) = h[\phi(x)]$ for all $x \in S$. As $h(x) = h_0(x)$ for every $x \in \phi(S)$, the function h is an extension of h_0. We conclude from Theorem 11 of Chapter 2, Section II that every subset of $\phi(S)$ is a restricted Baire set relative to $\phi(S)$.

---

[1]Eilenberg (1932, Théorème 3).

REMARK   Assuming CH, it follows from Theorem 6 of Section II that if $\mathscr{C}$ satisfies CCC, then condition (ii) can be replaced by the condition

(ii') $\phi(S)$ is a meager set.

Can one determine a fixed Lebesgue measurable function ψ such that every function f is representable in the form $f = \psi \circ h$ for some Lebesgue measurable function h? The response here is negative.[1]

THEOREM 19   For every Baire function ψ there exists a function f that is not representable in the form $f = \psi \circ h$ for any Baire function h.

Proof. We first determine a certain transfinite arrangement $\langle t_\alpha \rangle_{\alpha < \Lambda}$ of the set of all real numbers.
  Let

(1) $\qquad\qquad\qquad\qquad x_1, x_2, \ldots, x_\gamma, \ldots \qquad (\gamma < \Lambda)$

be an enumeration of all real numbers and let

(2) $\qquad\qquad\qquad\qquad Q_1, Q_2, \ldots, Q_\alpha, \ldots \qquad (\alpha < \Lambda)$

be an enumeration of all perfect subsets of $X \times Y$ that have at most one point on each line parallel to the Y-axis.
  Define t_1 to be the first term of the transfinite sequence (1) for which the line $\{\langle x, y \rangle : x = t_1 \text{ and } y \in Y\}$ has a nonempty intersection with Q_1. Assume that $\alpha < \Lambda$ and t_β has already been determined for all $\beta < \alpha$. The set Q_α being perfect and having at most one point on each vertical line, there are continuum many vertical lines that contain a point of Q_α. Define t_α to be the first term of the sequence (1) such that $t_\alpha \neq t_\beta$ for all $\beta < \alpha$ and the line $\{\langle x, y \rangle : x = t_\alpha \text{ and } y \in Y\}$ has a point in common with the set Q_α. We thus determine by transfinite induction the desired sequence $\langle t_\alpha \rangle_{\alpha < \Lambda}$. That every real number occurs as a term of this sequence is a consequence of the fact that the sequence (2) includes continuum many lines parallel to the X-axis.
  Suppose now that ψ is any given Baire function. Without loss of generality, we may assume that ψ is not a constant function. Let a and b be two different values of ψ and set

$$M = \{x \in X : \psi(x) = a\}$$

We define a function f in the following manner:
  Let x be any given real number. Then $x = t_\alpha$ for some $\alpha < \Lambda$ and there is a unique point $\langle t_\alpha, u_\alpha \rangle$ belonging to Q_α. We define

$$f(x) = \begin{cases} a, & u_\alpha \notin M \\ b, & u_\alpha \in M \end{cases}$$

[1]Sierpiński (1933g).

Assume that h is any Baire function. According to Theorem 5, there exists a perfect set P on which f is continuous. The set

$$Q = \{\langle x,y \rangle : x \in P \text{ and } y = h(x)\}$$

is a perfect set that is a term of the sequence (2), say $Q = Q_\alpha$. Let $u_\alpha = h(t_\alpha)$. If $u_\alpha \in M$, then $f(t_\alpha) = b$ and $\psi[h(t_\alpha)] = a$. If $u_\alpha \notin M$, then $f(t_\alpha) = a$ and $\psi[h(t_\alpha)] \neq a$. In any case we have $f(t_\alpha) \neq \psi[h(t_\alpha)]$.

In contrast to Theorem 19, we have the following result of Ruziewicz and Sierpiński.[1]

THEOREM 20   For any family $\mathcal{M}$ of functions that has the power of the continuum there exists a fixed Baire function ψ such that every function $f \in \mathcal{M}$ is expressible in the form $f = \psi \circ \phi$ for some increasing Baire function ϕ that is continuous on the set of irrational numbers if and only if $(X,\mathscr{C})$ is not equivalent to $(X,\mathscr{P})$.

Proof. If $(X,\mathscr{C})$ is equivalent to $(X,\mathscr{P})$, then it follows from Theorem 15 that the assertion fails to hold when $\mathcal{M}$ contains a non-Baire function.

Assume that $(X,\mathscr{C})$ is not equivalent to $(X,\mathscr{P})$ and let $\mathcal{M} = \{f_t : t \in \mathbb{R}\}$ be a family of functions that has the power of the continuum. We know from Theorem 1 of Section II that there exists a singular, perfect set P and, according to Theorem 20 of Section I, there is a family $\{P_t : t \in \mathbb{R}\}$ consisting of continuum many disjoint perfect subsets of P. Following the proof of Theorem 17, for each $t \in \mathbb{R}$, we determine the sets L_t, D_t and an isomorphism ϕ_t mapping X onto L_t such that $\phi_t(\mathbb{Q}) = D_t$. We define a Baire function ψ for each $y \in \mathbb{R}$ by

$$\psi(y) = \begin{cases} f_t[\phi_t^{-1}(y)] & \text{if } y \in L_t \text{ for some } t \in \mathbb{R} \\ 0 & \text{if } y \notin L_t \text{ for all } t \in \mathbb{R} \end{cases}$$

For each $t \in \mathbb{R}$ and every $x \in X$ we have $\phi_t(x) \in L_t$ and $\psi[\phi_t(x)] = f_t(x)$.

F.   Graphs

We next establish two theorems concerning graphs of Baire functions. We recall that for any sets X and Y, the cartesian product $X \times Y$ is the set of all ordered pairs $\langle x,y \rangle$ with $x \in X$ and $y \in Y$. The first term x of an ordered pair $\langle x,y \rangle$ is called the X-coordinate and the second term y is called the Y-coordinate of the ordered pair. The X-projection of a set $S \subset X \times Y$ is the set

$$\{x \in X : \langle xy \rangle \in S \text{ for some } y \in Y\}$$

---

[1]Ruziewicz and Sierpiński (1933); see also Sierpiński (1935e).

The graph of a function $f: X \to Y$, denoted by $\Gamma(f)$, is the set

$$\Gamma(f) = \{\langle x,y\rangle \in X \times Y : y = f(x)\}$$

We utilize Theorem 5 to establish[1]

THEOREM 21 There exists a function whose graph intersects the graph of each Baire function.

Proof. We have $X = \mathbb{R}^n$ and $Y = \mathbb{R}$. Let

(1) $p_1, p_2, \ldots, p_\alpha, \ldots$ $(\alpha < \Lambda)$

be a well-ordering of all points of $X \times Y$.

The family $\mathscr{H}$ of all perfect sets in $X \times Y$ whose X-projection is a perfect set in X has the power of the continuum. Accordingly, let

(2) $H_1, H_2, \ldots, H_\alpha, \ldots$ $(\alpha < \Lambda)$

be a well-ordering of all sets in $\mathscr{H}$.

We define by transfinite induction a transfinite sequence

(3) $q_1, q_2, \ldots, q_\alpha, \ldots$ $(\alpha < \Lambda)$

of points of $X \times Y$ in the following manner:

Define q_1 to be the first term of the sequence (1) that belongs to H_1. Assume that $1 < \alpha < \Lambda$ and the terms q_β have already been defined for all indices $\beta < \alpha$. The X-projection of H_α being a perfect set, hence of power of the continuum, there are points of H_α whose X-coordinate differs from the X-coordinates of all the points q_β with $\beta < \alpha$. We define q_α to be the first such point occurring in the enumeration (1).

The sequence (3) is thus defined by transfinite induction and, for each $\alpha < \Lambda$, we have

(4) $q_\alpha = \langle x_\alpha, y_\alpha\rangle$

where $x_\alpha \in X$, $y_\alpha \in Y$, and the elements

(5) $x_1, x_2, \ldots, x_\alpha, \ldots$ $(\alpha < \Lambda)$

are all different from one another. Moreover, it can be seen without difficulty that every element of X occurs as a term of the sequence (5). This is a consequence of the fact that each element of X occurs as the X-coordinate of a term p_γ of the sequence (1) which belongs to continuum many terms of the sequence (2).

[1]Cf. Sierpiński (1938a).

Let $g: X \to Y$ be the function defined by

(6) $\qquad\qquad\qquad g(x_\alpha) = y_\alpha \qquad$ for all $\alpha < \Lambda$

Suppose now that $f: X \to Y$ is any Baire function. According to Theorem 5, the function f is continuous on a perfect set $P \subset X$. The set

$$H = \{\langle x, y \rangle \in X \times Y : x \in P \text{ and } y = f(x)\}$$

is easily seen to be a perfect set in $X \times Y$ whose X-projection coincides with P. The set H thus belongs to the family $\mathscr{H}$ and consequently is a term of the sequence (2), say $H = H_\alpha$. By virtue of the definition of the sequence (3), we have $q_\alpha \in H_\alpha$. From (4) and the definition of the set H, we see that $x_\alpha \in P$ and $y_\alpha = f(x_\alpha)$. But in view of (6), this implies that $f(x_\alpha) = g(x_\alpha)$, so $\Gamma(f) \cap \Gamma(g) \neq \varnothing$.

REMARK Clearly, a function whose graph intersects the graph of each Baire function cannot itself be a Baire function.

In the case of a countable family of Baire functions, we have the following generalization of a theorem of Sierpiński.[1]

THEOREM 22 If $\mathscr{C}$ satisfies CCC and $\mathscr{M}$ is any countable family of Baire functions, then there exists a Baire function whose graph is disjoint from the graph of all the functions in $\mathscr{M}$.

Proof. We have $X = \mathbb{R}^n$ and $Y = \mathbb{R}$. Let $\mathscr{M} = \{f_i : i \in I\}$, where I is a countable set. For each index $i \in I$, the set $f_i^{-1}(y)$ is a Baire set for each $y \in Y$ and, because $\mathscr{C}$ satisfies CCC, the set

$$E_i = \{y \in Y : f_i^{-1}(y) \text{ is meager}\}$$

is a cocountable set. The set $E = \bigcap_{i \in I} E_i$ is also cocountable and hence nonempty. Let e be any point of E. For each element $x \in X$ we define $g(x) = e$ if $x \notin \bigcup_{i \in I} f_i^{-1}(e)$, and we define $g(x)$ to be an element of $Y - \{f_i(x) : i \in I\}$ if $x \in \bigcup_{i \in I} f_i^{-1}(e)$. As the function g is constant except for the meager set $\bigcup_{i \in I} f_i^{-1}(e)$, it is a Baire function. Clearly, $g(x) \neq f_i(x)$ for every $x \in X$ and every $i \in I$.

G. Sequences

In 1911, Egorov discovered the following measure-theoretic fact connecting pointwise convergence and uniform convergence:

(ε) If a sequence of Lebesgue measurable functions converges almost everywhere on an interval $[a, b]$, then one can always neglect a Lebesgue

[1]Cf. Sierpiński (1938a).

measurable set of arbitrarily small measure in $[a,b]$ so that the sequence will converge uniformly on the complement of that set.

This may be restated in the equivalent form:

(μ_2) If a sequence of Lebesgue measurable functions converges almost everywhere on $\mathbb{R}$, then every measurable set E of positive measure contains a measurable set F of positive measure on which the sequence converges uniformly.

The Baire category analogue of this statement is:

(τ_2) If a sequence of functions that have the Baire property converges on $\mathbb{R}$, with the possible exception of a set of the first category, then every second category set E that has the Baire property contains a second category set F that has the Baire property on which the sequence converges uniformly.

Now, this category analogue is not valid.[1] However, if we merely stipulate that the set F in statements (μ_2) and (τ_2) be a perfect set, then both of these modified statements are valid.

THEOREM 23 If a sequence of Baire functions converges essentially everywhere, then every abundant Baire set contains a perfect set on which the sequence converges uniformly.

Proof. Let $\langle f_n \rangle_{n\in\mathbb{N}}$ be a sequence of Baire functions converging essentially everywhere to a function f. For all $k,n \in \mathbb{N}$ the set

$$E_{k,n} = \bigcup_{m=n}^{\infty} \left\{ x \in X : |f_m(x) - f(x)| \geq \frac{1}{k} \right\}$$

is a Baire set. For each $k \in \mathbb{N}$ the sequence $\langle E_{k,n} \rangle_{n\in\mathbb{N}}$ is descending, and because $\langle f_n \rangle_{n\in\mathbb{N}}$ converges essentially everywhere to f, the set $\bigcap_{n=1}^{\infty} E_{k,n}$ is meager.

Let S be any abundant Baire set. According to Theorem 7 of Section II, there exists an increasing sequence $\langle n_k \rangle_{k\in\mathbb{N}}$ of natural numbers and a perfect set P contained in S which is disjoint from the set $\bigcup_{k=1}^{\infty} E_{k,n_k}$.

Suppose that ε is any given positive number. Choose a natural number k satisfying $0 < 1/k < \varepsilon$ and suppose that n is any natural number $\geq n_k$. For every point $x \in P$ we have $x \notin E_{k,n_k}$. From the inclusion $E_{k,n} \subset E_{k,n_k}$ we then obtain $|f_n(x) - f(x)| < \varepsilon$. Therefore, $\langle f_n \rangle_{n\in\mathbb{N}}$ converges uniformly to f on P.

___

[1] Cf. Oxtoby (1980, p. 38).

REMARK The equivalent form (μ_2) of Egorov's theorem can be obtained from the preceding proof merely by stipulating that the perfect set P have positive Lebesgue measure (see the Note following Theorem 7 of Section II).

The category analogue of Fréchet's theorem, discussed in Chapter 1, Section IV.F, is also invalid. In fact, Sierpiński has constructed a double sequence of continuous functions on $\mathbb{R}$ with an iterated limit converging pointwise everywhere to 0 from which it is impossible to extract a single sequence converging to 0 on any second category set having the Baire property.[1] However, we do have the following positive result of a general nature.

THEOREM 24 If the iterated limit of a double sequence of Baire functions converges essentially everywhere to a function f, then for every abundant Baire set S we can extract a single sequence from the double sequence that will converge pointwise to f on some perfect subset of S.

Proof. Let M be a meager set and let $\langle f_{k,n}\rangle_{k,n\in\mathbb{N}}$ be a double sequence of Baire functions whose iterated limit $\lim_{k\to\infty}(\lim_{n\to\infty}f_{k,n})$ converges to a function f on $X-M$. Let S be a given abundant Baire set.

For each $k\in\mathbb{N}$ define

$$f_k(x) = \lim_{n\to\infty} f_{k,n}(x)$$

for each $x\in X-M$ and define $f_k(x) = f(x)$ for each $x\in M$. Then

$$f(x) = \lim_{k\to\infty} f_k(x)$$

for all $x\in X$.

Consider the double sequence of sets $\langle E_{k,n}\rangle_{k,n\in\mathbb{N}}$, where

$$E_{k,n} = \bigcup_{m=n}^{\infty} \left\{x\in X-M : |f_{k,m}(x) - f_k(x)| \geq \frac{1}{k}\right\}$$

Since all the functions $f_{k,n}$ and f_k are Baire functions, the sets $E_{k,n}$ are Baire sets. Let $k\in\mathbb{N}$ be given. Obviously, $\langle E_{k,n}\rangle_{n\in\mathbb{N}}$ is a descending sequence. Suppose that $x\in X-M$. Because the sequence $\langle f_{k,n}(x)\rangle_{n\in\mathbb{N}}$ converges to $f_k(x)$, there is an index n such that $|f_{k,m}(x) - f_k(x)| < 1/k$ for all indices $m \geq n$, so that $x\notin E_{k,n}$. Therefore, $\bigcap_{n=1}^{\infty} E_{k,n} = \varnothing$. We thus see that the hypothesis of Theorem 7 of Section II is satisfied.

We apply that theorem to the abundant set $S-M$ to obtain an increasing sequence $\langle n_k\rangle_{k\in\mathbb{N}}$ of natural numbers and a perfect set P contained in the set $(S-M)-\bigcup_{k=1}^{\infty} E_{k,n_k}$.

[1]Sierpiński (1939a).

Suppose that x is any point of P and let ε be any given positive number. As $x \in X - M$, but $x \notin \bigcup_{k=1}^{\infty} E_{k,n_k}$, we have

$$|f_{k,n_k}(x) - f_k(x)| < \frac{1}{k}$$

for every $k \in \mathbb{N}$. Since $\langle f_k(x) \rangle_{k \in \mathbb{N}}$ converges to $f(x)$, we can find a natural number k_0 such that

$$|f_k(x) - f(x)| < \frac{\varepsilon}{2}$$

for all $k \geqslant k_0$ and $1/k_0 < \varepsilon/2$. Then for all $k \geqslant k_0$ we have

$$|f_{k,n_k}(x) - f(x)| < \varepsilon$$

Therefore, $\langle f_{k,n_k}(x) \rangle_{k \in \mathbb{N}}$ converges to $f(x)$ for every $x \in P$.

For further information concerning Egorov's theorem, Fréchet's theorem, and related matters, see Fréchet (1906, pp. 15–16), Egorov (1911), Hahn (1921, pp. 558–559, 561–563), Riesz (1922, 1928), Sierpiński (1928c; 1929c; 1932a; 1932d; 1933h; 1938c; 1938d; 1939a; 1939b; 1950; 1956, Chap. II, Secs. 5–7), Banach and Kuratowski (1929), Braun and Sierpiński (1932), Suckau (1935), Sierpiński and Szpilrajn (1936b), Tolstov (1939), Tsuchikura (1949), Bourbaki (1952, p. 198), Kvačko (1958), Neubrunn (1959), Weston (1959, 1960), Sindalovskiĭ (1960), Rozycki (1965), Walter (1977), Wilczyński (1977), Kuczma (1978), Bartle (1980), Oxtoby (1980), Wagner (1981), and Morgan (1986).

6

Translation Bases

I. DEFINITIONS AND BASIC PROPERTIES

A. Definitions and examples

Unless otherwise specified, all sets considered in this chapter are subsets of $\mathbb{R}$ and all functions are real-valued functions of a real variable.

If S is a set and $r \in \mathbb{R}$, then the translate of S by the amount r, denoted by $S(r)$, is defined by

$$S(r) = \{x + r : x \in S\}$$

A translate $S(r)$ is called a nonzero translate if $r \neq 0$. The translate $S(r)$ is called a proper translate if $S(r) \neq S$.

A set S is translation invariant if $S(r) = S$ for every $r \in \mathbb{R}$, that is, if S is invariant under all the translation mappings $\phi(x) = x + r$ for $r \in \mathbb{R}$. If $S(r) = S$ for every rational number r, then S is said to be invariant under rational translations. Sets S and T are called congruent if each is a translate of the other. Equivalently, S and T are congruent if $T = S(r)$ for some $r \in \mathbb{R}$. A set S is reflection invariant if it is invariant under the reflection mapping $\phi(x) = -x$. We say that S is linearly invariant if it is invariant under all linear mappings $\phi(x) = ax + b$, with $a, b \in \mathbb{R}$ and $a \neq 0$. A family $\mathscr{S}$ of sets is translation (resp., reflection, linearly) invariant if every set in $\mathscr{S}$ is translation (resp., reflection, linearly) invariant. The notion of essentially invariant sets or families and essentially congruent sets are defined in the obvious manner.

DEFINITION A category base $(\mathbb{R},\mathscr{C})$ is called a translation base if the following conditions are satisfied:
   (1) $\mathscr{C}$ is translation invariant.
   (2) If A is any region and D is any countable everywhere dense set, then the set $\bigcup_{r \in D} A(r)$ is abundant everywhere.[1]

REMARK Using the given condition (2), it is readily seen that the conclusion of this condition is also valid for any everywhere dense set D.

All bases considered in this chapter are assumed to be translation bases. The following are the most important translation bases.

EXAMPLE $\mathscr{C}$ is the family of all closed intervals $[a,b]$, where $a,b \in \mathbb{R}$ and $a < b$.

EXAMPLE $\mathscr{C}$ is the family of all intervals of the form $[a,b)$, where $a,b \in \mathbb{R}$ and $a < b$.

EXAMPLE $\mathscr{C}$ is the family of all compact sets in $\mathbb{R}$ of positive Lebesgue measure. Equivalently, we can take $\mathscr{C}$ to be the perfect base consisting of all compact sets that have positive Lebesgue measure in any open interval containing one of their points.

REMARK The verification of the condition (2) in this example is a consequence of the Lebesgue Density Theorem.[2]

EXAMPLE $\mathscr{S}$ is a translation invariant, proper σ-ideal of subsets of $\mathbb{R}$ and $\mathscr{C} = \{A : \mathbb{R}-A \in \mathscr{S}\}$. In this example, the singular and meager sets coincide with the sets in $\mathscr{S}$, while the Baire sets are the sets S such that either S or $\mathbb{R}-S$ belongs to $\mathscr{S}$.

All these examples satisfy CCC.

NOTE If $\mathscr{C}$ is the family of all sets of the form $\{x \in \mathbb{R} : x \leqslant a\}$, where $a \in \mathbb{R}$, then $(\mathbb{R},\mathscr{C})$ is a translation invariant category base satisfying CCC that does not satisfy the condition (2), and hence is not a translation base.

---

[1]Cf. Rademacher (1916, Sec. 1), Łomnicki (1918, Lemme I), Ostrowski (1929, p. 59), and Sierpiński (1929b, fn. p. 200).
[2]A proof of this theorem may be found in Goffman (1953).

B. General properties

We first note some simple consequences of the definition of a translation base.

(i) The families of singular sets, meager sets, and Baire sets are translation invariant.

(ii) Every region is an abundant set.

(iii) Every interval is an abundant set.

THEOREM 1   If A and B are any regions and D is any everywhere dense set, then there is an element $r \in D$ such that $A(r) \cap B$ contains a region.

Proof. Let E be a countable everywhere dense subset of D. According to condition (2) of the definition, the set $[\bigcup_{r \in E} A(r)] \cap B$ is abundant. Hence, there is an element $r_0 \in E$ for which the set $A(r_0) \cap B$ is abundant. By virtue of Theorem 2 of Chapter 1, Section II, the set $A(r_0) \cap B$ contains a region.

Utilizing the Fundamental Theorem we obtain a strengthening of condition (2).

THEOREM 2   If S is an abundant set and D is an everywhere dense set, then the set $\bigcup_{r \in D} S(r)$ is abundant everywhere.[1]

Proof. Let S be abundant everywhere in a region A. Suppose that B is any region. By Theorem 1 there is an element $r_0 \in D$ such that the set $A(r_0) \cap B$ contains a region C. According to Theorem 27 of Chapter 1, Section III, the set $S(r_0)$ is abundant everywhere in the region $A(r_0)$. Hence, $S(r_0) \cap C$ is an abundant set, as is also the set $S(r_0) \cap B$. Therefore, $\bigcup_{r \in D} S(r)$ is abundant in every region B.

The following consequence of this theorem is frequently used.

THEOREM 3   If S is an abundant Baire set and D is an everywhere dense set, then $\bigcup_{r \in D} S(r)$ is a comeager set.

NOTE   If E is any nowhere dense perfect set, then the set $\bigcup_{r \in Q} E(r)$ is a set of the first category whose complement, according to Theorem 8 of Chapter 5, Section II, contains a perfect set. This observation, in conjunction with Theorem 3, reveals the fact that there is no translation base which is equivalent to the category base of all perfect subsets of $\mathbb{R}$.

The following theorem is a generalization of a theorem of Scheeffer.[2]

THEOREM 4   If S is a meager set and D is a countable set, then there exists a comeager set T such that $S \cap D(t) = \varnothing$ for every $t \in T$.

[1] Sierpiński (1929, fn. p. 200; 1956, fn. p. 129) and Morgan (1975, Theorem 2).
[2] Scheeffer (1884, pp. 291–293) and Bagemihl (1954); see also Bagemihl (1959a).

Proof. Define $T = \mathbb{R} - \bigcup_{d \in D} S(-d)$. If $S \cap D(t) \neq \varnothing$, then there is an element $x \in S \cap D(t)$, say $x = d + t$, where $d \in D$. From $x \in S$ we obtain $t \in S(-d)$ and $t \notin \mathbb{R} - S(-d)$, whence $t \notin T$. Therefore, $S \cap D(t) = \varnothing$ for every $t \in T$.

COROLLARY 5 If S is a meager set and D is any countable set, then there exists a translate of S that is disjoint from D.

The next three theorems elucidate the nature of translation bases.

THEOREM 6 Every region has the power of the continuum.

Proof. Suppose that there is a region A which has power $< 2^{\aleph_0}$. The set of all elements representable in the form $x - y + r$, where $x, y \in A$ and $r \in \mathbb{Q}$, will then have power $< 2^{\aleph_0}$. Consequently, there exists an element $z \in \mathbb{R}$ which is not representable in this form. The set $A(z)$ is a region with

$$\left[\bigcup_{r \in \mathbb{Q}} A(r) \right] \cap A(z) = \varnothing$$

But this contradicts condition (2). We conclude that every region A must have power $2^{\aleph_0}$.

THEOREM 7 Every countable set is a meager set if and only if $\mathscr{C}$ is nontrivial.

Proof. If $\mathscr{C}$ is trivial, then every nonempty set is abundant, so there are countable sets which are not meager. Conversely, suppose that $\mathscr{C}$ is nontrivial. We shall show that every singleton set is singular. For this, it suffices to show that if A is any region containing x, then there is a subregion of A that does not contain x.

If $A = \mathbb{R}$, then, since $\mathscr{C}$ is nontrivial, there is a subregion B of A which is a proper subset of A. Choose an element $y \in A - B$. Then $B(x - y)$ is a subregion of A that does not contain x. Thus, we may henceforth assume that $A \neq \mathbb{R}$.

The remainder of the proof involves the consideration of three cases, in each of which we show there exists an element $y \notin A$ for which the set $A \cap A(x - y)$ is abundant. It then follows from Theorem 2 of Chapter 1, Section II that there is a region $B \subset A \cap A(x - y)$, and since $x \notin A(x - y)$, B is a subregion of A that does not contain x.

Case 1. *A contains no interval.*

Let E be a countable subset of $\mathbb{R} - A$ that is everywhere dense. Then $D = \{x - y : y \in E\}$ is also an everywhere dense set. From condition (2) we know $\bigcup_{r \in D} A(r)$ is abundant everywhere. Hence, $A \cap A(x - y)$ is an abundant set for some $y \in E$.

Case 2. *A contains an interval I and every interval containing x contains elements that do not belong to A.*

We choose an element y sufficiently close to x so that $y \notin A$ and $I \cap I(x - y)$ contains an interval. Every interval being an abundant set, this implies that $A \cap A(x - y)$ is an abundant set.

Case 3. There is an interval containing x that is entirely contained in A.

From $A \neq \mathbb{R}$ we obtain the existence of an element $z \notin A$ such that either $z < x$ or $x < z$. We consider only the possibility $x < z$, the other possibility being treated in a similar manner. Let U be the union of all intervals contained in A that contain x. From $x < z$ and $z \notin A$ it follows that $b = \sup U$ is an element of $\mathbb{R}$.

If $x = \inf U$, then we choose an element $y \notin A$ such that $y < x$ and $x - y < b - x$. The set $A \cap A(x - y)$ will then contain the interval $(2x - y, b)$ and, accordingly, is an abundant set.

If $x \neq \inf U$, then there is an element $a \in U$ such that $a < x$. Choose an element $y \notin A$ such that $b < y$ and $y - b < x - a$. The set $A \cap A(x - y)$ contains the interval $(a, b + x - y)$ and hence is an abundant set.

COROLLARY 8   If S is any countable set, then every subset of S which is a Baire set is meager.

Combining this theorem with an observation above, we obtain the following fact.

THEOREM 9   Every nontrivial translation base is a point-meager, Baire base.

THEOREM 10   If every open interval is a Baire set, then there exists a meager perfect set.

Proof. From the assumption that every open interval is a Baire set it follows that every perfect set is a Baire set. Let P be a nowhere dense, perfect set. If P is a meager set, then the proof is complete. Thus, we assume that P is abundant.

By Theorem 3, the set $S = \mathbb{R} - \bigcup_{r \in \mathbb{Q}} P(r)$ is a meager set. But this set is also a residual set which, according to Theorem 8 of Chapter 5, Section II, contains a perfect set Q. Therefore, Q is a meager perfect set.

REMARK   This theorem is also valid if we merely assume that there exists one open interval which is a Baire set, since this implies that every open interval is a Baire set.

C.   Invariant sets

Theorem 1, in conjunction with Theorem 28 of Chapter 1, Section III, yields the following fact.

THEOREM 11   If S is a Baire set that is essentially invariant under all rational translations, then S is either a meager set or a comeager set.

COROLLARY 12   Assume that every Baire set of power $< 2^{\aleph_0}$ is a meager set. If S is a Baire set such that $S\Delta S(r)$ has power less than $2^{\aleph_0}$ for every rational number r, then S is either a meager set or a comeager set.[1]

D.   Complementary bases

We give here two theorems of Sander.[2]

THEOREM 13   Two perfect, translation bases are nonequivalent if and only if they are complementary bases.

Proof. Suppose that $(\mathbb{R},\mathscr{C})$ and $(\mathbb{R},\mathscr{D})$ are perfect, translation bases that are not equivalent. Then, by Theorem 1 of Chapter 1, Section III, either

(a) there is a $\mathscr{C}$-region that contains no $\mathscr{D}$-region, or

(b) there is a $\mathscr{D}$-region that contains no $\mathscr{C}$-region.

By virtue of the symmetry involved, it suffices to assume that (a) holds.

Let A be a $\mathscr{C}$-region that contains no $\mathscr{D}$-region. By Corollary 6 of Chapter 5, Section I, the set A is $\mathscr{D}$-singular. Hence, the set

$$U = \bigcup_{r \in \mathbb{Q}} A(r)$$

is $\mathscr{D}$-meager. On the other hand, we know that the $\mathscr{C}$-region A is a $\mathscr{C}$-abundant set in $\mathfrak{B}(\mathscr{C})$. In view of Theorem 3, the set $V = \mathbb{R}—U$ is $\mathscr{C}$-meager. Thus,

$$\mathbb{R} = U \cup V$$

is a decomposition of $\mathbb{R}$ into disjoint sets, one of which is $\mathscr{C}$-meager and the other is $\mathscr{D}$-meager.

As for the converse, any two equivalent category bases for which $\mathbb{R}$ is an abundant set obviously cannot be complementary.

THEOREM 14   Assume that $2^{\aleph_0} = \aleph_1$. If $(\mathbb{R},\mathscr{C})$, $(\mathbb{R},\mathscr{C}')$ are perfect, translation bases satisfying CCC, then there exists a one-to-one mapping ϕ of $\mathbb{R}$ onto itself, with $\phi = \phi^{-1}$, such that a set S is $\mathscr{C}$-meager if and only if $\phi(S)$ is $\mathscr{C}'$-meager.

Proof. Use Theorem 13; Theorem 7 of Chapter 2, Section IV; Theorem 3 of Chapter 5, Section I; Theorem 1 of Chapter 2, Section I; Theorem 10 of Chapter 5, Section II; and Theorem 17 of Chapter 5, Section I.

---

[1]Sierpiński (1932b, p. 24).
[2]Sander (1981b, 1982).

II. ARITHMETIC OPERATIONS

A. Theorems of Steinhaus

Sierpiński proved that if S and T are any Lebesgue measurable sets of positive Lebesgue measure, then there always exists two different elements $x \in S$, $y \in T$ whose distance $x - y$ is a rational number.[1] Continuing further, Steinhaus proved the following:[2]

(i) If S and T are Lebesgue measurable sets of positive measure and D is any everywhere dense set, then there exists two elements $x \in S$ and $y \in T$ whose distance apart is a number in D.

(ii) If $\langle S_n \rangle_{n \in \mathbb{N}}$ is any sequence of Lebesgue measurable sets of positive measure, then there exists a sequence $\langle x_n \rangle_{n \in \mathbb{N}}$ of different elements such that $x_n \in S_n$ for each n and the distance between any pair of these elements is a rational number.

(iii) For every Lebesgue measurable set S, there exists a countable set P, which may be empty, whose elements are rational distances from one another, and there exists a set N of Lebesgue measure zero such that

$$P \subset S \subset P' \cup N$$

where P' is the set of limit points for P with respect to the usual topology for $\mathbb{R}$.

NOTATION   The distance set $\delta(S,T)$ for sets S and T is defined by

$$\delta(S,T) = \{|x - y| : x \in S \text{ and } y \in T\}$$

If $S = T$, then we write $\delta(S)$ in lieu of $\delta(S,S)$.

Steinhaus established the following facts about distance sets.[3]

(iv) The distance set for any two Lebesgue measurable sets of positive measure contains an interval.

(v) The distance set for any Lebesgue measurable set of positive measure contains an interval whose left endpoint is 0.

We now give general theorems that contain Steinhaus's theorems (i)–(iv)[4] and their category analogues.[5]

---

[1] Sierpiński (1917a, 1920b).
[2] Steinhaus (1920).
[3] See Piccard (1939, pp. 25–28); see also Rademacher (1921).
[4] For generalizations of Steinhaus's theorem (v), see Smith (1977).
[5] Cf. Piccard (1939, Chap. I; 1940).

THEOREM 1   If S and T are abundant sets, T is a Baire set, and D is an everywhere dense set, then there exists a denumerable subset D_0 of D such that $S(r) \cap T$ is an abundant set for every $r \in D_0$.[1]

Proof. Let E be a denumerable everywhere dense subset of D, let $U = \bigcup_{r \in E} S(r)$, and let T be abundant everywhere in a region A.

Since U is abundant everywhere, $U \cap A$ is an abundant set. Because T is a Baire set, the set $A—T$ is meager. Hence, $U \cap (A \cap T)$ and $U \cap T$ are abundant sets. Accordingly, there exists an element $r_1 \in E$ such that $S(r_1) \cap T$ is an abundant set.

Having defined $r_1, \ldots, r_n$ for a given natural number n, repeat the reasoning, with E replaced by the everywhere dense set $E - \{r_1, \ldots, r_n\}$, to obtain an element $r_{n+1} \in E - \{r_1, \ldots, r_n\}$ such that $S(r_{n+1}) \cap T$ is an abundant set. We thus obtain by mathematical induction an appropriate denumerable set $D_0 = \{r_n : n \in \mathbb{N}\}$.

COROLLARY 2   If S is an abundant Baire set, then there exist elements $x, y \in S$ such that $x - y$ is a nonzero rational number.

THEOREM 3   If $\langle S_n \rangle_{n \in \mathbb{N}}$ is a sequence of abundant Baire sets, then there exists a sequence $\langle x_n \rangle_{n \in \mathbb{N}}$ of different elements such that $x_n \in S_n$ for each n and the distance between any pair of these elements is a rational number.

Proof. Let $\langle p_n \rangle_{n \in \mathbb{N}}$ denote the sequence of all prime natural numbers, and for each $n \in \mathbb{N}$, let D_n be the set of all rational numbers r of the form

$$r = \frac{k}{p_n^m}$$

where k is any nonzero integer not divisible by p_n and $m \in \mathbb{N}$. Each set D_n being everywhere dense, the set $\bigcup_{r \in D_n} S_n(r)$ is a comeager set for every n. Therefore, the set $\bigcap_{n=1}^{\infty} \bigcup_{r \in D_n} S_n(r)$ is also a comeager set. Choose an element x in the latter set. Then there exists a sequence $\langle r_n \rangle_{n \in \mathbb{N}}$ of rational numbers and a sequence $\langle x_n \rangle_{n \in \mathbb{N}}$ of elements such that $r_n \in D_n$, $x_n \in S_n$, and $x = x_n + r_n$ for each $n \in \mathbb{N}$. If $m, n \in \mathbb{N}$ and $m \neq n$, then it follows from the equality $x_m + r_m = x_n + r_n$ that $x_m - x_n = r_n - r_m \neq 0$, so that the distance between x_m and x_n is a nonzero rational number.

COROLLARY 4   In any abundant Baire set there exists a sequence of different elements the distance between any two elements of which is a rational number.

---

[1] Morgan (1975, Theorem 3). This theorem also contains extensions of Steinhaus's theorem (i) given in Fukamiya (1935, p. 334) and Ray and Lahiri (1964); see also Kemperman (1957).

THEOREM 5 Assume that $\mathscr{C}$ is also a perfect base. If S is a Baire set, then there exists a countable set P, any two elements of which are a rational distance apart, and there exists a meager set N such that

$$P \subset S \subset P' \cup N$$

Proof. If S is a meager set, then we take $P = \varnothing$ and $N = S$. We thus assume that S is an abundant set.

Let S^* denote the set of all elements of S at which S is an abundant set and set $N = S{-}S^*$. We know from Theorem 12 of Chapter 1, Section II that N is a meager set.

Let $\langle I_n \rangle_{n\in\mathbb{N}}$ denote the sequence of all open intervals with rational endpoints whose intersection with S^* is an abundant set. For each $n \in \mathbb{N}$, set $S_n = S^* \cap I_n$ and apply Theorem 3 to obtain a denumerable subset P of S^*, containing at least one element in each set S_n, such that the distance between any two elements of P is a rational number. We have $P \subset S$ and, in order to establish $S \subset P' \cup N$, it suffices to show that $S^* \subset P'$.

Suppose that $x \in S^*$. Then there is a neighborhood A of x for every subneighborhood B of which $B \cap S^*$ is an abundant set. As $\mathscr{C}$ is a perfect base, there exists a descending sequence $\langle B_n \rangle_{n\in\mathbb{N}}$ of neighborhoods of x such that $B_n \subset A$ for every n and $\lim_{n\to\infty} \mathrm{diam}(B_n) = 0$. Consequently, there is a descending subsequence $\langle I_{n_k} \rangle_{k\in\mathbb{N}}$ of $\langle I_n \rangle_{n\in\mathbb{N}}$ such that $x \in I_{n_k}$ for every k and $\lim_{k\to\infty} \mathrm{diam}(I_{n_k}) = 0$. Choose an element $x_{n_k} \in P \cap I_{n_k}$ for each $k \in \mathbb{N}$. Then x is a limit point for the subset $\{x_{n_k} : k \in \mathbb{N}\}$ of P. Therefore, $x \in P'$.

NOTATION The arithmetic difference set $\Delta(S,T)$ of sets S and T is defined by

$$\Delta(S,T) = \{x - y : x \in S \text{ and } y \in T\}$$

THEOREM 6 If S, T are abundant sets and S is a Baire set, then the difference set $\Delta(S,T)$ contains an interval.[1]

Proof. Assume to the contrary that $\Delta(S,T)$ contains no interval. Then the set $D = \mathbb{R}{-}\Delta(S,T)$ is everywhere dense, as is also the set $\{-r : r \in D\}$. According to Theorem 3 of Section I, the set $U = \bigcup_{r\in D} S(-r)$ is comeager. This implies that $T \cap U$ is an abundant set and consequently, that there is an element $y \in T \cap U$. Hence, there is an element $r \in D$ and an element $x \in S$ such that $y = x - r$. This means that $r = x - y \in \Delta(S,T)$, contradicting the fact that $r \in D$.

COROLLARY 7 If S, T are abundant sets and S is a Baire set, then the distance set $\delta(S,T)$ contains an interval.

[1]Sander (1981a).

NOTATION   The arithmetic sum $S + T$ of two sets S and T is defined by

$$S + T = \{x + y : x \in S \text{ and } y \in T\}$$

As a further consequence of Theorem 6, we have

COROLLARY 8   Assume that $\mathscr{C}$ is reflection invariant. If S, T are abundant sets and S is a Baire set, then the arithmetic sum $S + T$ contains an interval.

REMARK   Kemperman has proved:[1] If S and T are Lebesgue measurable sets of positive measure, then upon deleting certain subsets of measure zero from S and T we obtain sets S_0 and T_0 such that $\Delta(S_0, T_0)$ and $S_0 + T_0$ are open sets.

   We note that there exists a set S having the Baire property which is of the second category and a Lebesgue measurable set T with infinite measure for which the arithmetic sum $S + T$ contains no interval.[2]

   Steinhaus also established the following two theorems[3] concerning the Cantor set C.

THEOREM 9   The difference set of the Cantor set coincides with the closed interval $[-1,1]$, so that $\delta(C) = [0,1]$.

Proof. Let E denote the set $C \times C$. The set E is also obtained by means of the following procedure. Divide the unit square $[0,1] \times [0,1]$ into 9 equal squares (by means of the lines $x = 1/3$, $x = 2/3$, $y = 1/3$, $y = 2/3$), delete the 5 squares whose interiors contain no point of E, and let U_1 denote the union of the 4 closed squares that remain. Divide each of these 4 squares in turn into 9 equal squares, delete the squares whose interiors contain no point of E, and let U_2 denote the union of the 16 closed squares that remain. Continue this process indefinitely to determine a descending sequence $\langle U_n \rangle_{n \in \mathbb{N}}$ of sets, where each set U_n is a union of 4^n squares, and

$$E = \bigcap_{n=1}^{\infty} U_n$$

   As is readily verified, the orthogonal projection of each of the sets U_n on the diagonal line segment joining the points $\langle 0,1 \rangle$ and $\langle 1,0 \rangle$ actually coincides with this segment. In other words, if L is any line with an equation of the form $x - y = h$, where $-1 \leqslant h \leqslant 1$, then $L \cap U_n \neq \varnothing$ for every natural number n. Since the sets $L \cap U_n$ form a descending sequence of nonempty compact sets, we must have $L \cap E \neq \varnothing$. Thus, if h is any number satisfying $-1 \leqslant h \leqslant 1$, then there exist elements $x, y \in C$ such that $x - y = h$.

[1]Kemperman (1957); see also Mueller (1965).
[2]Kominek (1983).
[3]Steinhaus (1917).

COROLLARY 10   The distance set of a meager set can be an abundant set.

In connection with this theorem, we note the following examples:[1]

EXAMPLE   The set P of all real numbers in the unit interval $[0,1]$ which have a decimal expansion that does not contain the digit 5 is also a perfect set whose difference set is the closed interval $[-1,1]$. The latter assertion follows from the fact that if $z = \Sigma_{n=1}^{\infty} z_n/10^n$ is any number between 0 and 1, then $z = x - y$, where $x = \Sigma_{n=1}^{\infty} x_n/10^n$, $y = \Sigma_{n=1}^{\infty} y_n/10^n$ are elements of P defined by $x_n = 6$ and $y_n = 1$ whenever $z_n = 5$, while $x_n = z_n$ and $y_n = 0$ whenever $z_n \neq 5$.

EXAMPLE   The set P of all real numbers in $[0,1]$ having a decimal expansion in which only the digits 0 and 1 occur is a perfect set whose difference set contains no interval. This is seen from the fact that the difference between any two elements of P can never equal any nonzero number z of the form $z = \Sigma_{n=1}^{\infty} z_n/10^n$, where $z_n = 0$ or $z_n = 5$.

THEOREM 11   The set $C + C$ coincides with the interval $[0,2]$.

Proof. We utilize the construction and notation employed in proving Theorem 9. It is readily seen that the orthogonal projection of each of the sets U_n on the diagonal line segment joining the points $\langle 0,0 \rangle$ and $\langle 1,1 \rangle$ coincides with this segment. Hence, if L is any line with an equation of the form $x + y = h$, where $0 \leqslant h \leqslant 2$, then $L \cap U_n \neq \emptyset$ for every $n \in \mathbb{N}$. This implies that $L \cap E \neq \emptyset$. Thus, if h is any number satisfying $0 \leqslant h \leqslant 2$, then there exist elements $x, y \in C$ such that $x + y = h$.

B.   Distance sets

In addition to the results of the preceding section, the following facts have been established concerning distance sets.[2]

THEOREM 12   The distance set of an abundant set is an abundant set.

Proof. Let S be an abundant set. Let A be a region in which S is abundant everywhere and let $T = S \cap A$. Choose an element a in T. The distance set $\delta(S)$ contains the set $T(-a)$, which is an abundant set. Hence, $\delta(S)$ is an abundant set.

We note that the distance set of a Baire set is not necessarily a Baire set.[3]

[1]Kestleman (1947).
[2]For an extensive discussion of discussion of distance sets, see Piccard (1939); see also Sierpiński (1946).
[3]See Theorem 24 of Section III.

The following three theorems concern sets whose distance set is the interval $[0,\infty)$.[1]

THEOREM 13   If S is a meager set and r is any real number, then there exist numbers $x, y \notin S$ such that $x - y = r$, so that $\delta(\mathbb{R}-S) = [0,\infty)$.

Proof. Let $T = \mathbb{R}-S$. The sets T and $T(r)$ being comeager, so also is the set $T \cap T(r)$. Therefore, there exist numbers $x, y \in T$ such that $x = y + r$.

THEOREM 14   If every open interval is a Baire set, then there exists a meager set S with $\delta(S) = [0,\infty)$.

Proof. Assume that the Cantor set $\mathbb{C}$ is a meager set. Then $S = \bigcup_{n=0}^{\infty} \mathbb{C}(n)$ is also a meager set. If r is any nonnegative number, then the number $r - [r]$ (where $[r]$ denotes the largest integer $\leqslant r$) belongs to the unit interval $[0,1]$ and, according to Theorem 9, there exist numbers y, $z \in \mathbb{C}$ such that $z - y = r - [r]$. Taking $x = z + [r]$, we have $x, y \in S$ and $x - y = r$. Thus, $\delta(S) = [0,\infty)$.

Assume that $\mathbb{C}$ is an abundant set. Then the set $E = \bigcup_{r \in Q} \mathbb{C}(r)$ is abundant everywhere. From the assumption that every open interval is a Baire set it follows that $\mathbb{C}$ is a Baire set and consequently, $S = \mathbb{R}-E$ is a meager set. Applying Theorem 13 to E, which is a set of the first category, we obtain $\delta(S) = \delta(\mathbb{R}-E) = [0,\infty)$.

THEOREM 15   Assume CH. If $\mathscr{C}$ satisfies CCC and has power at most $2^{\aleph_0}$, then there exists an uncountable rare set S with $\delta(S) = [0,\infty)$.

Proof. If $\mathscr{C}$ is a trivial base, then we can take $S = \mathbb{R}$. We assume therefore that $\mathscr{C}$ is nontrivial.

According to Theorem 1 of Chapter 2, Section I, the family of all meager $\mathscr{K}_{\delta\sigma}$-sets has power $2^{\aleph_0}$. Let

$$y_1, y_2, \ldots, y_\alpha, \ldots \qquad (\alpha < \Omega)$$

$$R_1, R_2, \ldots, R_\alpha, \ldots \qquad (\alpha < \Omega)$$

be transfinite enumerations of all positive real numbers and all meager $\mathscr{K}_{\delta\sigma}$-sets, respectively. We define by transfinite induction a pair p_α, q_α of numbers for each $\alpha < \Omega$.

Place $p_1 = y_1$ and $q_1 = 0$. Suppose that $1 < \alpha < \Omega$ and p_β, q_β have already been defined for all ordinal numbers $\beta < \alpha$. The set $U_\alpha = \bigcup_{\beta < \alpha} R_\beta$ being meager, we apply Theorem 13 to obtain elements $p_\alpha, q_\alpha \in \mathbb{R}-U_\alpha$ for which $|p_\alpha - q_\alpha| = y_\alpha$.

The set $S = \{p_\alpha, q_\alpha : \alpha < \Omega\}$ is then an uncountable rare set with $\delta(S) = [0,\infty)$.

[1] Sierpiński (1935d, 1935g).

THEOREM 16 Assume CH and that $\mathscr{C}$ is reflection invariant. For any meager set M of positive real numbers, there exists a set S for which $\delta(S) = [0,\infty) - M$.[1]

Proof. Denote by M^* the set symmetric to M with respect to 0; i.e.,

$$M^* = \{-x : x \in M\}$$

Let

(1) $\qquad\qquad\qquad x_1, x_2, \ldots, x_\alpha, \ldots \qquad (\alpha < \Omega)$

(2) $\qquad\qquad\qquad y_1, y_2, \ldots, y_\alpha, \ldots \qquad (\alpha < \Omega)$

be well-orderings of $\mathbb{R}$ and of all positive real numbers that do not belong to M, respectively. We define by transfinite induction a pair of numbers p_α, q_α for each $\alpha < \Omega$.

Place $p_1 = x_1$ and $q_1 = x_1 + y_1$. Assume that $1 < \alpha < \Omega$ and p_β, q_β have been defined for all $\beta < \alpha$. Let $S_\alpha = \{p_\beta, q_\beta : \beta < \alpha\}$. The set

$$U_\alpha = S_\alpha \cup \left[\bigcup_{z \in M} S_\alpha(z)\right] \cup \left[\bigcup_{z \in M} S_\alpha(-z)\right] = S_\alpha \cup \left[\bigcup_{t \in S_\alpha} M(t)\right] \cup \left[\bigcup_{t \in S_\alpha} M^*(t)\right]$$

is a meager set. Define p_α to be the first element in the enumeration (1) that does not belong to the meager set $U_\alpha \cup U_\alpha(-y_\alpha)$, so that $p_\alpha \notin U_\alpha$ and $p_\alpha + y_\alpha \notin U_\alpha$, and define $q_\alpha = p_\alpha + y_\alpha$.

The set $S = \{p_\alpha, q_\alpha : \alpha < \Omega\}$ thus determined is uncountable and every number in the enumeration (2) belongs to $\delta(S)$. It remains to show that $\delta(S)$ contains no element of M.

Assume to the contrary that there is a number $z \in \delta(S) \cap M$. Then there exist two numbers $r_1, r_2 \in S$ such that $|r_1 - r_2| = z$. The following three cases arise:

(I) $r_1 = p_\alpha$ and $r_2 = p_\beta$

(II) $r_1 = q_\alpha$ and $r_2 = q_\beta$

(III) $r_1 = p_\alpha$ and $r_2 = q_\beta$

where α, β are countable ordinal numbers. We show each of these cases leads to a contradiction.

As $z > 0$, we have $\alpha \neq \beta$ in cases (I) and (II). Without loss of generality we may assume in these two cases that $\beta < \alpha$. In case (I), because $p_\alpha \notin U_\alpha$ and $p_\beta \pm z \in U_\alpha$, it follows that $|p_\alpha - p_\beta| \neq z$. Similarly, in case (II), because $q_\alpha \notin U_\alpha$ and $q_\beta \pm z \in U_\alpha$, we must have $|q_\alpha - q_\beta| \neq z$.

Consider finally case (III). If $\alpha < \beta$, then, since $q_\beta \notin U_\beta$ and $p_\alpha \pm z \in U_\beta$, we cannot have $|p_\alpha - q_\beta| = z$. If $\alpha > \beta$, then $q_\alpha \notin U_\alpha$ and $p_\beta \pm z \in U_\alpha$, so we cannot

[1] Piccard (1939, pp. 199–200).

have $|p_\alpha - q_\beta| = z$. If $\alpha = \beta$, then because $|p_\alpha - q_\alpha| = y_\alpha \notin M$, we cannot have $|p_\alpha - q_\alpha| = z$.

In particular, this theorem implies the existence (under CH) of sets whose distance set contains all nonnegative numbers except one. One such set is the set

$$S = [0,1) \cup [2,3) \cup [4,5) \cup \cdots \cup [2n, 2n+1) \cup \cdots \cup \{-3, -5, -7, \ldots\}$$

whose distance set contains all nonnegative numbers with the exception of the number 1.

Combining the arguments used to prove Theorems 15 and 16, one obtains

THEOREM 17   Assume CH and that $\mathscr{C}$ is reflection invariant. If $\mathscr{C}$ satisfies CCC and has power at most $2^{\aleph_0}$, then for every meager set M of positive real numbers, there exists an uncountable rare set S for which $\delta(S) = [0,\infty) - M$.

C. Rationally independent sets

Theorem 9 implies that the Cantor set has the property that there exist two different elements of this set whose distance from one another is a rational number. The question arises whether this property is true of all perfect sets.

The property is true of all Lebesgue measurable sets (perfect or not) that have positive measure.[1] However, it is not true of all perfect sets. A simple example, due to Sierpiński,[2] of a perfect set such that the distance between any two different elements is an irrational number is the set of all real numbers x of the form

$$x = \sqrt{2} \sum_{n=1}^{\infty} \frac{x_n}{10^{n!}}$$

where the numbers x_n are any nonzero decimal digits $1, 2, \ldots, 9$.

The first example of a perfect set, any two different elements of which are an irrational distance from one another was given by Groß[3] and is the set of all numbers y of the form

$$y = \frac{x_1}{3!} + \frac{x_2}{(3 + x_1)!} + \frac{x_3}{(3 + 10x_1 + x_2)!} + \cdots$$

$$+ \frac{x_n}{(3 + 10^{n-2}x_1 + 10^{n-3}x_2 + \cdots + x_{n-1})!} + \cdots$$

[1]Cf. Burstin (1915, pp. 242–243).
[2]Sierpiński (1917a).
[3]Cf. Burstin (1915, pp. 239–240); see also Wolff (1924), Brouwer (1924), and Jones (1942, pp. 477–478).

where the numbers x_n are either 1 or 2. We show that a similarly constructed set has this same feature and is also a rationally independent set.

DEFINITION   A set S is called rationally independent if for every nonempty finite set of different nonzero elements $x_1, \ldots, x_k$ of S, the equation $\sum_{i=1}^{k} r_i x_i = 0$, where $r_1, \ldots, r_k \in \mathbb{Q}$, implies that $r_i = 0$ for all $i = 1, \ldots, k$.

We utilize a number-theoretic fact.[1]

LEMMA 18   Every real number z is uniquely representable in the form

$$(1) \qquad z = \sum_{n=1}^{\infty} \frac{z_n}{n!}$$

where the numbers z_n are integers satisfying the conditions

(a) $0 \leqslant z_n \leqslant n - 1$ for all indices $n \geqslant 2$.
(b) $z_n \leqslant n - 2$ for infinitely many indices n.

A number z is rational if and only if, in its representation (1), all terms from a certain index on are 0.

THEOREM 19   There exists a rationally independent, perfect set.[2]

Proof. Let Q be the set of all real numbers x of the form

$$(2) \qquad x = \frac{x_1}{10^1} + \frac{x_2}{10^2} + \cdots + \frac{x_n}{10^n} + \cdots$$

where each number x_n is either 1 or 2. As can be readily verified, Q is a perfect set.

With each number $x \in Q$, represented in the form (2), we correspond the number

$$(3) \qquad y = \frac{x_1}{(10x_1)!} + \frac{x_2}{(10^2 x_2 + 10 x_1)!} + \cdots + \frac{x_n}{(10^n x_n + 10^{n-1} x_{n-1} + \cdots + 10 x_1)!} + \cdots$$

and let P denote the set of all such corresponding numbers y. Because this correspondence between Q and P is one-to-one and continuous in both directions, the set P is also a perfect set.

It is also clear from the correspondence between Q and P that if y and y' are two different elements of Q, then there is a natural number n_0 such that each denominator

$$(10^n x_n + 10^{n-1} x_{n-1} + \cdots + 10 x_1)!$$

---

[1] A proof of this lemma can be found in Perron (1960, Sec. 33).
[2] Kormes (1924; 1926, p. 691; 1928, pp. 19–20) and Jones (1942, pp. 477–478). For additional examples of rationally independent sets, see von Neumann (1928), Erdös and Marcus (1957, pp. 448–449), and Mycielski (1964, 1967).

of a term of y whose index n is larger than n_0 differs from every one of the denominators of terms of y' with index larger than n_0. More generally, for any finite number of elements of Q, there is a natural number n_0 such that all the denominators of all the terms of these elements whose indices are larger than n_0 are different. We proceed to show that the set P is rationally independent.

Assume to the contrary that there exist different nonzero elements $y_1, \ldots, y_k$ of P and nonzero rational numbers $r_1, \ldots, r_k$ such that

$$\sum_{i=1}^{k} r_i y_i = 0$$

This relationship can also be expressed in the form

$$\sum_{i=1}^{k} n_i y_i = 0$$

where $n_1, \ldots, n_k$ are integers, or in the form

$$(4) \qquad \sum_{i=1}^{j} n_i' y_i' = \sum_{i=j+1}^{k} n_i' y_i'$$

where $n_1', \ldots, n_k'$ are natural numbers and $y_1', \ldots, y_k'$ are elements of P.

Let n_0 be a natural number such that all the denominators of all the terms of all the elements y_i' with indices larger than n_0 are different. Let m be a natural number larger than all the natural numbers $n_1', \ldots, n_k'$ and let p be a natural number larger than $2m$ and n_0.

Suppose that y' is any one of the elements $y_1', \ldots, y_k'$ and that in the representation (3), we have

$$y' = \frac{x_1'}{(10x_1')!} + \frac{x_2'}{(10^2 x_2' + 10x_1')!} + \cdots$$

$$+ \frac{x_n'}{(10^n x_n' + 10^{n-1} x_{n-1}' + \cdots + 10x_1')!} + \cdots$$

If we multiply y' by a natural number $n' \leqslant m$, then we obtain

$$(5) \qquad n'y' = \frac{n' x_1'}{(10x_1')!} + \frac{n' x_2'}{(10^2 x_2' + 10x_1')!} + \cdots$$

$$+ \frac{n' x_n'}{(10^n x_n' + 10^{n-1} x_{n-1}' + \cdots + 10x_1')!} + \cdots$$

For every index $n \geqslant p$, we have

$$n' x_n' \leqslant 2m < p < 10^p x_p' + 10^{p-1} x_{p-1}' + \cdots + 10x_1'$$

$$\leqslant 10^n x_n' + 10^{n-1} x_{n-1}' + \cdots + 10x_1'$$

which implies that

$$n'x_n' \leqslant (10^n x_n' + 10^{n-1} x_{n-1}' + \cdots + 10 x_1') - 2$$

for all indices $n \geqslant p$.

Now, the numbers $\Sigma_{i=1}^j n_i' y_i'$ can be arranged as a sum of two parts:

$$\sum_{i=1}^j n_i' y_i' = s_1 + t_1$$

where s_1 is the sum of all the terms of the numbers $n_1' y_1', \ldots, n_j' y_j'$ whose indices, in the representation (5), are smaller than p and t_1 is the sum of all the terms whose indices are larger than or equal to p. Similarly, we obtain the decomposition

$$\sum_{i=j+1}^k n_i' y_i' = s_2 + t_2$$

By (4), we have

$$s_1 + t_1 = s_2 + t_2$$

Writing the numbers $\Sigma_{i=1}^j n_i' y_i'$ and $\Sigma_{i=j+1}^k n_i' y_i'$ in the form (1), the initial terms of the two series (1) are determined by s_1 and s_2, respectively, and the sums of the remaining terms coincide with t_1 and t_2, respectively (both of which are already in the required form). From the uniqueness of the representation (1) we obtain $t_1 = t_2$. But this is impossible, since all the denominators of the terms of t_1 and t_2 are different from one another. Therefore, P must be a rationally independent set.

We now verify the assertion made above.

THEOREM 20   There exists a perfect set such that the distance between any two different elements of the set is an irrational number.

Proof. Let P be the perfect set defined in the proof of Theorem 19.

Assume to the contrary that there exist two different elements y, y' in P and a positive rational number r such that $y - y' = r$. Let n_0 be an index such that in the representations of y and y' in the form (3), all the denominators of the terms of y whose index n is larger than n_0 differ from all the denominators of the terms of y' whose indices are larger than n_0.

The number r, being rational, is representable in the form (1) with only a finite number of nonzero terms, so there exists a smallest index m_0 such that $z_n = 0$ for all indices $n \geqslant m_0$. Let p denote the maximum of the two numbers m_0 and n_0.

In the representation of the numbers y and $y' + r$ in the form (1), the terms of y with indices larger than p are different from the terms of $y' + r$

whose indices are larger than p. For if $y' + r$ is not already in the form (1), then it can be placed in that form and all terms of y' with indices larger than p will be unaffected. But all terms of y and $y' + r$ whose indices are larger than p in the representation (1) being different, we have contradicted the uniqueness of that representation. Therefore, the distance between any two different elements of P must be an irrational number.

NOTATION   For each set S and each $r \in \mathbb{R}$ we denote

$$rS = \{rx : x \in S\}$$

The next theorem is based upon the following lemma.

LEMMA 21   The set of all rational linear combinations of elements of a closed set is an $\mathcal{F}_\sigma$-set.

Proof. Let E be a given closed set. For each $n \in \mathbb{N}$, let L_n denote the set of all elements

$$x = r_1 e_1 + \cdots + r_n e_n$$

where $e_1, \ldots, e_n \in E$ and $r_1, \ldots, r_n \in \mathbb{Q}$. As $\bigcup_{n=1}^{\infty} L_n$ is the set of all rational linear combinations of elements of E, it suffices to show that each set L_n is an $\mathcal{F}_\sigma$-set.

It is clear that the set $L_1 = \bigcup_{r \in \mathbb{Q}} rE$ is an $\mathcal{F}_\sigma$-set.

For any elements $r, s \in \mathbb{Q}$ the set

$$rE \times sE = \{\langle u, v \rangle : u \in rE \text{ and } v \in sE\}$$

is a closed set in $\mathbb{R}^2$ and is representable as a countable union of compact sets in $\mathbb{R}^2$. The function $f: \mathbb{R}^2 \to \mathbb{R}$ defined by $f(u,v) = u + v$ being continuous, the set

$$E_{r,s} = f(rE \times sE) = \{rx + sy : x, y \in E\}$$

is an $\mathcal{F}_\sigma$-set. Consequently, the set $L_2 = \bigcup_{r \in \mathbb{Q}} \bigcup_{s \in \mathbb{Q}} E_{r,s}$ is an $\mathcal{F}_\sigma$-set.

Continuing inductively, it is seen that L_n is an $\mathcal{F}_\sigma$-set for each $n \in \mathbb{N}$.

THEOREM 22   Assume that every open interval is a Baire set. If $\mathscr{C}$ satisfies CCC, then there exists a rationally independent, perfect set P such that the set of all rational linear combinations of elements of P is a meager set.[1]

Proof. According to Theorem 19, there exists a rationally independent, perfect set S. By virtue of Theorem 20 of Chapter 5, Section I, the set S contains continuum many disjoint perfect sets P_α ($\alpha < \Lambda$) and each of these sets is also rationally independent. For each $\alpha < \Lambda$, let Q_α denote the set of all

---

[1] Erdös and Marcus (1957, pp. 448–449).

rational linear combinations of elements of P_α. It follows from our assumption and Lemma 21 that each set Q_α is a Baire set. Due to the rational independence, for $\alpha < \beta < \Lambda$, the sets R_α and R_β have only the number 0 in common. Applying Theorem 9 of Section I and Theorem 3 of Chapter 2, Section I, at least one set R_{α_0} is a meager set. The set P_{α_0} then has the required properties.

REMARK It has been shown[1] that the Continuum Hypothesis is equivalent to the following proposition:

(P) The set of all real numbers can be decomposed into a countable number of rationally independent sets.

D. Intersections of translates

From the existence of a rationally independent, perfect set, we obtain

THEOREM 23 There exists a perfect set that has at most one point in common with each of its nonzero translates.[2]

Proof. Let P be a rationally independent, perfect set. Assume that $t \neq 0$ and $x, y \in P \cap P(t)$. We show that $x = y$.

From $x, y \in P(t)$ we obtain the existence of $u, v \in P$ such that

$$x = u + t \qquad \text{and} \qquad y = v + t$$

This yields the relationship

(1) $x - u + v - y = 0$

Because P is rationally independent, this equality can hold only if either $x = u$, $x = v$, $x = y$, $u = v$, $u = y$, or $v = y$. However, since $t \neq 0$, we cannot have $x = u$ or $v = y$. It remains to consider the two possibilities $x = v$ and $u = y$.

Suppose that $x = v$. Then (1) becomes

(2) $2x - u - y = 0$

Again applying the rational independence of P, we must have either $x = u$, $x = y$, or $u = y$. The case $x = u$ has already been ruled out, while the case $x = y$ is the desired conclusion. In the case that $u = y$, equation (2) reduces to

(3) $2x - 2y = 0$

from which we conclude that $x = y$.

[1] Erdős and Kakutani (1943).
[2] Ruziewicz and Sierpiński (1932, Théorème I).

If we suppose that $u = y$, then a similar argument leads to the conclusion $x = y$.

In the event that $(\mathbb{R}, \mathscr{C})$ is also a perfect base, any perfect set having the foregoing property is necessarily a meager set, as is evident from the next theorem.

THEOREM 24 If S is any set that has at most a countable number of elements in common with each of its nonzero translates, then S contains no abundant Baire set.

Proof. Assume first that $\mathscr{C}$ is a trivial base. If S satisfies the given condition, then we cannot have $S = \mathbb{R}$, and since $\mathbb{R}$ is the only abundant Baire set, S contains no abundant Baire set. We can thus assume that $\mathscr{C}$ is a nontrivial base.

Suppose that S is an arbitrary set which contains an abundant Baire set E. Let D be a denumerable everywhere dense set of nonzero real numbers. We know from Theorem 3 of Section I that the set $\mathbb{R} - \bigcup_{r \in D} E(r)$ is meager. Consequently, $S - \bigcup_{r \in D} S(r)$ is also a meager set. From the equality

$$S = \left[S \cap \left(\bigcup_{r \in D} S(r) \right) \right] \cup \left[S - \bigcup_{r \in D} S(r) \right]$$

and the fact that S is an abundant set, it follows that there is an element $r_0 \in D$ for which the set $S \cap S(r_0)$ is abundant. In light of Theorem 7 of Section I, the set $S \cap S(r_0)$ must be uncountable.

From Theorem 23 we further derive

THEOREM 25 There exists a family $\mathscr{M}$ of power of the continuum, consisting of disjoint perfect sets, such that for every pair of different sets in $\mathscr{M}$, each translate of one has at most one element in common with each translate of the other.[1]

Proof. Let P be the perfect set given by Theorem 23 and let $\mathscr{M}$ be any family consisting of continuum many, disjoint perfect subsets of P.

Suppose that Q, R are two different sets in $\mathscr{M}$ and a, b are any real numbers such that $Q(a) \cap R(b) \neq \varnothing$. Since $Q(a) \cap R(a) = \varnothing$, we must have $a \neq b$. Accordingly, $P(b - a)$ is a nonzero translate of P. From the inclusion $Q \cap R(b - a) \subset P \cap P(b - a)$ we see that $Q \cap R(b - a)$ consists of at most one element. Translating the latter set by a, we conclude that $Q(a) \cap R(b)$ contains at most one element.

[1]Ruziewicz and Sierpiński (1932, Théorème II).

THEOREM 26 If $2^{\aleph_0} = \aleph_1$, then there exists a family of power $2^{2^{\aleph_0}}$ consisting of sets of power $2^{\aleph_0}$ such that for any pair of differents sets in the family, each translation of one set has at most countably many elements in common with each translate of the other set.[1]

Proof. We first show that if S and T are any two uncountable subsets of the perfect set P given by Theorem 23 such that $S \cap T$ is countable, then each translate of S has at most countably many elements in common with each translate of T.

Suppose that a, b are any real numbers. If $a = b$, then the intersection $S(a) \cap T(b)$ is a translate of $S \cap T$ and, accordingly, is a countable set. If $a \neq b$, then $P(b - a)$ is a nonzero translate of P and the inclusion $S \cap T(b - a) \subset P \cap P(b - a)$ implies that the set $S \cap T(b - a)$, hence also the set $S(a) \cap T(b)$, contains at most one element.

We now apply Proposition 14 of Chapter 2, Section III.

Turning to decomposition theorems, we first establish a theorem of Trzeciakiewicz.[2]

THEOREM 27 If the real line is decomposed into two disjoint infinite sets S and T, then there always exists a translate of S having infinitely many elements in common with T.

Proof. If S is a denumerable set, then there is a number a that is not representable in the form $x - y$, where $x, y \in S$. The translate $S(a)$ of S will then be disjoint from S and consequently contained in T. On the other hand, if T is denumerable, then there is a number b such that $T(b)$ is contained in S, so $S(-b)$ will contain T. We thus have only to consider the case in which both S and T are uncountable.

Following Vitali, we partition $\mathbb{R}$ into continuum many equivalence classes, each of which is denumerable, by calling two numbers x and y equivalent if their difference $x - y$ is a rational number.[3]

Suppose that there are only countably many equivalence classes which contain simultaneously an element of S and an element of T. Since each equivalence class is countable and S is uncountable, there is an equivalence class A that contains only elements of S. Similarly, there is an equivalence class B that contains only elements of T. Choose elements $a \in A$, $b \in B$ and place $c = b - a$. Then $A(c) = B$ and consequently, $S(c) \cap T$ is infinite.

Suppose on the other hand that there are uncountably many equivalence classes which contain simultaneously an element of S and an element of T. In

[1] Ruziewicz and Sierpiński (1932).
[2] Trzeciakiewicz (1932).
[3] Vitali's partition is discussed in detail in Section III.A.

each such equivalence class determine a pair of elements $\langle x, y \rangle$ with $x \in S$ and $y \in T$. For each such pair the difference $x - y$ is a rational number and Q being denumerable, there exists a rational number a such that $x - y = a$ for uncountably many pairs $\langle x, y \rangle$. All the elements y in these pairs of numbers belong to the set $S(-a) \cap T$. Consequently, $S(-a) \cap T$ is an uncountable set.

The real line cannot be decomposed into two disjoint sets, each having more than one element, such that each translate of one set has at most one element in common with each translate of the other set. This is obvious in the case that one of the two sets is finite and is an elementary consequence of Theorem 21 if both sets are infinite. However, we do have the following result.

THEOREM 28 For each cardinal number m, with $2 \leqslant m \leqslant 2^{\aleph_0}$, the real line can be decomposed into m disjoint Bernstein sets such that each translate of one of these sets has fewer than $2^{\aleph_0}$ elements in common which each translate of any other one of these sets.

Proof. This is a consequence of Theorem 23 of Chapter 5, Section II and the fact that if ϕ denotes any translation, then for any two disjoint Bernstein sets S, T given by that theorem, we have the inclusions

$$S \cap \phi(T) \subset S \cap \phi(\mathbb{R}-S) \subset S \bigtriangleup \phi(S)$$

COROLLARY 29 If $2^{\aleph_0} = \aleph_1$, then the real line can be decomposed into two disjoint uncountable sets such that each translate of one set has at most countably many elements in common with each translate of the other set.

The preceding theorems concern sets that have few points in common with their nonzero translates.[1] As for sets that have many points in common with their nonzero translates, Theorem 23 of Chapter 5, Section II guarantees the existence of proper subsets of $\mathbb{R}$ which differ from each of their nonzero translates by fewer than continuum many elements. In this instance, the sets do not have the Baire property and are not Lebesgue measurable. On the other hand, using the fact that any Sierpiński set is a set of the first category and any Mahlo-Luzin set is a set of Lebesgue measure zero, we obtain from Theorem 12 of Chapter 2, Section III:

(1) If $2^{\aleph_0} = \aleph_1$, then there exists an uncountable set of the first category which differs from each of its nonzero translates by at most countably many elements.
(2) If $2^{\aleph_0} = \aleph_1$, then there exists an uncountable set of Lebesgue measure zero that differs from each of its nonzero translates by at most countably many elements.

[1]See also Theorem 21 of Section III.

Theorem 23 of Chapter 5, Section II yields the existence of a set containing no perfect set whose complement also contains no perfect set and which differs from each of its nonzero translates by fewer than continuum many elements. One can also obtain such sets, which have perfect subsets.

THEOREM 30 There exists a set containing a perfect set whose complement also contains a perfect set and which differs from each of its nonzero translates by fewer than continuum many elements.[1]

Proof. As a consequence of Theorem 25 we obtain the existence of two disjoint perfect sets P and Q such that each translate of one of them has at most one element in common with each translate of the other.

Let

$$x_1, x_2, \ldots, x_\alpha, \ldots \qquad (\alpha < \Lambda)$$

be a well-ordering of $\mathbb{R}$ with $x_1 = 0$.

For a given set E and each ordinal number $\alpha < \Lambda$, we denote by E_α the union of all the translates $E(x_{\xi_1} \pm \cdots \pm x_{\xi_n})$, where $n \in \mathbb{N}$ and $\xi_1, \ldots, \xi_n$ are ordinal numbers $\leqslant \alpha$. It is clear that $E_\alpha \subset E_\beta$ whenever $\alpha \leqslant \beta < \Lambda$ and that $E_\alpha(x_\alpha) = E_\alpha$ for each $\alpha < \Lambda$. This notation is specialized to the sets P and Q. We show the set

$$S = \bigcup_{\alpha < \Lambda} (P_\alpha - Q_\alpha)$$

has the stated properties.

Since $x_1 = 0$ we have $P_1 = P(0) = P$ and $Q_1 = Q(0) = Q$. The sets P and Q being disjoint, the set $P_1 - Q_1 = P - Q = P$ is thus a perfect subset of S. On the other hand, $Q = Q_1 \subset Q_\alpha$ for every $\alpha < \Lambda$ implies that $Q \cap S = \varnothing$. Hence, Q is a perfect subset of the complement of S.

Suppose that r is any given real number. Then $r = x_\lambda$ for some $\lambda < \Lambda$. We have

$$S(r) = \bigcup_{\alpha < \Lambda} [P_\alpha(x_\lambda) - Q_\alpha(x_\lambda)]$$

For $\alpha \geqslant \lambda$, $P_\alpha(x_\lambda) = P_\alpha$ and $Q_\alpha(x_\lambda) = Q_\alpha$. Hence, the foregoing equalities yield

$$S(r) - S \subset \bigcup_{\alpha < \lambda} P_\alpha(x_\lambda) \subset P_\lambda(x_\lambda) = P_\lambda$$

From $P_\lambda - Q_\lambda \subset S$ we obtain $X - S \subset [(X - P_\lambda) \cup Q_\lambda]$ and, consequently,

$$S(r) - S \subset P_\lambda \cap [(X - P_\lambda) \cup Q_\lambda] = P_\lambda \cap Q_\lambda$$

Now, P_λ is the union of fewer than $2^{\aleph_0}$ sets of the form $P(x)$ with $x \in \mathbb{R}$ and Q_λ is the union of fewer than $2^{\aleph_0}$ sets of the form $Q(y)$ with $y \in \mathbb{R}$. Hence,

[1]Sierpiński (1936d).

$P_\lambda \cap Q_\lambda$ is the union of fewer than $2^{\aleph_0}$ sets $P(x) \cap Q(y)$ with $x, y \in \mathbb{R}$. Each set $P(x) \cap Q(y)$ contains at most one element. Therefore, $P_\lambda \cap Q_\lambda$ contains fewer than $2^{\aleph_0}$ elements, as does also the set $S(r)$—S.

The number r being arbitrary, it follows that the set $S(-r)$—S has fewer than $2^{\aleph_0}$ elements for any $r \in \mathbb{R}$, as does its translate S—$S(r)$.

We conclude that $S \bigtriangleup S(r)$ has power $< 2^{\aleph_0}$ for every real number r.

III. CONSTRUCTIONS OF VITALI AND HAMEL

A. Vitali sets

Vitali was the first person to establish the existence of a set of real numbers that is not Lebesgue measurable.[1] The set he constructed also fails to have the Baire property.

Define two real numbers x, y to be equivalent if their difference $x - y$ is a rational number. This definition actually determines an equivalence relation that partitions $\mathbb{R}$ into continuum many, disjoint equivalence classes, called Vitali equivalence classes. Each equivalence class is denumerable and invariant under all rational translations. Choosing one element in each equivalence class, we form a set S, called a Vitali set. We shall show that any Vitali set is not a Baire set, using the following lemma.

LEMMA 1 If S is any set, D is an everywhere dense set, and

$$S(r) \cap S(t) = \varnothing \qquad \text{for all } r, t \in D, \quad r \neq t$$

then S contains no abundant Baire set.

Proof. Assume to the contrary that S contains an abundant Baire set T. Let d_0 be a fixed element of D and set $E = D$—$\{d_0\}$. In view of our hypothesis, we have $T(r) \cap T(d_0) = \varnothing$ for every $r \in E$, and consequently, $T(d_0)$ is contained in the complement of the set $\bigcup_{r \in E} T(r)$. According to Theorem 3 of Section I, the latter set is a comeager set. Hence, $T(d_0)$ is a meager set. But this implies that T is also a meager set, contradicting our assumption.

That a Vitali set is not a Baire set is established in the course of proving the following decomposition theorem.[2]

THEOREM 2 The real line can be decomposed into denumerably many disjoint, congruent sets, none of which is a Baire set.

[1]Vitali (1905a).
[2]For an extension of this theorem, see Kellerer (1973).

Proof. Let S be a Vitali set. Then we have

$$S(r) \cap S(t) = \varnothing \qquad \text{for all } r,t \in \mathbb{Q}, \quad r \neq t$$

Hence, we obtain the decomposition

$$\mathbb{R} = \bigcup_{r \in \mathbb{Q}} S(r)$$

of $\mathbb{R}$ into denumerably many, disjoint, congruent sets.

If S were a Baire set, then, by virtue of Lemma 1, the set S would be meager. For each $r \in \mathbb{Q}$, the set $S(r)$ would also be meager. But this implies that $\mathbb{R}$ is a meager set, which is a contradiction. Thus, S is not a Baire set, nor are any of the sets $S(r)$ Baire sets.

From the existence of one set that is not a Baire set, we derive a stronger result.[1]

THEOREM 3 Every abundant set contains a set that is not a Baire set.

Proof. Suppose that S is any abundant set. The theorem being trivally true if S is not a Baire set, we assume that S is a Baire set.

Let D be a countable everywhere dense set, let

$$U = \bigcup_{r \in D} S(r)$$

and let T be a set that is not a Baire set. According to Theorem 3 in Section I, the set $\mathbb{R}-U$ is meager. This fact and the equality

$$T = (T \cap U) \cup (T-U)$$

implies the existence of an element $r \in D$ such that the set $T \cap S(r)$ is not a Baire set. Hence, $T(-r) \cap S$ is a subset of S that is not a Baire set.

THEOREM 4 If S is a Vitali set and E is a Baire set that has a nonempty intersection with at most finitely many of the sets $S(r)$, for $r \in \mathbb{Q}$, then E is a meager set.[2]

Proof. Assume that the set

$$D = \{r \in \mathbb{Q} : E \cap S(r) = \varnothing\}$$

is cofinite. If E were an abundant set, then, by Theorem 3 in Section I, the complement of the set $\bigcup_{r \in D} E(-r)$ is a meager set that contains S. This implies that S is a meager set, contradicting the fact that a Vitali set is not a Baire set. We conclude that E must be a meager set.

[1] Rademacher (1916, Satz VI) and Wolff (1923).
[2] Weston (1959).

Using a Vitali set, we next establish a simple generalization of a theorem of Sierpiński which states: If $2^{\aleph_0} = \aleph_1$, then there exists a decomposition of the unit interval into $2^{2^{\aleph_0}}$ sets, each of Lebesgue measure 1 and of the second category in every interval, such that the intersection of any two different sets is countable.[1]

THEOREM 5 Let $\langle \mathscr{C}_n \rangle_{n \in \mathbb{N}}$ be a sequence of translation bases, each of which has power at most $2^{\aleph_0}$ and satisfies CCC. If $2^{\aleph_0} = \aleph_1$, then $\mathbb{R}$ is representable as the union of $2^{2^{\aleph_0}}$ sets, each $\mathscr{C}_n$-abundant for every n, such that the intersection of any two different sets is a countable set.

Proof. Let S be a Vitali set and let the set Q of all rational numbers be decomposed into a sequence $\langle Q_n \rangle_{n \in \mathbb{N}}$ of disjoint everywhere dense sets $Q_n = \{r_{nk} : k \in \mathbb{N}\}$.

The set S being $\mathscr{C}_n$-abundant for each n, we apply Theorem 1 of Chapter 2, Section III to obtain a $\mathscr{C}_n$-rare set $S_n \subset S$ that has the power of the continuum. According to Proposition 14 of Chapter 2, Section III, each set S_n is representable as the union of $2^{2^{\aleph_0}}$ uncountable sets, the intersection of any two different sets of which is countable, say,

$$S_n = \bigcup_{\alpha < \Theta} S_{n\alpha}$$

where Θ denotes the smallest ordinal number of power $2^{2^{\aleph_0}}$.

For each ordinal number $\alpha < \Theta$, set

$$T_\alpha = \bigcup_{n=1}^{\infty} \bigcup_{k=1}^{\infty} S_{n\alpha}(r_{nk})$$

We show the sets T_α are $\mathscr{C}_n$-abundant everywhere for every n and $T_\alpha \cap T_\beta$ is a countable set whenever $\alpha < \beta < \Theta$.

Suppose that $\alpha < \Theta$ and $n \in \mathbb{N}$ are given. Since $S_{n\alpha}$ is an uncountable subset of the $\mathscr{C}_n$-rare set S_n, the set $S_{n\alpha}$ is $\mathscr{C}_n$-abundant. The set Q_n being everywhere dense, the set $\bigcup_{k=1}^{\infty} S_{n\alpha}(r_{nk})$ is $\mathscr{C}_n$-abundant everywhere. Hence, T_α is $\mathscr{C}_n$-abundant everywhere for every n.

Suppose that $\alpha < \beta < \Theta$ is given. Then

$$T_\alpha \cap T_\beta = \left[\bigcup_{n=1}^{\infty} \bigcup_{k=1}^{\infty} S_{n\alpha}(r_{nk}) \right] \cap \left[\bigcup_{m=1}^{\infty} \bigcup_{p=1}^{\infty} S_{m\beta}(r_{mp}) \right]$$

$$= \bigcup_{n=1}^{\infty} \bigcup_{k=1}^{\infty} \bigcup_{m=1}^{\infty} \bigcup_{p=1}^{\infty} [S_{n\alpha}(r_{nk}) \cap S_{m\beta}(r_{mp})]$$

[1]Sierpiński (1929b); see also Theorem 15 of Chapter 2, Section III and Theorem 28 of Chapter 5, Section I.

According to the definition of S, if r and r' are two different rational numbers, then $S(r) \cap S(r') = \varnothing$ and consequently $S_{n\alpha}(r) \cap S_{m\beta}(r')$, which is a subset of $S(r) \cap S(r')$, is also empty. Hence,

$$T_\alpha \cap T_\beta = \bigcup_{n=1}^{\infty} \bigcup_{k=1}^{\infty} [S_{n\alpha}(r_{nk}) \cap S_{n\beta}(r_{nk})]$$

Since $S_{n\alpha} \cap S_{n\beta}$ is a countable set, so also is the set $S_{n\alpha}(r_{nk}) \cap S_{n\beta}(r_{nk})$ for each n and each k. Therefore, $T_\alpha \cap T_\beta$ is a countable set.

Finally, upon adjoining the set $\mathbf{R} - \bigcup_{\alpha < \theta} T_\alpha$ to one of the sets T_α, we have the desired representation.

We note that one can sometimes form Vitali sets with special properties. For example, we have

THEOREM 6   There exists a Vitali set that contains a perfect set.[1]

Proof. If all elements of a rationally independent, perfect set are multiplied by a fixed nonzero number, then the resulting set is also a rationally independent, perfect set. Accordingly, we can obtain from the proof of Theorem 19 of Section II a rationally independent, perfect set P of nonzero numbers that contains the number 1.

The intersection of P with any given Vitali equivalence class must contain at most one element. For suppose that $x, y \in P$, $r \in \mathbb{Q}$, $x \neq y$, and $x - y = r$. Then $x - y - r = 0$. Due to the rational independence we must have either $x = y$, $x = 1$, or $y = 1$. However, we do not have $x = y$. Suppose then that $x = 1$. This gives $(1 - r) - y = 0$ with $1 - r \neq 0$ and, again invoking the rational independence, we obtain $y = 1$. But this yields the impossibility $y = x$. Similarly, assuming that $y = 1$ leads to the same impossibility.

Choosing an element of P in each equivalence class having a nonempty intersection with P and an arbitrary element in each other equivalence class, we obtain a Vitali set containing P.

B.   Hamel's basis

DEFINITION   A set B of nonzero real numbers is called a Hamel basis if every nonzero real number x is uniquely representable as a rational linear combination of elements of B; that is, there is a unique representation of x of the form

$$x = r_1 b_1 + \cdots + r_k b_k$$

where $k \in \mathbb{N}$, $b_1, \dots, b_k \in B$, and $r_1, \dots, r_k \in \mathbb{Q} - \{0\}$.

[1]Cf. Kuratowski (1973, p. 71).

The existence of such a set B was first established by Hamel and is a consequence of the following general principle (with $E = \mathbb{R}$).[1]

THEOREM 7 If E is a set of real numbers such that every real number of some interval I can be represented as a rational linear combination of elements of E, then E contains a Hamel basis.

Proof. We first note that every real number is representable as a rational linear combination of elements of E, since every real number is a rational multiple of some element of I. Next, we note that the set E must have the power of the continuum. This is a consequence of the facts that E is an infinite set and the set of all rational linear combinations of elements of any infinite set has the same power as the set itself.[2]

We shall define a Hamel basis contained in E by transfinite induction. Let

(*) $\qquad\qquad\qquad x_1, x_2, \ldots, x_\alpha, \ldots \qquad (\alpha < \Lambda)$

be a well-ordering of all nonzero elements of E. Place $b_1 = x_1$. Assume that $\alpha < \Lambda$ and the elements b_ξ have already been defined for all ordinal numbers $\xi < \alpha$. The set of all rational linear combinations of the form $r_1 b_{\xi_1} + \cdots + r_k b_{\xi_k}$, where $k \in \mathbb{N}$, $r_1, \ldots, r_k \in \mathbb{Q}$, and $\xi_1, \ldots, \xi_k$ are ordinal numbers less than α, has power less than the power of the continuum. Hence, there exist elements of (*) that are not representable in this form. We define b_α to be the first such element of (*).

The set $B = \{b_\alpha : \alpha < \Lambda\}$ of elements thus defined is a Hamel basis contained in E.

It is a simple matter to verify the following fact:[3]

THEOREM 8 Every Hamel basis has the power of the continuum.

In place of the field $\mathbb{Q}$ of rational numbers one can utilize other fields of real numbers. A field is a set F of real numbers having the following properties for all elements $x, y \in \mathbb{R}$:

(i) $0, 1 \in F$.
(ii) If $x, y \in F$, then $x + y \in F$.
(iii) If $x \in F$, then $-x \in F$.
(iv) If $x, y \in F$, then $xy \in F$.
(v) If $x \in F$ and $x \neq 0$, then $x^{-1} \in F$.

---

[1]Hamel (1905) and Jones (1942, Lemma 1); see also Broggi (1907).
[2]Cf. Burstin (1916b, Satz 1).
[3]Burstin (1916b).

Any set E of real numbers generates a field that is defined in an external manner as the intersection of all fields containing the given set. This field is characterized in an internal manner as the set of all real numbers x representable in the form of a quotient

$$x = \frac{r_0 + r_1 p_1 + \cdots + r_m p_m}{s_0 + s_1 q_1 + \cdots + s_n q_n}$$

with nonzero denominator, where $m,n \in \mathbb{N}$, $r_0, r_1, \ldots, r_m, s_0, s_1, \ldots, s_n \in \mathbb{Q}$, and each of the elements $p_1, \ldots, p_m, q_1, \ldots, q_n$ is either 0 or the product of a finite number of elements of E. From this characterization we obtain the following fact: The field generated by an infinite set E has the same power as the set E.

DEFINITION   A set B of nonzero real numbers is a Hamel basis relative to a field F if every nonzero real number x is uniquely representable in the form

$$x = r_1 b_1 + \cdots + r_k b_k$$

where $k \in \mathbb{N}$, $b_1, \ldots, b_k \in B$, and $r_1, \ldots, r_k$ are nonzero elements of F. When $F = \mathbb{Q}$ the phrase "relative to F" is deleted.

A simple generalization of the argument used to prove Theorem 7 yields

THEOREM 9   If F is a field that has power less than the power of the continuum, then there exists a Hamel basis relative to F.

We note that any Hamel basis is a rationally independent set.

THEOREM 10   Every abundant Baire set contains a Hamel basis.

Proof. Use Theorem 6 of Section II and Theorem 7.

THEOREM 11   There exists a Hamel basis that contains a perfect set.[1]

Proof. As seen in the proof of Theorem 19 of Section II, there exists a rationally independent, perfect set P consisting of nonzero real numbers. Assuming that P is not already a Hamel basis, we proceed by transfinite induction to determine a Hamel basis containing P.

Let

(+) $x_1, x_2, \ldots, x_\alpha, \ldots$ $(\alpha < \Lambda)$

be an enumeration of all elements of $\mathbb{R}$ that do not belong to P. Define y_1 to be the first element of (+) that is not representable as a rational linear combination of elements of P. Assume that $\alpha < \Lambda$ and that we have already defined elements y_β for all ordinal numbers $\beta < \alpha$. If the set

---

[1]See the references to Theorem 13 of Section II and Marcus (1957).

$R_\alpha = P \cup \{y_\beta : \beta < \alpha\}$ is a Hamel basis, then we are finished. If not, then we define y_α to be the first element of $(+)$ that is not representable as a rational linear combination of elements of R_α. Either this procedure terminates at a certain ordinal number $\alpha < \Lambda$ and R_α is a Hamel basis or we ultimately obtain the set $P \cup \{y_\alpha : \alpha < \Lambda\}$ as a Hamel basis.

Burstin established the existence of a Hamel basis having at least one element in common with each perfect set and employed this basis to show that the real line can be decomposed into continuum many, disjoint sets, none of which is Lebesgue measurable or has the Baire property.[1] Sierpiński observed that the complement of any Hamel basis having at least one element in each perfect set also has this property.[2]

THEOREM 12 There exists a Hamel basis that is a Bernstein set.[3]

Proof. Let

(1) $$x_1, x_2, \ldots, x_\alpha, \ldots \qquad (\alpha < \Lambda)$$

(2) $$P_1, P_2, \ldots, P_\alpha, \ldots \qquad (\alpha < \Lambda)$$

be well-orderings of $\mathbb{R}$ and of all perfect sets, respectively.

Define b_1 to be the first nonzero element of (1) that belongs to P_1. Assume that $\alpha < \Lambda$ and nonzero elements $b_\xi \in P_\xi$ have been defined for all $\xi < \alpha$. The set of all rational linear combinations of elements b_{ξ_i} with $\xi_i < \alpha$ having power less than the power of the continuum, the set

$$E_\alpha = P_\alpha - \{r_1 b_{\xi_1} + \cdots + r_k b_{\xi_k} : k \in \mathbb{N}, \ r_1, \ldots, r_k \in \mathbb{Q}, \ \xi_1, \ldots, \xi_k < \alpha\}$$

has the power of the continuum. We define b_α to be the first nonzero element of (1) that belongs to E_α. The set $B = \{b_\alpha : \alpha < \Lambda\}$ thus determined is a Hamel basis having at least one element in each perfect set.

Suppose that B is any Hamel basis and $b \in B$. Then we must have $B \cap B(b) = \emptyset$. For otherwise, there is an element $b' \in B \cap B(b)$ and $b'' \in B$ such that $b' = b'' + b$, contradicting the fact that b' has a unique representation as a rational linear combination of elements of B.

The Hamel basis B we have constructed has at least one element in each perfect set. The translate $B(b)$ of B also has this property. But since $B(b)$ is contained in the complement of B, both B and $\mathbb{R} - B$ must have at least one element in each perfect set. Thus, B is a Bernstein set.

---

[1]Burstin (1916b).
[2]Sierpiński (1935g).
[3]See also Kuczma (1985a, Theorem 6).

COROLLARY 13   If $\mathscr{C}$ is also a perfect base, then there exists a Hamel basis that is not a Baire set.[1]

The existence of a Hamel basis implies the existence of a set that is not a Baire set.

THEOREM 14   The set S of all real numbers whose representation as a rational linear combination of elements of a given Hamel basis B does not involve a given fixed element $b \in B$ is a set that is not a Baire set. In fact, both S and $\mathbb{R}—S$ are abundant everywhere.[2]

Proof. Let c be a fixed element of $B—\{b\}$ and let T be the set of all real numbers representable as a rational linear combination of elements of $B—\{b,c\}$. From the equalities

$$\mathfrak{r} = \bigcup_{r \in \mathbb{Q}} S(br) \qquad \text{and} \qquad S = \bigcup_{r \in \mathbb{Q}} T(cr)$$

it is seen that both S and T are abundant sets.

The set $\{cr : r \in \mathbb{Q}\}$ being everywhere dense and T being an abundant set, it follows from Theorem 2 of Section I that the set $S = \bigcup_{r \in \mathbb{Q}} T(cr)$ is abundant everywhere. The set $\{br : r \in \mathbb{Q}\}$ is also everywhere dense and $S(br) \cap S(bt) = \varnothing$ for all $r, t \in \mathbb{Q}$, $r \neq t$. Applying Lemma 1 and Theorem 4 of Chapter 2, Section I, we see that the complement of S is also abundant everywhere. Therefore, S is not a Baire set.

In view of the fact that the sets $S(br)$, with $r \in \mathbb{Q}$, utilized in proving the foregoing theorem are disjoint and

$$\mathbb{R} = \bigcup_{r \in \mathbb{Q}} S(br)$$

we immediately obtain the following strengthening of Theorem 2, which generalizes two results of Sierpiński.[3]

THEOREM 15   The real line can be decomposed into denumerably many, disjoint congruent sets, each of which is abundant everywhere.

Sierpiński further proved that any Hamel basis has inner Lebesgue measure zero[4] and later[5] stated the Baire category analog that the complement of any Hamel basis is everywhere of the second category. In the

[1] Sierpiński (1920b, 1935g).
[2] Sierpiński (1920b), Abian (1970), and Hewitt and Stromberg (1974); see also Sierpiński (1935i) for a special property of this set.
[3] Sierpiński (1917b).
[4] [1] Sierpiński (1920b).
[5] [2] Sierpiński (1935g).

measure-theoretic direction, Ruziewicz obtained a more general result which is a version of the following theorem.[1]

THEOREM 16 A union of fewer than continuum many translates of a Hamel basis contains no abundant Baire set.

Proof. Let B be a Hamel basis, let E be a set of real numbers of power less than the power of the continuum, and let

$$U = \bigcup_{t \in E} B(t)$$

Let D be an everywhere dense set of nonzero rational numbers such that $|r - s| \neq 1$ for all $r, s \in D$ and let b be an element of B that does not occur in the representation (with respect to the Hamel basis B) of any element of E.

Assume that $r, s \in D$ and $U(br) \cap U(bs) \neq \varnothing$. Then there exist elements $t_1, t_2 \in E$ and $b_1, b_2 \in B$ such that

$$b_1 + t_1 + br = b_2 + t_2 + bs$$

Because the element b does not occur in the representations of t_1 and t_2, the element b occurs in the representation of $b_1 + t_1 + br$ either with coefficient r or with coefficient $1 + r$ and occurs in the representation of $b_2 + t_2 + bs$ either with coefficient s or with coefficient $1 + s$. In view of the uniqueness of representations, one of the following equalities must hold:

$$r = s \qquad r = 1 + s \qquad 1 + r = s \qquad 1 + r = 1 + s$$

However, by virtue of the definition of D, we must have $r = s$. Summarizing, if $r, s \in D$ and $r \neq s$, then $U(br) \cap U(bs) = \varnothing$. The set $\{br : r \in \mathbb{Q}\}$ being everywhere dense, it follows from Lemma 1 that U contains no abundant Baire set.

COROLLARY 17 No Hamel basis contains an abundant Baire set.

COROLLARY 18 A Hamel basis that is a Baire set is a meager set.[2]

In the following example we see that there is a Hamel basis which is simultaneously of Lebesgue measure zero and of the first category.[3]

EXAMPLE Let S be the set of all real numbers x having a representation of the form

$$x = c_0 + \frac{c_2}{2^2} + \frac{c_4}{2^4} + \frac{c_6}{2^6} + \cdots$$

[1] Ruziewicz (1936); see also Sierpiński (1935f, 1936c).
[2] Sierpiński (1920b) and Mehdi (1964).
[3] Sierpiński (1920b).

and let T be the set of all real numbers y that have a representation of the form

$$y = \frac{c_1}{2^1} + \frac{c_3}{2^3} + \frac{c_5}{2^5} + \cdots$$

where c_0 is an integer and $c_n = 0$ or 1 for each $n \in \mathbb{N}$. It can be shown that both of the sets S and T have Lebesgue measure zero and are nowhere dense. The set $E = S \cup T$ also has Lebesgue measure zero and is nowhere dense. Now, every real number z is representable in the form $z = x + y$, where $x \in S$ and $y \in T$. According to Theorem 7, the set E contains a Hamel basis B.

NOTE One can also establish the existence of a Hamel basis that is a subset of the Cantor set by means of Theorem 7 and either Theorem 9 of Section II or Theorem 11 in Section II.

THEOREM 19 If every open interval is a Baire set, then there exists a Hamel basis that is a meager set.

Proof. The Cantor set $\mathbb{C}$ will then be a Baire set. If $\mathbb{C}$ is a meager set, then, as just noted, $\mathbb{C}$ contains a Hamel basis that is a meager set.

Suppose that $\mathbb{C}$ is an abundant set. Then the set $S = \bigcup_{r \in \mathbb{Q}} \mathbb{C}(r)$ is, according to Theorem 3 of Section I, a comeager set. Since $\mathbb{C}$ is a nowhere dense set, the complement of S is a set of the second category that has the Baire property. According to the Baire category case of Theorem 10, the complement of S contains a Hamel basis and this Hamel basis must be a meager set.

THEOREM 20 Assume CH. If $\mathscr{C}$ satisfies CCC and has power at most $2^{\aleph_0}$, then there exists a Hamel basis that is a rare set.[1]

Proof. Apply Theorem 15 of Section II and Theorem 7.

We utilize this theorem to establish a further result concerning intersections of translates.

THEOREM 21 Assume CH. If $\mathscr{C}$ satisfies CCC and has power at most $2^{\aleph_0}$, then for every cardinal number $\mathfrak{m}$, with $2 \leqslant \mathfrak{m} \leqslant 2^{\aleph_0}$, there exists a family of power $\mathfrak{m}$ consisting of disjoint abundant sets such that for any pair of different sets in the family, each translation of one set has at most one element in common with the other set.[2]

Proof. Let B be a Hamel basis that is a rare set and let $\mathscr{S}$ be a family consisting of $\mathfrak{m}$ disjoint uncountable subsets of B. Let S and T be two

[1] Cf. Sierpiński (1935g) and Darst (1965).
[2] Cf. Eames (1977).

different sets in $\mathscr{S}$. Suppose that r is any real number and $x, y \in S \cap T(r)$. Then there exist $u, v \in T$ such that $x = u + r$ and $y = v + r$ so that $x = y + u - v$. Due to the uniqueness of representation for x we must have either $x = y$, $x = u$, or $x = v$. If we assume that $x = u$, then we would obtain $r = 0$ and $x \in S \cap T$, contradicting the disjointness of S and T. Hence, we cannot have $x = u$. Suppose, then, that $x = v$. This implies that $y = 2x - u$ and consequently, by virtue of the uniqueness of representation for y, either $y = x$ or $y = u$. But $y = u$ implies that $2x = 2y$ and $x = y$. We are thus led inevitably to the conclusion that $x = y$. Thus, $S \cap T(r)$ contains at most one element.

We note the following additional existence theorems:[1]

(1)   There exists a Hamel basis that is not a projective set.
(2)   For any number $\alpha \geqslant 0$, there exists a Hamel basis that has Lebesgue outer measure α and contains no perfect set.

We next consider arithmetic combinations of sets.

THEOREM 22   Assume that $\mathscr{C}$ is linearly invariant. If $\mathscr{A}$ is a family of Baire sets satisfying the conditions
   (i) $S + T \in \mathscr{A}$ for all sets $S, T \in \mathscr{A}$,
   (ii) $rS \in \mathscr{A}$ for all sets $S \in \mathscr{A}$ and all $r \in \mathbb{Q}$,
then no set in $\mathscr{A}$ is a Hamel basis.[2]

Proof. Assume to the contrary that there exists a set B in $\mathscr{A}$ that is a Hamel basis. According to Corollary 18, the set B must be a meager set. For each $n \in \mathbb{N}$ define

$$B_n = \left\{ \sum_{i=1}^{n} r_i b_i : r_1, \ldots, r_n \in \mathbb{Q} \text{ and } b_1, \ldots, b_n \in B \right\};$$

i.e., B_n consists of all elements x whose expansion with respect to the Hamel basis B has at most n nonzero coefficients. We show by induction that B_n is a meager set for every n.

Clearly,

$$B_1 = \bigcup_{r \in \mathbb{Q}} \{rb : b \in B\} = \bigcup_{r \in \mathbb{Q}} rB$$

is a meager set. Suppose that $n \in \mathbb{N}$ and that the set B_n has been shown to be a meager set. Let b be a fixed element of B, and for each nonzero rational number r, let C_r denote the set of all elements in B_{n+1} whose expansion

---

[1] Marcus (1958, Satz 2 and 3, resp.).
[2] Cf. Sierpiński (1920b).

contains b with the coefficient r; that is,

$$C_r = \left\{ rb + \sum_{i=1}^{n} r_i b_i : r_1, \ldots, r_n \in \mathbb{Q} \text{ and } b_1, \ldots, b_n \in B \text{---} \{b\} \right\}$$

The set

$$C = \bigcup \{C_r : r \in \mathbb{Q} \text{ and } r \neq 0\}$$

consisting of all elements of B_{n+1} whose expansion contains the element b with a nonzero coefficient, will then be a meager set, since we have $C_r \subset B_n(rb)$ for each r.

From the assumption that B belongs to $\mathscr{A}$ and the conditions (i) and (ii), it follows that B_{n+1} also belongs to $\mathscr{A}$, so B_{n+1} is a Baire set. If B_{n+1} were an abundant set, then the set B_{n+1}---C, consisting of all elements in B_{n+1} whose expansion does not contain the element b, would be an abundant Baire set. The set $\{x/b : x \in B_{n+1}$---$C\}$ would also be an abundant Baire set and, by Corollary 2 of Section II, there would exist elements $x, y \in B_{n+1}$---C such that $x/b - y/b = r$, for some nonzero rational number r. Then $b = (x - y)/r$ and the element b is thus expressible as a rational linear combination of a finite number of elements of B different from b, which is impossible. Therefore, B_{n+1} must be a meager set. Hence, all the sets B_n are meager sets.

From the equality $\mathbb{R} = \bigcup_{n=1}^{\infty} B_n$, we then obtain the contradictory result that $\mathbb{R}$ is a meager set. We conclude that no set B in $\mathscr{A}$ can be a Hamel basis.

COROLLARY 23   Assume that $\mathscr{C}$ is linearly invariant. If every open interval is a Baire set, then there exist Baire sets S and T whose arithmetic sum $S + T$ is not a Baire set.

Proof. Let $\mathscr{A}$ be the family of all Baire sets. Because $\mathscr{C}$ is linearly invariant, condition (ii) of Theorem 22 is satisfied. According to Theorem 19, the conclusion of Theorem 22 is not valid. Therefore, condition (i) is not satisfied.

THEOREM 24   Assume that $\mathscr{C}$ is reflection invariant. If every open interval is a Baire set, then there is a Baire set whose distance set is not a Baire set.[1]

Proof. Assume to the contrary that the distance set of every Baire set is a Baire set. Since the difference set of a Baire set consists of the distance set and its reflection, this assumption implies the difference set of every Baire set is a Baire set.

Now, by Theorem 19, there exists a Hamel basis B that is a meager set. Let b_0 be a fixed element of B and let $B_0 = B$---$\{b_0\}$. Since B_0 is a Baire set, the set

$$E_1 = \{r_1 b_1 : r_1 \in \mathbb{Q} \text{ and } b_1 \in B_0\} = \bigcup_{r \in \mathbb{Q}} r B_0$$

[1]Sierpiński (1925).

of all rational multiples of elements of B_0 is a Baire set. By assumption, the difference set of E_1, which we shall denote by E_2, is also a Baire set. We have

$$E_2 = \{q_1 b_1 - q_2 b_2 : q_1, q_2 \in \mathbb{Q} \text{ and } b_1, b_2 \in B_0\}$$
$$= \{r_1 b_1 + r_2 b_2 : r_1, r_2 \in \mathbb{Q} \text{ and } b_1, b_2 \in B_0\}$$

Continuing inductively, we define E_{n+1} to be the difference set of E_n. The set E_{n+1} is a Baire set and

$$E_{n+1} = \left\{ \sum_{i=1}^{2^n} r_i b_i : r_i \in \mathbb{Q} \text{ and } b_i \in B_0 \text{ for all } i = 1, 2, \dots, 2^n \right\}$$

We note that E_n contains all the numbers expressible in the form $r_1 b_1 + \cdots + r_n b_n$, where $r_1, \dots, r_n \in \mathbb{Q}$ and $b_1, \dots, b_n \in B_0$. Consequently, the set $S = \bigcup_{n=1}^{\infty} E_n$ is the set of all rational linear combinations of elements of B_0. Being a countable union of Baire sets, S is a Baire set. This contradicts Theorem 14.

C. Cauchy's functional equation

A function f is called an additive function if it satisfies Cauchy's functional equation,[1]

(1) $$f(x + y) = f(x) + f(y)$$

for all real numbers x and y.

It follows immediately from (1) that for any real number x,

$$f(2x) = 2f(x)$$

and continuing inductively, that

(2) $$f(nx) = nf(x)$$

for all natural numbers n. If $x = 0$, then we have $f(0) = 2f(0)$, so

$$f(0) = 0$$

This implies that $f(x) + f(-x) = f(x - x) = f(0) = 0$, and consequently,

$$f(-x) = -f(x)$$

for every real number x. This fact, in conjunction with (2), reveals that the equality (2) is valid for every integer n. If $r = m/n$ is any rational number,

[1]For an extensive discussion of Cauchy's functional equation and Hamel bases, see Aczél (1966) and Kuczma (1985b).

where m and n are integers with $n \neq 0$, then we have

$$mf(1) = f(m) = f(nr) = nf(r)$$

which yields

(3) $$f(r) = rf(1)$$

Setting $c = f(1)$, we thus have

(4) $$f(r) = cr$$

for every rational number r.

    If the function f is continuous, then using (4) and taking limits, we see that

$$f(x) = cx$$

for every real number x. Summarizing, we have established

THEOREM 25   If f is a continuous additive function, then $f(x) = cx$ for some constant c and every $x \in \mathbb{R}$.

    Any linear function of the form $f(x) = cx$ is obviously additive. Therefore, within the class of continuous functions, such linear functions constitute the solution set of the equation (1). The question of whether there are any other additive functions was unresolved for a long time. The existence of discontinuous solutions of the equation (1) was resolved in the affirmative by Hamel, utilizing his basis for the set of all real numbers.

    Let B be a Hamel basis for $\mathbb{R}$ and let f_0 be any one-to-one mapping of B onto B. If x is any real number, then x is uniquely representable as a rational linear combination of elements of B, say,

$$x = \sum_{i=1}^{n} r_i b_i$$

Then defining

(*) $$f(x) = \sum_{i=1}^{n} r_i f_0(b_i)$$

we obtain an extension of the function f_0 to a function f defined for all real numbers x. It is easily seen that the function f so-defined satisfies Cauchy's functional equation and f will be continuous if and only if there is a constant c such that $f(b) = cb$ for every $b \in B$. Since every Hamel basis has the power of the continuum, it is clear from this construction that there exist $2^{2^{\aleph_0}}$ discontinuous solutions of Cauchy's functional equation.

    One can also determine a discontinuous solution of (1) as follows:[1] Let b

---

[1] Cf. Kormes (1926) and Sander (1981a).

be a fixed element of a Hamel basis B, define

$$f_0(x) = \begin{cases} 1, & x = b \\ 0, & x \in B-\{b\} \end{cases}$$

and extend f_0 to a function f on $\mathbb{R}$ by (*).

There are various conditions which guarantee that an additive function f satisfying the given condition must be a continuous function.[1] Of special importance is the following condition:

THEOREM 26 If f is an additive function that is bounded above (or below) on some interval, then $f(x) = cx$ for some constant c and every $x \in \mathbb{R}$.

Proof. Suppose that f is bounded above by a number M on an interval $[a,b]$. The function

$$g(x) = f(x) - xf(1)$$

satisfies the equation $g(x + y) = g(x) + g(y)$ and is also bounded above on $[a,b]$ by $M-af(1)$. Moreover, by (3) above, we have $g(r) = 0$ for every rational number r, and consequently,

$$g(y + r) = g(y)$$

for all real numbers y and all rational numbers r.

If x is any real number, then there is a rational number r such that $x - r$ is in the interval $[a,b]$. From the equality

$$g(x) = g[(x - r) + r] = g(x - r)$$

we see that g is bounded above on $\mathbb{R}$.

If there were a number x_0 such that $g(x_0) \neq 0$, then, in view of the fact that $g(nx_0) = ng(x_0)$ holds for all integers n, we would have to conclude that g is not bounded above. Therefore, we must have $g(x) = 0$ for all real numbers x, or

$$f(x) = cx$$

for all real numbers x, with $c = f(1)$.

The validity of the theorem when f is only assumed to be bounded below reduces to the preceding case by considering, instead of f, the function $f'(x) = -f(x)$, which is bounded above and also additive.

COROLLARY 27 If f is an additive function that is bounded above (or below) on a set S whose arithmetic sum $S + S$ contains an interval, then $f(x) = cx$ for some constant c and every $x \in \mathbb{R}$.[2]

[1] See Kuczma (1985b).
[2] See Kemperman (1957, Theorem 3.1).

Proof. Suppose that f is bounded above on S by a number M. For each element x of the arithmetic sum $S + S$ we have $f(x) \leqslant 2M$. As $S + S$ contains an interval, f is bounded above on an interval.

The case where f is bounded below is treated as before.

THEOREM 28 Assume that $\mathscr{C}$ is reflection invariant. If f is an additive Baire function, then f is a continuous function.[1]

Proof. Setting $S_n = \{x \in \mathbb{R} : f(x) \leqslant n\}$ for each $n \in \mathbb{N}$, we have $\mathbb{R} = \bigcup_{n=1}^{\infty} S_n$. Because $\mathbb{R}$ is abundant, there is a natural number n_0 such that S_{n_0} is an abundant set. As S_{n_0} is also a Baire set, it follows from Corollary 8 of Section II that the arithmetic sum $S_{n_0} + S_{n_0}$ contains an interval. The conclusion now follows from Corollary 27.

IV. GROUPS AND PERIODIC FUNCTIONS

A. Groups considered

By a group we shall mean below an additive group of real numbers; i.e., a subset G of $\mathbb{R}$ having the following properties:

(i) $0 \in G$.
(ii) If $x, y \in G$, then $x + y \in G$.
(iii) If $x \in G$, then $-x \in G$.

How many groups are there?

THEOREM 1 There are $2^{2^{\aleph_0}}$ different groups.

Proof.[2] Let E be a rationally independent set of real numbers having the power of the continuum. Let $\mathscr{S}$ be the family of all sets of nonzero elements in E which have the power of the continuum. Obviously, $\mathscr{S}$ has power $2^{2^{\aleph_0}}$.

For each set $S \in \mathscr{S}$ we define G_S to be the set of all real numbers x representable in the form

$$(*) \qquad\qquad x = r_1 x_1 + \cdots + r_n x_n$$

where $n \in \mathbb{N}$, $x_1, \ldots, x_n \in S$, and $r_1, \ldots, r_n \in \mathbb{Q}$. It is readily verified that G_S is a group.

---

[1] Cf. Lebesgue (1907a), Fréchet (1913, 1914), Banach (1920), Sierpiński (1920d, 1920e, 1924c), Kormes (1926), Ostrowski (1929), Kac (1936), Braun, Kuratowski, and Szpilrajn (1937, p. 240), Alexiewicz and Orlicz (1945), Kestelman (1947), Mehdi (1964), Letac (1978), and Sander (1981a); see also Aczél (1966) and Kuczma (1985b).
[2] Cf. Mycielski and Sierpiński (1966, Théorème 3).

Suppose that $S, T \in \mathscr{S}$ and $S \neq T$. Without loss of generality, we assume that there is an element $x_0 \in T$ with $x_0 \notin S$. If $x_0 \in G_S$, then x_0 has a representation of the form (*)

$$x_0 = r_1 x_1 + \cdots + r_n x_n$$

However, since $x_0 \neq x_i$ for all $i = 1, \ldots, n$, this representation contradicts the rational independence of E. Hence we must have $x_0 \notin G_S$. As $x_0 \in G_T$, we must thus have $G_S \neq G_T$.

We conclude that there exist $2^{2^{\aleph_0}}$ groups G_S.

B. Periods and periodic functions

DEFINITION   A number p is a period of a function $f : \mathbb{R} \to \mathbb{R}$ if

$$f(x + p) = f(x)$$

for every $x \in \mathbb{R}$. The set of all periods of a function f is called the period set of f and is denoted by $P[f]$. The period set $P[f]$ is called nontrivial if it contains at least one nonzero period, in which case, f is called a periodic function.

The simplest periodic functions are the constant functions.

THEOREM 2   A function is constant if and only if its period set is $\mathbb{R}$.

THEOREM 3   A function having at least one point of continuity is constant if and only if its period set is everywhere dense.

Proof. Let f be a function continuous at a point x_0. If f is constant, then $P[f]$ is everywhere dense.

Suppose, on the other hand, that $P[f]$ is everywhere dense. Let x be any element of $\mathbb{R}$. The translate $\{p + x_0 : p \in P[f]\}$ of $P[f]$ by x_0 being also everywhere dense, there exists a sequence $\langle p_n \rangle_{n \in \mathbb{N}}$ of elements of $P[f]$ such that the sequence $\langle p_n + x_0 \rangle_{n \in \mathbb{N}}$ converges to x. The sequence $\langle x - p_n \rangle_{n \in \mathbb{N}}$ will then converge to x_0. For each $n \in \mathbb{N}$,

$$f(x) = f[(x - p_n) + p_n] = f(x - p_n)$$

Since f is continuous at x_0, this implies that

$$f(x) = \lim_{n \to \infty} f(x - p_n) = f(x_0)$$

Thus, f is a constant function.

The investigation of periodic functions inevitably leads to a study of groups, as is evident from the following two theorems, whose proofs are elementary.

THEOREM 4 The period set of any function is a group.

COROLLARY 5 Any integral linear combination of periods is also a period. That is, if $k \in \mathbb{N}$, $p_1, \ldots, p_k$ are periods of f, and $n_1, \ldots, n_k$ are integers, then $p = n_1 p_1 + \cdots + n_k p_k$ is also a period of f.

THEOREM 6 If G is any group, then there exists a function f whose period set is precisely G; e.g., take $f = \chi_G$, the characteristic function of G.

Combining Theorems 4 and 6, we obtain

THEOREM 7 A necessary and sufficient condition that a set be the period set of some function is that the set be a group.[1]

Any nonempty set S of real numbers generates a group, e.g., the set of all real numbers x representable in the form

$$x = n_1 x_1 + \cdots + n_k x_k$$

where $k \in \mathbb{N}$, $x_1, \ldots, x_k \in S$, and $n_1, \ldots, n_k \in \mathbb{I}$. For a given number p we denote by $G\langle p \rangle$ the group generated by $\{p\}$; that is,

$$G\langle p \rangle = \{np : n \in \mathbb{I}\}$$

More generally, we denote by $G\langle p_1, \ldots, p_k \rangle$ the group generated by k numbers $p_1, \ldots, p_k$; i.e.,

$$G\langle p_1, \ldots, p_k \rangle = \{n_1 p_1 + \cdots + n_k p_k : n_1, \ldots, n_k \in \mathbb{I}\}$$

We note that the group generated by an infinite set S of power $\mathfrak{m}$ also has the power $\mathfrak{m}$.

DEFINITION Two periods p_1, p_2 of a function f are called distinct if they are not both integral multiples of one period p. In other words, two periods p_1, p_2 of a function f are distinct if there is no period p of f such that $G\langle p_1, p_2 \rangle = G\langle p \rangle$.[2]

More generally, k periods $p_1, \ldots, p_k$ of a function f are distinct if the group $G\langle p_1, \ldots, p_k \rangle$ cannot be generated by fewer than k periods of f.

Note that if $p_1, \ldots, p_k$ are k distinct periods, for $k \geqslant 2$, then they must all be nonzero.

DEFINITION A periodic function is called singly periodic, doubly periodic, or k-ply periodic if its period set is generated by one period, two distinct periods, or k distinct periods, respectively. A periodic function that is not singly periodic is called multiperiodic.

---

[1] Łomnicki (1918, Twierdzenie VIII).
[2] Cf. Tannery and Molk (1893, pp. 141–142) and Jordan (1894, p. 322).

EXAMPLE The functions $f(x) = \sin x$ and $f(x) = x - [x]$, where $[x]$ denotes the largest integer $\leqslant x$, are both singly periodic functions.

EXAMPLE The characteristic function of the group

$$G\langle 1, \sqrt{2} \rangle = \{m + n\sqrt{2} : m,n \in \mathbb{I}\}$$

is doubly periodic.

EXAMPLE If k is a natural number greater than 1 and $p_1, \ldots, p_{k-1}$ are the first $k - 1$ prime numbers, then the characteristic function of the group

$$G\langle 1, p_1, \ldots, p_{k-1} \rangle = \{n_0 + n_1 p_1 + \cdots + n_{k-1} p_{k-1} : n_0, n_1, \ldots, n_k \in \mathbb{I}\}$$

is k-ply periodic.

EXAMPLE The Dirichlet function $f = \chi_\mathbb{Q}$ is a periodic function whose period set $\mathbb{Q}$ has no two distinct periods, and hence f is not k-ply periodic for any natural number k.

In connection with this last example, we have

THEOREM 8 The period set of a periodic function that has no pair of distinct periods is countable.

Proof. Let p be a given nonzero period of such a periodic function. If q is any period, then, since p and q are not distinct, there exist integers m and n, with $n \neq 0$, such that $q/p = m/n$ and hence $q = (m/n)p$. Accordingly, every period must be a rational multiple of p. Therefore, the period set is countable.

As seen in this proof, the ratio of any two nonzero periods that are not distinct is a rational number. We now establish the converse.[1]

THEOREM 9 The ratio of two distinct periods of a periodic function is an irrational number.

Proof. Suppose that p_1, p_2 are two nonzero periods and $p_1/p_2 = m/n$, where m, n are integers that are relatively prime. From number theory we know that there exist integers r, s such that $rm + sn = 1$. Since p_1 and p_2 are periods, the number $p = rp_1 + sp_2$ is also a period and we have

$$p = rp_1 + sp_2 = rp_1 + s\left(\frac{np_1}{m}\right) = (rm + sn)\frac{p_1}{m} = \frac{p_1}{m}$$

so that $p_1 = mp$. Similarly, it is seen that $p_2 = np$. Therefore, p_1 and p_2 are not distinct periods.

[1]Tannery and Molk (1893, p. 144, Note) and Łomnicki (1918, Twierdzenie I).

Burstin investigated periodic functions whose period set is everywhere dense, while Łomnicki investigated multiperiodic functions.[1] Using the next lemma, we show that these classes of periodic functions are identical.[2]

LEMMA 10 If z is any irrational number, then the group

$$G\langle 1,z \rangle = \{m + nz : m,n \in \mathbb{I}\}$$

is an everywhere dense set.

Proof. Let $S = \{m + nz : m,n \in \mathbb{I}\}$. For each natural number n, the half-open interval

$$[nz, nz + 1) = \{x \in \mathbb{R} : nz \leqslant x < nz + 1\}$$

contains exactly one integer k_n. Define $x_n = k_n - nz$ for each $n \in \mathbb{N}$ and set $T = \{x_n : n \in \mathbb{N}\}$. This set is a subset of the half-open interval $[0,1)$ and is an infinite set. For, if $m \neq n$ and we had $x_m = x_n$, then $z = (k_m - k_n)/(m - n)$ is a rational number, contrary to our hypothesis.

Now, according to the Bolzano-Weierstrass Theorem, the set T has at least one limit point. Hence, given any number $\varepsilon > 0$, there exist elements $x_m, x_n \in T$ such that $0 < |x_m - x_n| < \varepsilon$.

Suppose now that a, b are any two positive real numbers with $a < b$. Place $\varepsilon = b - a$ and let x_m, x_n be elements of T such that $0 < |x_m - x_n| < \varepsilon$ and $x_m < x_n$. Choosing N to be the first natural number satisfying $a < N(x_n - x_m)$, we have $a < N(x_n - x_m) < b$. The number $N(x_n - x_m)$ is expressible in the form $p + qz$, where $p,q \in \mathbb{I}$. Therefore, the open interval (a,b) contains an element of the set S.

From this special case one easily obtains the general result that any open interval (a,b) contains an element of S. Thus, S is an everywhere dense set.

THEOREM 11 Every multiperiodic function has arbitrarily small nonzero periods.[3]

Proof. First, consider the case in which the period set contains two nonzero periods p, q whose ratio $z = p/q$ is an irrational number. By Lemma 10, we know that the set

$$E = \{m + nz : m,n \in \mathbb{I}\}$$

is everywhere dense. This implies that the set

$$F = \{qx : x \in E\} = \{mq + np : m,n \in \mathbb{I}\}$$

---

[1] See Burstin (1915) and Łomnicki (1918).
[2] See also Kern (1977).
[3] Jordan (1894, pp. 323–325) and Łomnicki (1918, Twierdzenie II).

is also everywhere dense. But every element of F is a period. Hence, there exist arbitrarily small nonzero periods of the form $mq + np$, with $m,n \in I$. We thus have only to consider the case in which the ratio of every two nonzero periods is a rational number.

Let p_0 be any nonzero period. Then there must exist a period p such that the ratio

$$r = \frac{p}{p_0}$$

is a rational number that is not an integer, since otherwise every period p would be an integral multiple of p_0 and the function would not be multiperiodic. Now, we can place

$$r = k + x$$

where k is an integer (either less than or greater than r) and x is a fraction satisfying

$$0 < |x| \leqslant \frac{1}{2}$$

We then obtain

$$p = kp_0 + xp_0$$

and the number

$$p_1 = p - kp_0 = xp_0$$

as an integral linear combination of two periods, is also a period. Clearly,

$$0 < |p_1| \leqslant \frac{1}{2}|p_0|$$

Similarly, starting with the period p_1 we obtain a period p_2 such that

$$0 < |p_2| \leqslant \frac{1}{2}|p_1|$$

so that

$$0 < |p_2| \leqslant \frac{1}{2^2}|p_0|$$

In general, for each natural number n, there is a period p_n such that

$$0 < |p_n| \leqslant \frac{1}{2^n}|p_0|$$

Hence, there are arbitrarily small periods p_n.

THEOREM 12 A function is multiperiodic if and only if its period set is everywhere dense.

Proof. It follows from Theorem 11 that a multiperiodic function has an everywhere dense period set. On the other hand, the period set of any singly periodic function consists of all integral multiples of one nonzero period and hence is not everywhere dense.

REMARK The fact that multiperiodic functions coincide with the functions having arbitrarily small nonzero periods prompted Łomnicki to call such functions "microperiodic."

As a further consequence of Theorem 11, we have

THEOREM 13 Any group that is not generated by a single element is everywhere dense. In particular, any uncountable group is an everywhere dense set.

We note the following fact concerning the range of singly periodic, continuous functions.[1]

THEOREM 14 If f is a nonconstant, continuous, periodic function that has a least positive period p_0, then the range of f is a closed interval $[a,b]$ and $f(\mathbf{I})$ is everywhere dense in $[a,b]^2$ if and only if p_0 is an irrational number.

This theorem implies, for instance, that the set $\{\sin n : n \in \mathbf{I}\}$ is everywhere dense in $[-1,1]$.

C. The nature of multiperiodic functions

THEOREM 15 A multiperiodic function having at least one point of continuity is a constant function.[3]

Proof. Apply Theorems 3 and 12.

THEOREM 16 A multiperiodic function assumes each of its values on an everywhere dense set.[4]

Proof. Let y be an element of the range of a multiperiodic function f and let x_0 be an element such that $f(x_0) = y$. As noted in the proof of Theorem 3, the

[1]See Mullikin (1976).
[2]A set is everywhere dense in an interval if the set has at least one element in each subinterval of that interval.
[3]See Burstin (1915, Sec. 2, Satz 2), Łomnicki (1918, Twierdzenie III), Mullikin (1976), and Kern (1977, Theorem 1.8).
[4]Łomnicki (1918, Twierdzenie IV).

set $\{p + x_0 : p \in P[f]\}$ is everywhere dense. For every element x of this set we have $f(x) = y$.

We now establish the most important theorem concerning multiperiodic Baire functions. The Lebesgue measure-theoretic version of this theorem and a category version are due to Burstin.[1]

THEOREM 17   Every multiperiodic Baire function is essentially constant.

Proof. From the equality

$$R = \bigcup_{n=1}^{\infty} \{x \in \mathbb{R} : -n \leqslant f(x) < n\}$$

it follows that there exist real numbers a_1 and b_1 such that the set

$$E_1 = \{x \in \mathbb{R} : a_1 \leqslant f(x) < b_1\}$$

is an abundant Baire set. This set E_1 is actually a comeager set. For if p is any period of f, then it is easily seen that $E_1(p) = E_1$. The period set $P[f]$ being everywhere dense by Theorem 12, this equality implies that the set $E_1 = \bigcup_{p \in P[f]} E_1(p)$ is abundant everywhere. Since E_1 is a Baire set, it must thus be a comeager set.

We continue inductively to define a descending sequence $\langle E_n \rangle_{n \in \mathbb{N}}$ of comeager sets in the following manner: Assuming that we have already defined the sets $E_1, \ldots, E_n$, where

$$E_n = \{x \in \mathbb{R} : a_n \leqslant f(x) < b_n\}$$

we choose an increasing sequence $\langle a_{n,k} \rangle_{k \in \mathbb{N}}$ of numbers with

$$a_n = a_{n,1} < a_{n,2} < \cdots < a_{n,k} < \cdots < b_n$$

$$|a_{n,k+1} - a_{n,k}| \leqslant \frac{1}{2}|b_n - a_n|$$

for each $k \in \mathbb{N}$, and

$$b_n = \sup_k a_{n,k}$$

Among the sets

$$E_{n,k} = \{x \in \mathbb{R} : a_{n,k} \leqslant f(x) < a_{n,k+1}\}$$

[1]Burstin (1915, Sec. 2, Satz 2, Satz 4; 1916); see also Łomnicki (1918, Twierdzenie V Wniosek), Hartman and Kershner (1937, fn. p. 815), Agnew (1940), Boas (1953, 1957), Marcus (1959), Semadeni (1964), Dodziuk (1969), Kern (1977, Theorem 1.9), Cignoli and Hounie (1978), and Henle (1980).

for $k \in \mathbb{N}$, we choose the one with smallest index k_0 for which E_{n,k_0} is an abundant set. Such a set necessarily exists, because E_n is an abundant set and $E_n = \bigcup_{k=1}^{\infty} E_{n,k}$. Now, place $a_{n+1} = a_{n,k_0}$, $b_{n+1} = a_{n+1,k_0}$, and define $E_{n+1} = E_{n,k_0}$ so that

$$E_{n+1} = \{x \in \mathbb{R} : a_{n+1} \leqslant f(x) < b_{n+1}\}$$

By an argument similar to that above, we see that E_{n+1} is a comeager set. Because

$$a_1 \leqslant a_2 \leqslant \cdots \leqslant a_n \leqslant a_{n+1} \leqslant \cdots < b_{n+1} < b_n < \cdots < b_2 < b_1$$

and $\lim_{n \to \infty} |b_n - a_n| = 0$, the set $\bigcap_{n=1}^{\infty} [a_n, b_n)$ consists of a single element c. The set $E = \bigcap_{n=1}^{\infty} E_n$ is a comeager set and

$$f(E) = f\left(\bigcap_{n=1}^{\infty} E_n\right) \subset \bigcap_{n=1}^{\infty} f(E_n) \subset \bigcap_{n=1}^{\infty} [a_n, b_n) = \{c\}$$

Therefore, $f(x) = c$ for all $x \in E$.

If f is a multiperiodic function that has the Baire property and is Lebesgue measurable, it follows from Theorem 17 that f has the following two properties:

(i) f is constant, with the possible exception of a set M of the first category.
(ii) f is constant, with the possible exception of a set N of Lebesgue measure zero.

We give an example[1] to show that we need not have $M = N$.

EXAMPLE Let E be any nowhere dense, perfect set of positive Lebesgue measure and let $M = \bigcup_{r \in \mathbb{Q}} E(r)$ be the union of all rational translates of E. The set M is of the first category and is invariant under all rational translations, while its complement N is a set of Lebesgue measure zero that is also invariant under all rational translations. The characteristic function χ_M of M is a multiperiodic function whose period set contains the set $\mathbb{Q}$ of all rational numbers and which has the Baire property and is Lebesgue measurable. This function is constant and equal to 0 except for the set M of the first category. It is also constant and equal to 1 except for the set N of Lebesgue measure zero. However, $M \neq N$.

It follows from Theorems 1, 6, 12, and 13 that there exist $2^{2^{\aleph_0}}$ different multiperiodic functions. As for the power of the set of all multiperiodic Baire functions, we have[2]

---

[1] Burstin (1915, p. 237).
[2] Cf. Burstin (1915, Sec. 3, Satz 1).

THEOREM 18   If there exists a meager set of power of the continuum, then the set of all multiperiodic Baire functions has power $2^{2^{\aleph_0}}$.

Proof. Let E be a meager set of power of the continuum. The set $M = \bigcup_{r \in Q} E(r)$ is then a meager set of power of the continuum that is invariant under all rational translations. Well-ordering M and proceeding by transfinite induction, determine a subset S of M having the power of the continuum such that the difference between any two different elements of S is an irrational number.

If T is any subset of S, then the characteristic function χ_U of the set $U = \bigcup_{r \in Q} T(r)$, as the characteristic function of a meager set, is a Baire function. Furthermore, χ_U is a multiperiodic function whose period set contains the set Q of all rational numbers.

Now, there are $2^{2^{\aleph_0}}$ different subsets T of S. Moreover, if T_1 and T_2 are different subsets of S, then the sets $U_1 = \bigcup_{r \in Q} T_1(r)$ and $U_2 = \bigcup_{r \in Q} T_2(r)$ are also different, since $T_1 - T_2 \subset U_1 - U_2$ and $T_2 - T_1 \subset U_2 - U_1$. Therefore, there are $2^{2^{\aleph_0}}$ different functions χ_U. Since the family of all functions has power $2^{2^{\aleph_0}}$, we conclude that there are precisely $2^{2^{\aleph_0}}$ multiperiodic Baire functions.

In the converse direction we have the following result:

THEOREM 19   Assume CH. If the set of all multiperiodic Baire functions has power $2^{2^{\aleph_0}}$, then there exists a meager set having the power of the continuum.

Proof. If not, then every meager set is countable. The family of all meager sets then has power at most $2^{\aleph_0}$. By virtue of Theorem 17, the set of multiperiodic Baire functions must have power at most $2^{\aleph_0} \cdot 2^{\aleph_0} = 2^{\aleph_0}$.

D.   Groups and Baire sets

It can be seen from Theorems 6, 12, and 17 that if G is a group which is everywhere dense and is a Baire set, then either G or $\mathbb{R} - G$ is a meager set. In fact, a stronger result holds.[1]

THEOREM 20   A group that is a Baire set is either a meager set or identical to $\mathbb{R}$.

Proof. Let G be a group that is a Baire set. If $\mathscr{C}$ is a trivial base, then the only Baire sets are $\varnothing$ and $\mathbb{R}$, so $G = \mathbb{R}$. We thus assume that $\mathscr{C}$ is nontrivial.

[1]Cf. Mazur and Sternbach (1930), Banach (1931; 1932b, Chap. I, Sec. 2), Kuratowski (1933b; 1966, pp. 114–115), and Cignoli and Hounie (1978).

Suppose that G is an abundant set. According to Theorem 7 of Section I, the set G must be an uncountable set and we know by Theorem 13 that G is everywhere dense.

Let x be any element of $\mathbb{R}$. Theorem 2 of Section I implies that $\bigcup_{t \in G} G(t - x)$ is abundant everywhere. It follows from Theorem 3 of Section I that the set $G \cap [\bigcup_{t \in G} G(t - x)]$ is abundant. Let t be an element of G for which $G \cap G(t - x) \neq 0$ and let y be an element of $G \cap G(t - x)$. Then $x + y = z + t$ for some element $z \in G$. This implies that $x = z + t - y$ is an element of G. We conclude that $G = \mathbb{R}$.

As for the existence of uncountable groups that are meager sets, we have

THEOREM 21   Assume that every open interval is a Baire set. If $\mathscr{C}$ satisfies CCC, then there exists an uncountable group which is a meager set.

Proof. The set of all rational linear combinations of elements of the perfect set P of Theorem 22 of Section II is such a group.

COROLLARY 22   Assume that every open interval is a Baire set. If $\mathscr{C}$ satisfies CCC, then there exists a multiperiodic Baire function whose period set is an uncountable meager set.[1]

The first example of an uncountable group that is a set of the first category and has Lebesgue measure zero is apparently the following example of Łomnicki.[2]

EXAMPLE   Let S be the set of all real numbers x representable in the form

$$x = [x] + \sum_{n=1}^{\infty} \frac{a_n}{10^{n!}}$$

where the numbers a_n are decimal digits $0, 1, 2, \ldots, 9$. In addition to all integers and certain rational numbers, the set S contains an uncountable set of transcendental numbers defined by Liouville. The set G of all real numbers that are integral linear combinations of element of S is an uncountable group which is of the first category and has Lebesgue measure zero.

Using the set S of Theorem 14 in Section III, which is a group, we obtain

THEOREM 23   There exists a group that is not a Baire set.[3]

We also have the following existence theorem.[4]

[1]Cf. Łomnicki (1918, Sec. 9).
[2]Łomnicki (1918, Sec. 9).
[3]Cf. Łomnicki (1918, Sec. 8), Sierpiński (1920b), Korovkin (1969, Teorema 4), and Korec (1972).
[4]Smítal (1968, Theorem 9) and Erdös (1979).

THEOREM 24 Assume CH and that $\mathscr{C}$ is linearly invariant. If $\mathscr{C}$ satisfies CCC and has power at most $2^{\aleph_0}$, then there exists an uncountable group which is a rare set.

Proof. If $\mathscr{C}$ is trivial, then $\mathbb{R}$ itself will be an uncountable rare set. Assume therefore that $\mathscr{C}$ is nontrivial.

According to Theorem 9 of Section I and Theorem 1 of Chapter 2, Section I, the family of all meager $\mathscr{K}_{\delta\sigma}$-sets has power $2^{\aleph_0}$. Let

$$E_1, E_2, \ldots, E_\beta, \ldots \qquad (\beta < \Omega)$$

be well-ordering of these sets. We define by transfinite induction an ascending transfinite sequence $\langle G_\alpha \rangle_{\alpha < \Omega}$ of denumberable groups.

Choose a nonzero element $x_1 \in \mathbb{R}$ and let G_1 be the group generated by x_1. Assume that for $1 < \alpha < \Omega$, the elements x_β and groups G_β have been defined for all ordinal numbers $\beta < \alpha$. The set $H_\alpha = \bigcup_{\beta < \alpha} G_\beta$ is a denumerable set. For each $\beta < \alpha$, each $h \in H_\alpha$, and each nonzero integer m, the set

$$T_{\beta h m} = \frac{1}{m} E_\beta(-h)$$

is meager. As there are only countably many such sets, we can choose an element $x_\alpha \in \mathbb{R} - H_\alpha$ that belongs to none of the sets $T_{\beta h m}$. We then define G_α to be the group generated by the denumerable set $H_\alpha \cup \{x_\alpha\}$. We note that because H_α is itself a group, G_α is comprised of all elements x of the form $x = h + m x_\alpha$, where $h \in H_\alpha$ and m is an integer.

Having thus determined the transfinite sequence $\langle G_\alpha \rangle_{\alpha < \Omega}$, we set $G = \bigcup_{\alpha < \Omega} G_\alpha$. It is a simple matter to show G is an uncountable group.

Let us assume that G has an uncountable meager subset S. Then, according to Theorem 5 of Chapter 1, Section II, there is an ordinal number $\beta < \Omega$ for which $S \subset E_\beta$. Since the groups G_α are denumerable sets and S is uncountable, there is an index $\alpha_0 > \beta$ such that $S \cap (G_{\alpha_0} - H_{\alpha_0}) \neq \varnothing$. Choose $x \in S \cap (G_{\alpha_0} - H_{\alpha_0})$. Then $x = h + m x_{\alpha_0}$, where $h \in H_{\alpha_0}$ and m is an integer. As $x \notin H_{\alpha_0}$, we must have $m \neq 0$. Now, from $x \in S$ we obtain $x \in E_\beta$ and $x_{\alpha_0} = (1/m)(x - h) \in (1/m) E_\beta(-h) = T_{\beta h m}$. This contradicts the choice of x_{α_0}. We conclude that G has no uncountable meager sets; i.e., G is a rare set.

NOTE In addition to the foregoing results, we note: For $0 \leqslant \alpha \leqslant 1$, there exists a group that has Hausdorff dimension α.[1]

THEOREM 25 If G is a group that is not a Baire set, then neither G nor $\mathbb{R} - G$ contains any abundant Baire set.

[1] See Erdös and Volkmann (1966) and Foran (1974).

242   Chapter 6

Proof. The assertion being obvious when $\mathscr{C}$ is trivial, we assume that $\mathscr{C}$ is nontrivial. By Theorem 7 of Section I and Theorem 13, the set G is uncountable and everywhere dense.

Let S be an abundant Baire set. According to Theorem 3 of Section I, the set $E = \bigcup_{t\in G} S(t)$ is a comeager set. If $S \subset G$, then $E \subset G$ and G must be a comeager set. If $S \subset \mathbb{R}$—G, then $E \subset \mathbb{R}$—G and G must be a meager set. Since G is not a Baire set, neither the inclusion $S \subset G$ nor the inclusion $S \subset \mathbb{R}$—G is valid.

THEOREM 26 If G is a group that is a proper subset of $\mathbb{R}$, then every subset of G that is a Baire set is a meager set.

Proof. Apply Theorem 20 and the preceding theorem.

The period set of a function being a group, this theorem and Theorem 2 yield

THEOREM 27 A function is constant if and only if its period set contains an abundant Baire set.[1]

THEOREM 28 The period set of any periodic Baire function is a Baire set.

Proof. If f is a constant periodic function, then the assertion follows, because the period set will then be $\mathbb{R}$. Assume therefore that f is a nonconstant, periodic Baire function. This assumption implies that $\mathscr{C}$ is nontrivial, since the constant functions are the only Baire functions when $\mathscr{C}$ is trivial.

If f is singly periodic, then the period set is a countable set which, according to Theorem 7 of Section I, is a meager set.

Suppose that f is multiperiodic. By Theorem 17, there is a constant c such that the set

$$E = \{x \in \mathbb{R} : f(x) \neq c\}$$

is meager. Let P denote the period set of f and let e be an element of E. Using the periodicity of f we obtain $P(e) \subset E$. This implies that P is a meager set.

COROLLARY 29 The period set of any nonconstant, periodic Baire function is a meager set.[2]

We note that the characteristic function of the set S of Theorem 14 of Section III is a multiperiodic function that is not a Baire function and whose period set is an uncountable abundant set that is not a Baire set.[3]

---

[1] Cf. Łomnicki (1918, Twierdzenie VI) and Korovkin (1969, Teorema 1).
[2] Cf. Łomnicki (1918, Twierdzenie VI).
[3] Cf. Łomnicki (1918, Sec. 8).

We now give an example[1] of a multiperiodic function that is not a Baire function, but whose period set is a Baire set for any nontrivial base $\mathscr{C}$. Thus, the converse of Theorem 28 is not true.

EXAMPLE Let ϕ be a one-to-one mapping of the family of all Vitali equivalence classes onto the set of all real numbers. For each $x \in \mathbb{R}$ define

$$f(x) = \phi(E_x)$$

where E_x is the Vitali equivalence class containing x. If x and y are any two real numbers, then we have $f(x) = f(y)$ if and only if x and y are equivalent; i.e., $y = x + r$ for some rational number r. The function f is a multiperiodic function that is constant on each Vitali equivalence class and whose period set is the set $\mathbb{Q}$ of all rational numbers. The period set is thus a meager set for any nontrivial base $\mathscr{C}$. Because f is not constant on any comeager set, it follows from Theorem 17 that f is not a Baire function for any base $\mathscr{C}$.

REMARK There being $(2^{\aleph_0})^{2^{\aleph_0}} = 2^{2^{\aleph_0}}$ possible choices for the function ϕ and different choices yielding different functions f, there are $2^{2^{\aleph_0}}$ such functions f.

E. Essential periodicity

DEFINITION A number p is said to be essentially a period, or an essential period, for a function f if there is a meager set M such that

$$f(x + p) = f(x)$$

for all $x \notin M$.

The set $P^*[f]$ of all essential periods for a function f is a group. If $P^*[f]$ is not generated by a single element, then we say that f is essentially multiperiodic.

We have the following simple extension of Theorem 17.

THEOREM 30 Every essentially multiperiodic Baire function is essentially constant.[2]

Proof. Repeat the argument used to prove Theorem 17 with the following alteration to show that the abundant sets

$$E = \{x \in \mathbb{R} : a \leqslant f(x) < b\}$$

are comeager:

[1] Cf. Burstin (1915, p. 234 fn.) and Kern (1977, Example 2.2).
[2] Cf. Agnew (1940), Boas (1953, 1957), and Marcus (1959).

Choose a countable, everywhere dense set $D \subset P^*[f]$. For each $p \in D$ we have $E(p) \prec E$. Hence, $\bigcup_{p \in D} E(p) \prec E$. This implies that E is a comeager set.

If f is a function with period set $P[f] = \mathbb{R}$, then we know by Theorem 2 that f is a Baire function. In general, a similar result for an essential period set $P^*[f]$ is not valid.[1]

THEOREM 31   Assume CH. If $\mathscr{C}$ is also a perfect base, then there exists a function f such that $P^*[f] = \mathbb{R}$ and f is not a Baire function.

Proof. It follows from Theorems 23 and 11 of Chapter 5, Section II that there exists a set S that is not a Baire set and which is essentially transformed into itself by each translation. The characteristic function of S is a suitable function f.

V.   CONGRUENT SETS

A.   Translation homogeneous sets

DEFINITION   A set S is called a translation homogeneous set[2] if for any two elements $x, y \in S$, the set S remains invariant under the translation taking y into x; i.e., $S(x - y) = S$.

This notion of homogeneity may be characterized in various alternative ways.[3]

THEOREM 1   The following statements are equivalent for a set S:
   (i) For all elements $x, y \in S$, we have $S(x - y) = S$.
   (ii) For all elements $x, y, z \in S$, we have $x + (y - z) \in S$.
   (iii) S is a translate of some group.
   (iv) Each translate of S either coincides with S or is disjoint from S.

Proof. (i) $\Rightarrow$ (ii). From $S(z - y) = S$ and $x \in S$ we obtain $x = t + (z - y)$ for some element $t \in S$. In fact, $t = x + (y - z)$.

   (ii) $\Rightarrow$ (iii). Let t be any given element of S and set $G = \{x - t : x \in S\}$. Using condition (ii) it is readily verified that G is a group and $S = G(t)$.

   (iii) $\Rightarrow$ (iv). Assume that $S = G(t)$, where G is a group and $t \in \mathbb{R}$. Suppose that $r \in \mathbb{R}$. We have $G(r) = G$ if $r \in G$ and $G(r) \cap G = \varnothing$ if $r \notin G$. If $G(r) = G$, then $S(r) = G(t + r) = G(r + t) = [G(r)](t) = G(t) = S$. If $G(r) \cap G = \varnothing$, then $S(r) \cap S = G(t + r) \cap G(t) = G(r + t) \cap G(t) = [G(r) \cap G](t) = \varnothing(t) = \varnothing$.

   (iv) $\Rightarrow$ (i). If x, y are any elements of S, then, by condition (iv), either

---

[1] Sierpiński (1932b).
[2] Cf. Borel (1946).
[3] Cf. Mycielski and Sierpiński (1966).

$S(x - y) = S$ or $S(x - y) \cap S = \varnothing$. However, the latter equality is ruled out since x is an element of $S(x - y) \cap S$.

REMARK It follows from condition (iv) that if S is a translation homogeneous set, then any two translates of S are either identical or disjoint.

From Theorems 20 and 26 in Section IV, and condition (iii), we obtain

(1) If S is a translation homogeneous Baire set that is a proper subset of $\mathbb{R}$, then S is a meager set.
(2) If S is a translation homogeneous set that is a proper subset of $\mathbb{R}$, then every subset of S that is a Baire set is a meager set.

B. Decomposition theorems

Ruziewicz proved that if $\mathfrak{m}$ is a cardinal number and $\aleph_0 \leqslant \mathfrak{m} \leqslant 2^{\aleph_0}$, then $\mathbb{R}$ can be decomposed into $\mathfrak{m}$ disjoint congruent sets that are not Lebesgue measurable.[1] More generally, we have[2]

THEOREM 2 For any infinite cardinal number $\mathfrak{m} \leqslant 2^{\aleph_0}$, the real line can be decomposed into $\mathfrak{m}$ disjoint congruent, translation homogeneous, Bernstein sets.

Proof. Let

$$x_1, x_2, \ldots, x_\alpha, \ldots \qquad (\alpha < \Lambda)$$

$$P_1, P_2, \ldots, P_\alpha, \ldots \qquad (\alpha < \Lambda)$$

be well-orderings of $\mathbb{R}$ and of all perfect subsets of $\mathbb{R}$, respectively. Proceeding by transfinite induction, determine a Hamel basis B having the property that for each ordinal number $\xi < \Lambda$, there are two different elements a_ξ, b_ξ, in B both of which belong to P_ξ.

Let A be a subset of the set $\{a_\xi : \xi < \Lambda\}$ of power $\mathfrak{m}$, let E be the set of all rational linear combinations of elements in A, and let S be the set of all rational linear combinations of elements in $B-A$. The set E has power $\mathfrak{m}$ and we have the decomposition

$$\mathbb{R} = \bigcup_{x \in E} S(x)$$

of $\mathbb{R}$ into $\mathfrak{m}$ disjoint congruent sets. In view of the fact that the set S is a group, the sets $S(x)$ are all translation homogeneous sets.

---

[1] See Ruziewicz (1924), Rindung (1950), Erdös and Marcus (1957), Mycielski and Sierpiński (1966). Compare this result with Theorems 2 and 15 of Section III.
[2] Cf. Halperin (1951) and Erdös and Marcus (1957).

Since S contains all the elements b_ξ with $\xi < \Lambda$, the set S has at least one element in common with each perfect set. Consequently, the sets $S(x)$ also have at least one element in common with each perfect set. Therefore, the sets $S(x)$ are Bernstein sets.

In regard to finite decompositions, Erdös and Marcus have proved that $\mathbb{R}$ cannot be decomposed into a finite number of nonempty, disjoint congruent, translation homogeneous sets.[1]

None of the sets occurring in the decomposition given by Theorem 2 contains a perfect set. However, we can also obtain a decomposition into sets that do have perfect subsets.[2]

THEOREM 3   For any infinite cardinal number $\mathfrak{m} \leqslant 2^{\aleph_0}$, the real line can be decomposed into $\mathfrak{m}$ disjoint congruent, translation homogeneous sets, each of which contains a perfect set.

Proof. Let B be a Hamel basis that contains a perfect set, the existence of which was established in Theorem 11 in Section III, and let Q and R be two disjoint perfect subsets of P. Let M be a subset of R of power $\mathfrak{m}$, let S be the set of all rational linear combinations of elements of $B-M$, and let T be the set of all rational linear combinations of elements of M. Then

$$\mathbb{R} = \bigcup_{t \in T} S(t)$$

is a decomposition of the desired form.

In addition, we have the following theorem[3] whose proof appears to require the assumption of the Continuum Hypothesis.

THEOREM 4   Assume CH and that every open interval is a Baire set. If $\mathscr{C}$ satisfies CCC and has power at most $2^{\aleph_0}$, then for every infinite cardinal number $\mathfrak{m} \leqslant 2^{\aleph_0}$, the real line can be decomposed into $\mathfrak{m}$ disjoint congruent, translation homogeneous sets, each of which contains a perfect set and is abundant everywhere.

Proof. By Theorem 22 in Section II, there is a rationally independent, perfect set P for which the set R_1 of all rational linear combinations of elements of P is a meager set.

Let

(1) $E_1, E_2, \ldots, E_\alpha, \ldots$ $(\alpha < \Lambda)$

---

[1]See Erdös and Marcus (1957, Théorème 2) and Mycielski & Sierpiński (1966, Théorème 2).
[2]Cf. Erdös and Marcus (1957, Théorème 4).
[3] [1]Cf. Erdös and Marcus (1957, Théorème 5).

be a well-ordering of all abundant $\mathscr{C}_{a\delta}$-sets. (That the family of all abundant $\mathscr{C}_{a\delta}$-sets has the power of the continuum is a consequence of Theorem 1 of Chapter 2, Section I.) In addition, let

(2) $x_1, x_2, \ldots, x_{a_1}, \ldots$ $(\alpha < \Lambda)$

be all well-ordering of all elements of $\mathbb{R}-P$. We proceed by transfinite induction to determine a Hamel basis B such that $P \subset B$, and for each ordinal number $\alpha < \Lambda$, the basis B contains at least two elements belonging to the set E_a that do not belong to P.

The set R_1 being meager, E_1-R_1 is an abundant set. We define y_{11} to be the first term of the transfinite sequence (2) that is an element of E_1-R_1. The set S_1 of all rational linear combinations of elements of the set $P \cup \{y_{11}\}$ is representable in the form

$$S_1 = \bigcup_{t \in Q} R_1(ty_{11})$$

and is thus a meager set. Hence, E_1-S_1 is an abundant set. Define y_{21} to be the first term of (2) which is an element of E_1-S_1, and let T_1 denote the set of all rational linear combinations of elements of $P \cup \{y_{11}, y_{21}\}$. By virtue of the equality

$$T_1 = \bigcup_{t \in Q} S_1(ty_{21})$$

the set T_1 is a meager set.

Assume that $\alpha < \Lambda$, that the elements $y_{1\beta}, y_{2\beta}$ have already been defined for all ordinal numbers $\beta < \alpha$, and that the set T_β consisting of all rational linear combinations of elements of the set $P \cup \{y_{i\xi} : i \in \{1,2\} \text{ and } \xi \leqslant \beta\}$ is meager. The set R_α of all rational linear combinations of elements of $P \cup \{y_{i\beta} : i \in \{1,2\} \text{ and } \beta < \alpha\}$ is representable as

$$R_\alpha = \bigcup_{\beta < \alpha} T_\beta$$

and since we are assuming CH, it follows that R_α is a meager set. In the manner described in the preceding paragraph, we define elements $y_{1\alpha}, y_{2\alpha}$ belonging to the set $E_\alpha-R_\alpha$, which is a subset of $E_\alpha-P$, and let T_α be the set consisting of all rational linear combinations of elements of the set $P \cup \{y_{i\beta} : i \in \{1,2\} \text{ and } \beta \leqslant \alpha\}$. The set

$$B = P \cup \{y_{i\alpha} : i \in \{1,2\} \text{ and } \alpha < \Lambda\}$$

will then be a Hamel basis with the desired properties.

Let M be a subset of the set $\{y_{2\alpha} : \alpha < \Lambda\}$ that has power $\mathfrak{m}$, let S be the set of all rational linear combinations of elements belonging to the set $B-M$, and let T be the set of all rational linear combinations of elements of M.

Clearly, T has power $\mathfrak{m}$. We proceed to show that the set S, which contains the perfect set P, is abundant everywhere.

Suppose, to the contrary, that S is not abundant everywhere. Then there is a region A such that $A \cap S$ is a meager set. The set $A—(A \cap S)$, which is a subset of $\mathbb{R}—S$, will then be an abundant Baire set. According to condition (iv) of Theorem 7 of Chapter 1, Section III, the set $A—(A \cap S)$, hence also $\mathbb{R}—S$, must contain an abundant $\mathscr{C}_{\sigma\delta}$-set. But this is impossible, since S contains at least one element $y_{1\alpha}$ belonging to E_{α} for each $\alpha < \Lambda$, and all abundant $\mathscr{C}_{\sigma\delta}$-sets occur among the terms of the transfinite sequence (1). Therefore, S must be abundant everywhere.

The decomposition

$$\mathbb{R} = \bigcup_{t \in T} S(t)$$

will then have the required properties.

C.   The number of sets congruent to a given set

For any set S we denote by $\tau(S)$ the family of all subsets of $\mathbb{R}$ that are congruent to S. As is readily seen,

$$\tau(S) = \{S(t) : t \in \mathbb{R}\}$$

It is clear that $\tau(\mathbb{R}) = \{\mathbb{R}\}$, and, in fact, $S = \mathbb{R}$ is the only nonempty set for which $\tau(S)$ consists of a single set. For any nonempty, proper subset of $\mathbb{R}$, Sierpiński has shown that $\tau(S)$ must always be infinite.[1] We also have the following theorems concerning the power of the family $\tau(S)$.[2]

THEOREM 5   For every infinite cardinal number $\mathfrak{m} \leqslant 2^{\aleph_0}$, there exists a set S for which the family $\tau(S)$ has the power $\mathfrak{m}$.

Proof. The theorem being obviously true for $\mathfrak{m} = 2^{\aleph_0}$, we assume that $\mathfrak{m} < 2^{\aleph_0}$. Let F be any field of real numbers of power $\mathfrak{m}$. For example, if E is any set of power $\mathfrak{m}$, then we can take F to be the subfield of $\mathbb{R}$ generated by E.

According to Theorem 9 of Section III, there exists a Hamel basis B relative to F. Let b be any fixed element of B. Define S to be the set of all real numbers x representable in the form

$$x = r_1 b_1 + \cdots + r_k b_k$$

where $k \in \mathbb{N}$, $b_1, \ldots, b_k \in B—\{b\}$, and $r_1, \ldots, r_k \in F$; that is, S consists of all

---

[1] Sierpiński (1948, Théorème 1).
[2] Cf. Sierpiński (1948, Théorèmes 2–6; 1954, Sec. 2).

numbers whose representation with respect to the base B does not contain the element b with a nonzero coefficient in F. The family

$$\mathcal{F} = \{S(rb):r \in F\}$$

has power $\mathfrak{m}$ and consists of disjoint sets, each of which is congruent to S.

Suppose now that T is any set congruent to S, say $T = S(t)$ for a given real number t. The number t is uniquely representable either in the form $t = r_0 b$, where $r_0 \in F$, or in the form

$$t = r_0 b + r_1 b_1' + \cdots + r_m b_m'$$

where $m \in \mathbb{N}$, $b_1', \ldots, b_m' \in B - \{b\}$, $r_0 \in F$, and $r_1, \ldots, r_m$ are nonzero elements of F. In either case it is seen that $S(t) = S(r_0 b)$, so that T belongs to the family $\mathcal{F}$. Therefore, $\mathcal{F} = \tau(S)$.

THEOREM 6 If S is a nonempty set such that $\tau(S)$ is countable, then S is an abundant set.

Proof. If $\tau(S)$ is countable, then there exists a sequence $\langle S_n \rangle_{n \in \mathbb{N}}$ of sets, not necessarily distinct, such that $\tau(S) = \{S_n : n \in \mathbb{N}\}$. In view of the fact that $\bigcup_{n=1}^{\infty} S_n = \mathbb{R}$, at least one of the sets S_n must be an abundant set. But each set S_n being congruent to S, this means that S is an abundant set.

THEOREM 7 If S is an abundant set and $\tau(S)$ has power $< 2^{\aleph_0}$, then S is abundant everywhere.

Proof. For each real number x, the set $S(x)$ belongs to the family $\tau(S)$. Since $\tau(S)$ has power $< 2^{\aleph_0}$, there exists a set $S_0 \in \tau(S)$ and an uncountable set E of real numbers such that $S(x) = S_0$ for every $x \in E$.

If x and y are any two elements of E, then it follows from the equality $S(x) = S_0 = S(y)$ that $S(x - y) = S$. That is, $S(t) = S$ for every number t belonging to the difference set of E. This difference set, being uncountable, contains arbitrarily small nonzero numbers t. In conjunction with the fact that $S(t) = S$ implies that $S(nt) = S$ for every integer n, this implies that $S(t) = S$ for an everywhere dense set D of numbers t. Applying Theorem 2 of Section I, we conclude that S is abundant everywhere.

THEOREM 8 If S is a set for which $\tau(S)$ is denumerable, then S is not a Baire set.

Proof. If $\tau(S)$ is denumerable, then S is a nonempty, proper subset of $\mathbb{R}$. Let T be the complement of S. From the equality $\mathbb{R} = S \cup T$ it is seen that for any real numbers x and y, we have $S(x) = S(y)$ if and only if $T(x) = T(y)$. This implies that $\tau(T)$ has the same power as $\tau(S)$. Hence, according to Theorem 7, both S and its complement T are abundant everywhere. Therefore, S is not a Baire set.

THEOREM 9 If $2^{\aleph_0} = \aleph_1$ and S is a nonempty, proper subset of $\mathbb{R}$ that is a Baire set, then there exist continuum many different sets congruent to S.

Proof. Assume to the contrary that $\tau(S)$ has power $< 2^{\aleph_0}$ and let T denote the complement of S. As seen in the proof of the preceding theorem, $\tau(T)$ has the same power as $\tau(S)$. From the Continuum Hypothesis, Theorem 6, and Theorem 8, we see that both S and T are abundant everywhere. Therefore, S is not a Baire set.

Concerning the nature of the sets occurring in a decomposition of $\mathbb{R}$ into congruent, translation homogeneous, abundant sets, we have[1]

THEOREM 10 If m is any infinite cardinal number $< 2^{\aleph_0}$ and $\mathbb{R}$ is decomposed into a family $\mathcal{M}$ consisting of m disjoint congruent, translation homogeneous, abundant sets, then each set in $\mathcal{M}$ is abundant everywhere.

Proof. Let S be any set in $\mathcal{M}$. We know from the homogeneity that any two translates of S are either identical or disjoint. According to the congruence assumption, each set in $\mathcal{M}$ is a translate of S. Combining these facts with the fact that the sets in $\mathcal{M}$ constitute a decomposition of $\mathbb{R}$, we see that the set

$$\tau(S) = \{S(t) : t \in \mathbb{R}\}$$

must have power m. Hence, according to Theorem 7, the set S must be abundant everywhere.

[1]Cf. Erdös and Marcus (1957, Théorème 3).

Bibliography

Aarts, J. M. and Lutzer, D. J. Completeness properties designed for recognizing Baire spaces. Dissertationes Math. (Rozprawy Mat.) 116, 1974, 48p.

Abian, A. The outer and inner measures of a nonmeasurable set. Boll. Un. Mat. Ital. (4) 3, 1970, 555–558.

Aczél, J. Lectures on Functional Equations and Their Applications. Academic Press, New York, 1966.

Agnew, R. P. On continuity and periodicity of measurable functions. Ann. of Math. 41, 1940, 727–733.

Alas, O. F. On Blumberg's theorem. Atti Accad. Naz. Lincei Rend. Cl. Sci. Fis. Mat. Natur. (8) 60, 1976, 579–582.

Albanese, B. Sugli insiemi di non-misurabilità secondo Carathéodory. Atti Acad. Naz. Lincei, Rend. Cl. Sci. Fis. Mat. Natur. (8) 57, 1974, 156–158.

Alexandrow, W. Elementare Grundlagen für die Theorie des Masses. Dissertation (1915) University of Zurich, Gerbrüder Leeman, Zurich, 1916, 85p.

Alexiewicz, A. and Orlicz, W. Remarque sur l'équation fonctionelle $f(x + y) = f(x) + f(y)$. Fund. Math. 33, 1945, 314–315.

Bagemihl, F. A note on Scheeffer's theorem. Michigan Math. J. 2, 1954, 149–150.

—— Concerning the continuum hypothesis and rectilinear sections of spatial sets. Arch. Math. (Basel) 10, 1959a, 360–362.

—— Some results connected with the continuum hypothesis. Z. Math. Logic Grundlag. Math. 5, 1959b, 97–116.

Baire, R. Sur les fonctions discontinues qui se rattachent aux fonctions continues. C. R. Acad. Sci. Paris 126, 1898, 1621–1623.

—— Sur les fonctions de variables réelles. Ann. Mat. Pura Appl. (3) 3, 1899a, 1–123.

—— Sur la théorie des ensembles. C. R. Acad. Sci. Paris 129, 1899b, 946–949.

Ballew, D. W. Not every function has a measurable range. Math. Student 42, 1974, 100.

Banach, S. Sur l'équation fonctionnelle $f(x + y) = f(x) + f(y)$. Fund. Math. 1, 1920, 123–124.

—— Théorème sur les ensembles de première catégorie. Fund. Math. 16, 1930, 395–398.

——Über metrische Gruppen. Studia Math. 3, 1931, 101–113.

—— Sur les transformations biunivoques. Fund. Math. 19, 1932a, 10–16.

—— Théorie des opérations linéaires, Monografie Matematyczne, Vol. 1, Hafner Press, New York, 1932b. (2nd. ed., Chelsea, New York, 1955.)

—— Stefan Banach, oeuvres (avec des commentaires). In Travaux sur les fonctions réelles et sur les séries orthogonales (ed. S. Hartman and E. Marczewski). Państwowe Wydawnictwo Naukowe, Warsaw, 1967.

Banach, S. and Kuratowski, K. Sur une généralisation du problème de la mesure. Fund. Math. 14, 1929, 127–131.

Bartle, R. G. An extension of Egorov's theorem. Amer. Math. Monthly 87, 1980, 628–633.

Bendixson, I. Quelques théorèmes de la théorie des ensembles de points. Acta Math. 2, 1883, 415–429.

Bernstein, F. Zur Theorie der trigonometrischen Reihen. Sitzungsber. Sächs. Akad. Wiss. Leipzig. Math.-Natur. Kl. 60, 1908, 325–338.

Besicovitch, A. S. Concentrated and rarified sets of points. Acta Math. 62, 1934, 289–300.

Bettazzi, R. Su una corrispondènza fra un gruppo di punti ed un continuo ambedùe lineari. Ann. Mat. Pura Appl. (2) 16, 1888, 49–60.

Bhaskara Rao, K. P. S. and Bhaskara Rao, M. A category analogue of the Hewitt-Savage zero-one law. Proc. Amer. Math. Soc. 44, 1974, 497–499.

Bhaskara Rao, K. P. S. and Pol, R. Topological zero-one law. Colloq. Math. 39, 1978, 13–23.

Billingsley, P. Probability and Measure, 2nd ed. Wiley, New York, 1986.

Birkhoff, G. Lattice Theory, rev. ed., Amer. Math. Soc. Colloq. Publ. 25. American Mathematical Society, Providence, R.I., 1948.

Blakney, S. S. Lusin's theorem in an abstract space. Mat. Časopis Sloven. Akad. Vied. 19, 1969, 249–251.

Blumberg, H. New properties of all real functions. Trans. Amer. Math. Soc. 24, 1922, 113–128.

———— The measurable boundaries of an arbitrary function. Acta Math. 65, 1935, 263–282.

Boas, R. P. Functions which are odd about several points. Nieuw Arch. Wisk. (3) 1, 1953, 27–32.

———— Functions which are odd about several points: addendum. Nieuw Arch. Wisk. (3) 5, 1957, 25.

Bois Reymond, P. du. Die allgemeine Functionentheorie. H. Laupp, Tübingen, Germany, 1882; translated by G. Milhaud and A. Girot as Théorie générale des fonctions, Imprimerie Niçoise, Nice, 1887.

Borel, E. Un théorème sur les ensembles mesurables. C. R. Acad. Sci. Paris 137, 1903, 966–967.

———— Les ensembles homogènes. C. R. Acad. Sci. Paris 222, 1946, 617–618.

Bourbaki, N. Eléments de mathématiques XIII. Première partie: Les structures fondamentales de l'analyse. Livre VI. Intégration, Actualités Scientifiques et Industrielles No. 1175. Hermann, Paris 1952.

Bradford, J. C. and Goffman, C. Metric spaces in which Blumberg's theorem holds. Proc. Amer. Math. Soc. 11, 1960, 667–670.

Braun, S., Kuratowski, K. and Szpilrajn, E. Annexe, Fund. Math. 1, 1920, éd. nouvelle 1937, 225–254.

Braun, S. and Sierpiński, W. Sur quelques propositions équivalentes à l'hypothèse du continu. Fund. Math. 19, 1932, 1–7. [Sierpiński O.C. III, 84–89.]

Broggi, U. Sur un théorème de M. Hamel. Enseignement Math. 9, 1907, 385–387.

Brouwer, L. E. J. Perfect sets of points with positively-irrational distances. Kon. Akad. van Weten. Amsterdam 27, 1924, 487. Perfecte punktverzamelingen met positief-irrationale afstanden. Amst. Ak. Versl. 33, 1924, 81.

Brown, J. B. Metric spaces in which a strengthened form of Blumberg's theorem holds. Fund. Math. 71, 1971, 243–253.

Brown, J.B. A measure theoretic variant of Blumberg's theorem. Proc. Amer. Math. Soc. 66, 1977, 266–268.

———— Variations on Blumberg's theorem. Real Anal. Exchange 9, 1983–1984, 123–137.

———— Continuous restrictions of Marczewski measurable functions. Real Anal. Exchange 11, 1985–1986, 64–71.

Brown, J. B. and Cox, G. V. Classical theory of totally imperfect spaces. Real Anal. Exchange 7, 1981–1982, 185–232.

Brown, J. B. and Prikry, K. Variations on Lusin's theorem. Trans. Amer. Math. Soc. 302, 1987, 77–86.

Burstin, C. Eigenschaften messbarer und nichtmessbarer Mengen. Sitzungsber. Kaiserlichen Akad. Wiss. Math.-Natur. Kl. Abteilung IIa., 123, 1914, 1525–1551.

———— Über eine spezielle Klasse reeller periodischer Funktionen. Monatsh. Math. 26, 1915, 229–262.

———— Berichtigung der Arbeit über eine spezielle Klasse reeller periodischer Funktionen. Monatsh. Math. 27, 1916a, 163–165.

———— Die Spaltung des Kontinuums in c im L. Sinne nichtmessbare Mengen. Sitzungsber. Kaiserlichen Akad. Wiss. Math.-Natur. Kl. Abteilung IIa, 125, 1916b, 209–217.

Cantor, G. Grundlagen einer allgemeinen Mannichfaltigkeitslehre. Ein mathematisch-philosophischer Versuch in der Lehre des Unendlichen. B.G. Teubner, Leipzig, 1883.

———— Gesammelte Abhandlungen. Mathematischen und philosophischen Inhalts (ed. E. Zermelo). Georg Olms, Hildesheim, West Germany, 1966.

Carathéodory, C. Über das lineare Mass von Punktmengen—Eine Verallgemeinerung des Längenbegriffs. Göttingen Nachr., 1914, 404–426.

Cholewa, P. Sierpiński-Erdös duality principle in the abstract Baire category theory. Manuscript, 1982, 12p.

Choquet, G. Ensembles singuliers et structures des ensembles mesurables pour les mesures de Hausdorff. Bull. Soc. Math. France 14, 1946, 1–14.

Christensen, J. P. R. Borel structures and a topological zero-one law. Math. Scand. 29, 1971, 245–255.

———— Borel structures and a topological zero-one law. Papers from the "Open House for Probabilists" (Math. Inst., Aarhus Univ., Aarhus, 1971), Various Publ. Ser., No. 21. Math. Inst., Aarhus University, Aarhus, Denmark, 1972, pp. 32–46.

Cignoli, R. and Hounie, J. Functions with arbitrarily small periods. Amer. Math. Monthly 85, 1978, 582–584.

Cohen, L. W. A new proof of Lusin's theorem. Fund. Math. 9, 1927, 122–123.

Covington, E. A case history of the evolution of mathematical ideas. M.S. degree project, California State Polytechnic University, Pomona, Calif., 1986.

Darst, R. B. On measure and other properties of a Hamel basis. Proc. Amer. Math. Soc. 16, 1965, 645–646.

Davies, R. O. Two counter-examples concerning Hausdorff dimensions of projections. Colloq. Math. 42, 1979, 53–58.

Denjoy, A. Mémoire sur les nombres dérivées des fonctions continues. J. Math. (7) 1, 1915, 105–240.

Dodziuk, J. O funkcjach mierzalnych okresowych [On measurable periodic functions]. Wiadom. Mat. 11, 1969, 13–14.

Douwen, E. K. Van, Tall, F. D., and Weiss, W. A. R. Nonmetrizable hereditarily Lindelöf spaces with point-countable bases from CH. Proc. Amer. Math. Soc. 64, 1977, 139–145.

Dressler, R. E. and Kirk, R. B. Non-measurable sets of reals whose measurable subsets are countable. Israel J. Math. 11, 1972, 265–270.

Eames, W. On the translates of sets of the second category. Math. Sem. Notes Kobe Univ. 5, 1977, 359–360.

Eggleston, H. G. A measureless one-dimensional set. Proc. Cambridge Philos. Soc. 50, 1954, 391–393.

——— On measureless sets. Proc. London Math. Soc. (3) 8, 1958, 631–640.

Egorov, D. F. Sur les suites de fonctions mesurables. C. R. Acad. Sci. Paris 152, 1911, 244–246.

Eilenberg, S. Remarques sur les ensembles et les fonctions relativement mesurables. Spraw. Tow. Nauk. Warszaw. Wyd. III. 25, 1932, 93–98.

Emeryk, A., Frankiewicz, R., and Kulpa, W. On functions having the Baire property. Bull. Acad. Polon. Sci. Sér. Sci. Math. 27, 1979, 489–491.

Erdös, P. Some remarks on set theory. Ann. of Math. (2) 44, 1943, 643–646.

——— Some remarks on set theory. Proc. Amer. Math. Soc. 1, 1950, 127–141.

——— Some remarks on set theory, IV. Michigan Math. J. 2, 1953–1954, 169–173.

——— Some remarks on subgroups of real numbers. Colloq. Math. 42, 1979, 119–120.

Erdös, P. and Kakutani, S. On non-denumerable graphs. Bull. Amer. Math. Soc. 49, 1943, 457–461.

Erdös, P. and Marcus, S. Sur la décomposition de l'espace Euclidien en ensembles homogènes. Acta Math. Acad. Sci. Hungar. 8, 1957, 443–452.

Erdös, P. and Volkman, B. Additive Gruppen mit vorgegebener Hausdorffscher Dimension. J. Reine Angew. Math. 221, 1966, 203–208.

Feldman, M. B. A proof of Lusin's theorem. Amer. Math. Monthly 88, 1981, 191–192.

Foran, J. Some relationships of subgroups of the real numbers to Hausdorff measures. J. London Math. Soc. (2) 7, 1974, 651–661.

Frankiewicz, R. On functions having the Baire property, II. Bull. Acad. Polon. Math. 30, 1982, 559–560.

Fréchet, M. Sur quelques points du calcul fonctionnel. Rend. Circ. Mat. Palermo 22, 1906, 1–74.

———— Les ensembles abstraits et le calcul fonctionnel. Rend. Circ. Mat. Palermo 30, 1910, 1–26.

———— Pri la funkcia ekvacio $f(x + y) = f(x) + f(y)$. Enseignement Math. 15, 1913, 390–393.

———— Pri la funkcia ekvacio $f(x + y) = f(x) + f(y)$. Enseignement Math. 16, 1914, 136.

———— Quelques propriétés des ensembles abstraits. Fund. Math. 10, 1927, 328–357.

———— Au sujet de deux citations contenues dans un mémoire précédent. Fund. Math. 14, 1929, 118–121.

Freud, G. Ein Beitrag zu dem Sätze von Cantor und Bendixson. Acta Math. Acad. Sci. Hungar. 9, 1958, 333–336.

Fukamiya, M. Sur une propriété des ensembles mesurables. Sci. Rep. Tôhoku Univ. 24, 1935, 332–334.

Galvin, F. Chain conditions and products. Fund. Math. 108, 1980, 33–48.

Goffman, C. Real Functions. Holt, Rinehart and Winston, New York, 1953.

Goffman, C. and Waterman, D. On upper and lower limits in measure. Fund. Math. 48, 1960, 127–133.

Goffman, C. and Zink, R. E. Concerning the measurable boundaries of a real function. Fund. Math. 48, 1960, 105–111.

Gomes, Ruy Luiz. Exemplo de conjunto não mensuravel à Lebesgue. Gaz. Mat. (Lisbon) 13, 1952, No. 51, 4–6.

Gottschalk, W. H. and Hedlund, G. A. Topological Dynamics. Amer. Math. Soc. Colloq. Publ. 36. American Mathematical Society, Providence, R.I., 1955.

Grattan-Guinness, I. The correspondence between Georg Cantor and Philip Jourdain. Jber. Deutsch. Math.-Verein. 73, 1971, 111–130.

Grzegorek, E. On a paper by Karel Prikry concerning Ulam's problem on families of measures. Colloq. Math. 42, 1979, 197–208.

——— On saturated sets of Boolean rings and Ulam's problem on sets of measures. Fund. Math. 110, 1980, 153–161.

Hahn, H. Theorie der reellen Funktionen, Vol. 1 Julius Springer, Berlin, 1921.

——— Reelle Funktionen. Erster Teil. Punkfunktionen. Akademische Verlagsgesellschaft M.B.H., Leipzig, 1932. (Reprinted by Chelsea, New York, 1948.)

Hahn, H. and Rosenthal, A. Set Functions. University of New Mexico Press, Albuquerque, N.Mex., 1948.

Halmos, P. R. Lectures on Boolean Algebras. Van Nostrand Math. Studies No. 1. Van Nostrand, Princeton, N.J., 1963.

Halperin, I. Non-measurable sets and the equation $f(x + y) = f(x) + f(y)$. Proc. Amer. Math. Soc. 2, 1951, 221–224.

Hamel, G. Eine Basis aller Zahlen und die unstetigen Lösungen der Funktionalgleichung $f(x + y) = f(x) + f(y)$. Math. Ann. 60, 1905, 459–462.

Hankel, H. Untersuchungen über die unendlich oft oscillerenden und unstetigen Functionen (Abdruck aus dem Gratulationsprogramm der Tübinger Universität vom 6 März 1870). F. Fues, Freiburg, Germany, 1870. (Reprinted in Math. Ann. 20, 1882, 63–112.)

Hansell, R. W. Borel measurable mappings for nonseparable metric spaces. Trans. Amer. Math. Soc. 161, 1971, 145–169.

Harazišvili, A. B. Ob odnom pokrytii kontinuuma [On a covering of the continuum]. Soobšč. Akad. Nauk Gruzin. SSR 96, 1979, 273–276.

Hardy, G. H. A theorem concerning the infinite cardinal numbers. Quart. J. Pure Appl. Math 35, 1903, 87–94.

——— The continuum and the second number class. Proc. London Math. Soc. (2) 4, 1905, 10–17.

Hartman, P. and Kershner, R. The structure of monotone functions. Amer. J. Math. 59, 1937, 809–822.

Hashimoto, H. On some local properties of spaces. Math. Japon. 2, 1952, 127–134.

——— On the resemblance of point sets. Math. Japon. 3, 1954, 53–56.

——— On the *topology and its application. Fund. Math. 91, 1976, 5–10.

Hausdorff, F. Grundzüge der Mengenlehre. Veit, Leipzig, 1914. (Reprinted by Chelsea, New York, 1965.)

——— Dimension und äusseres Mass. Math. Ann. 79, 1919, 157–179.

Hausdorff, F. Summen von $\aleph_1$ Mengen. Fund. Math. 26, 1936, 241–255.

———— Dualität zwischen Kategorie und Mass. (Übersicht verschiedener in der Literatur mitgeteilter Ergebnisse.) Nachgelassene Schriften, Band I: Studien und Referate. B.G. Teubner, Stuttgart, West Germany, 1969, pp. 19–22.

Haworth, R. C. and McCoy, R. A. Baire spaces. Dissertationes Math. (Rozprawy Mat.) 141, 1977, 77p.

Hayashi, E. Topologies defined by local properties. Math. Ann. 156, 1964, 205–215.

———— More on topologies defined by local properties. Bull. Nagoya Inst. Tech. 26, 1974, 129–133.

Henle, J. M. Functions with arbitrarily small periods. Amer. Math. Monthly 87, 1980, 816.

Hewitt, E. and Savage, L. J. Symmetric measures on cartesian products. Trans. Amer. Math. Soc. 80, 1955, 470–501.

Hewitt, E. and Stromberg, K. Some examples of nonmeasurable sets. J. Austral. Math. Soc. 18, 1974, 236–238.

Ho, S. M. F. and Naimpally, S. A. An abstract model for measure and category. Math. Japon. 24, 1979, 123–140.

Hobson, E. W. The Theory of Functions of a Real Variable and the Theory of Fourier Series, 3rd ed., Vol. 1. Cambridge, 1927. (Reprinted by Dover, New York, 1957.)

Hong, Y. and Tong, J. Decomposition of a function into measurable functions. Amer. Math. Monthly 90, 1983, 573.

Horn, S. and Schach, S. An extension of the Hewitt-Savage zero-one law. Ann. Math. Statist. 41, 1970, 2130–2131.

Horn, A. and Tarski, A. Measures in Boolean algebras. Trans. Amer. Math. Soc. 64, 1948, 467–497.

Hurewicz, W. Une remarque sur l'hypothèse du continu. Fund. Math. 19, 1932, 8–9.

Inagaki, T. Contribution à la topologie, III. Math. J. Okayama Univ. 4, 1954, 79–96.

Itzkowitz, G. L. Continuous measures, Baire category, and uniform continuity in topological groups. Pacific J. Math. 54, 1974, 115–125.

Jones, F. B. Measure and other properties of a Hamel basis. Bull. Amer. Math. Soc. 48, 1942, 472–481.

Jordan, C. Cours d'Analyse de l'École Polytechnique, 2nd ed., Vol. 2. Gauthier-Villars, Paris, 1894.

Kac, M. Une remarque sur les équations fonctionnelles. Comment. Math. Helv. 9, 1936–1937, 170–171.

Kechris, A. S. On a notion of smallness for subsets of the Baire space. Trans. Amer. Math. Soc. 229, 1977, 191–207.

Kellerer, H. G. Non measurable partitions of the real line. Adv. in Math. 10, 1973, 172–176.

Kemperman, J. H. B. A general functional equation. Trans. Amer. Math. Soc. 86, 1957, 28–56.

Kern, R. Multiply periodic real functions and their period sets. Manuscripta Math. 20, 1977, 153–175.

Kestelman, H. On the functional equation $f(x + y) = f(x) + f(y)$. Fund. Math. 34, 1947, 144–147.

Kingman, J. F. C. and Taylor, S. J. Introduction to Measure and Probability. Cambridge University Press, Cambridge, 1966.

Knaster, B. Sur une propriété caractéristique de l'ensemble des nombres réels. Mat. Sb. (N.S.) 16 (58), 1945, 281–290.

Kominek, Z. Measure, category, and the sums of sets. Amer. Math. Monthly 90, 1983, 561–562.

Kondô, M. Sur les notions de la catégorie et de la mesure dans la théorie des ensembles de points. J. Fac. Sci. Hokkaido Univ. Ser. 1, Math. 4, 1936, 123–180.

———— Sur l'hypothèse de M. B. Knaster dans la théorie des ensembles de points. J. Fac. Sci. Hokkaido Univ. Ser. 1, Math. 6, 1937, 1–20.

Korec, I. A class of Lebesgue non-measurable subfields of the field of real numbers. Acta Fac. Rerum Natur. Univ. Comenian. Math. 26, 1972, 29–32.

Kormes, M. Treatise on basis sets. Bull. Amer. Math. Soc. 30, 1924, 483.

———— On the functional equation $f(x + y) = f(x) + f(y)$. Bull. Amer. Math. Soc. 32, 1926, 689–693.

———— On basis-sets. Ph.D. thesis, Columbia University; printed by the Zincograph Company, January 1928.

Korovkin, P. P. O množestve periodov funkciĭ odnogo peremennogo [On the set of periods of a function of one variable]. Kalinin. Gos. Ped. Inst. Učen. Zap. 69, 1969, 70–76.

Králik, D. Bemerkungen über nicht-messbare Punktmengen. Publ. Math. Debrecen 2, 1951–1952, 229–231.

Krawczyk, A. and Pelc, A. On families of σ-complete ideals. Fund. Math. 109, 1980, 155–161.

Kuczma, M. A remark containing measure and category. Uniw. Śląski. w Katowicach Prace Mat. 3, 1973, 51–52.

Kuczma, M. Additive functions and the Egorov theorem. In General Inequalities 1 (ed. E. F. Beckenbach). Birkhäuser Verlag, Basel, 1978, pp. 169–173.

——— On some analogies between measure and category and their applications in the theory of additive functions. Ann. Math. Silesianae 13, 1985a, 155–162.

——— An Introduction to the Theory of Functional Equations and Inequalities. Cauchy's Equation and Jensen's Inequality. Państwowe Wydawnictwo Naukowe. Uniwersytet Ślaski, Warsaw, 1985b.

Kunugui, K. Sur les fonctions jouissant de la propriété de Baire. Japan. J. Math. 13, 1937, 431–433.

Kuratowski, K. Sur les fonctions représentables analytiquement et les ensembles de première catégorie. Fund. Math. 5, 1924, 75–86.

——— Sur la puissance de l'ensemble des "nombres de dimension" au sens de M. Fréchet. Fund. Math. 8, 1926, 201–208.

——— La propriété de Baire dans les espaces métriques. Fund. Math. 16, 1930, 390–394.

——— Topologie I. Monografje Matematyczne, Vol. 3. Z subwencji funduszu kultury narodowej, Warsawa-Lwów, 1933a.

——— Sur la propriété de Baire dans les groupes métriques. Studia Math. 4, 1933b, 38–40.

——— Sur le rapport des ensembles de M. Lusin à la théorie générale des ensembles. Fund. Math. 22, 1934, 315–318.

——— Quelques problèmes concernant les espaces métriques non-séparables. Fund. Math. 25, 1935, 534–545.

——— Topology, Vol. 1. Academic Press, New York, 1966.

——— Applications of the Baire category method to the problem of independent sets. Fund. Math. 81, 1973, 65–72.

——— On the concept of strongly transitive systems in topology. Ann. Mat. Pura Appl. (4) 98, 1974, 357–363.

Kuratowski, K. and Sierpiński, W. Sur l'existence des ensembles projectifs non mesurables. Spis. Bulgar. Akad. Nauk. 61, 1941, 207–212. [Sierpiński O.C. III, 417–421.]

Kvačko, M. Izmerimye otobraženija prostranstv [Measurable mappings of spaces]. Vestnik Leningrad. Univ. 13, 1958, 87–101.

Lebesgue, H. Sur une propriété des fonctions. C. R. Acad. Sci. Paris 137, 1903, 1228–1230.

——— Sur les fonctions représentables analytiquement. J. Math. (6) 1, 1905, 139–216.

Lebesgue, H. Sur les transformations ponctuelles transformant les plans en plans qu'on peut définir par des procédés analytiques. Atti Accad. Sci. Torino 42, 1907a, 532-539.

———— Contribution à l'étude des correspondances de M. Zermelo. Bull. Soc. Math. France 35, 1907b, 202-212.

Lennes, N. J. Concerning Van Vleck's non-measurable set. Trans. Amer. Math. Soc. 14, 1913, 109-112.

Letac, G. Cauchy functional equation again. Amer. Math. Monthly 85, 1978, 663-664.

Levy, R. A totally ordered Baire space for which Blumberg's theorem fails. Proc. Amer. Math. Soc. 41, 1973, 304. (Erratum, Proc. Amer. Math. Soc. 45, 1974, 469.)

———— Strongly non-Blumberg spaces. General Topology Appl. 4, 1974, 173-177.

Lindelöf, E. Sur quelques points de la théorie des ensembles. C. R. Acad. Sci. Paris 137, 1903, 697-700.

Lisagor, L. R. Igra Banaha-Mazura. Mat. Sb. 110 (152), 1979, 218-234. [See following for English translation.]

———— The Banach-Mazur game. Math. USSR-Sb. 38, 1981, 201-216.

Ljapunov, A. A. Ob R-množestvah [On R-sets]. Dokl. Akad. Nauk SSSR (N.S.) 58, 1947, 1887-1890.

———— O teoretiko-množestvennyh operacijah, sohranjajuščih izmerimost' [On set-theoretical operations which preserve measurability]. Dokl. Akad. Nauk SSSR (N.S.) 65, 1949a, 609-612.

———— O δs-operacijah, sohranjajuščih izmerimost' i sboĭstvo Bèra [On δs-operations preserving measurability and the Baire property]. Mat. Sb. (N.S.) 24 (66), 1949b, 119-127.

———— R-množestva [R-sets]. Trudy Mat. Inst. Steklov. 40, 1953, 1-68.

———— O metode transfinitnyh indeksov v teorii operaciĭ nad množestvami. Trudy Mat. Inst. Steklov. 133, 1973, 132-148. On the method of transfinite indices in the theory of operations on sets. Proc. Steklov Inst. Math. 133, 1973, 133-149.

Łomnicki, A. O wielookresowych funkcjach jednoznacznych zmiennej rzeczywistej [On uniform multiperiodic functions of a real variable]. Spraw. Tow. Nauk. Warszaw. Wyd. III. 11, 1918, 808-846.

Lukeš, J. and Zajíček, L. Fine topologies as examples of non-Blumberg spaces. Comment. Math. Univ. Carolinae 17, 1976, 683-688.

Luzin, N. K″ osnovnoĭ teoremje integral'nago isčislenija. Mat. Sb. 28, 1911, 266-294.

———— Sur les propriétés des fonctions mesurables. C. R. Acad. Sci. Paris 154, 1912, 1688-1690.

———— Sur un problème de M. Baire. C. R. Acad. Sci. Paris 158, 1914, 1258-1261.

Luzin, N. Sur la classification de M. Baire. C. R. Acad. Sci. Paris 164, 1917, 91–94.

———— Sur l'existence d'un ensemble non dénombrable qui est de première catégorie dans tout ensemble parfait. Fund. Math. 2, 1921, 155–157.

Luzin, N. and Sierpiński, W. Sur une décomposition d'un intervalle en une infinité non dénombrable d'ensembles non mesurables. C. R. Acad. Sci. Paris 165, 1917a, 422–424. [Sierpiński O.C. II, 177–179]

———— Sur une propriété du continu. C. R. Acad. Sci. Paris 165, 1917b, 498–500. [Sierpiński O.C. II, 180–182]

———— Sur quelques propriétés des ensembles (A). Bull. Internat. Cracovie, 1918, 35–48. [Sierpiński O.C. II, 192–204]

———— Sur un ensemble non mesurable B. J. Math. (9) 2, 1923, 53–72. [Sierpiński O.C. II, 504–519]

Mahlo, P. Über Teilmengen des Kontinuums von dessen Mächtigkeit. Sitzungsber. Sächs. Akad. Wiss. Leipzig. Math.-Natur. Kl. 65, 1913, 283–315.

Marcus, S. Sur un théorème de F. B. Jones. Sur un théorème de S. Kurepa. Bull. Math. R.P.R. 1 (49), 1957, 433–434.

———— Hamelsche Basis und projektive Mengen. Math. Nachr. 17, 1958–1959, 143–150.

———— Conditions d'équivalence à une constante pour les fonctions intégrables Riemann et pour les fonctions jouissant de la propriété de Baire. Rev. Roumaine Math. Pures Appl. 4, 1959, 283–285.

Marczewski, E. Remarques sur l'équivalence des classes d'ensembles. Ann. Soc. Polon. Math. 19, 1946, 228.

———— Séparabilité et multiplication cartésienne des espaces topologiques. Fund. Math. 34, 1947, 127–143.

———— Sur l'oeuvre scientifique de Stefan Banach, II. Théorie des fonctions réelles et théorie de la mesure. Colloq. Math. 1, 1948, 93–102.

Marczewski, E. and Sikorski, R. Measures in non-separable metric spaces. Colloq. Math. 1, 1948, 133–139.

———— Remarks on measure and category. Colloq. Math. 2, 1949, 13–19.

Martin, N. F. G. Generalized condensation points. Duke Math. J. 28, 1961, 507–514.

Mazur, S. and Sternbach, L. Über Mannigfaltigkeiten in Funktionalräumen. Ann. Soc. Polon. Math. 9, 1930, 195.

Mehdi, M. R. On convex functions. J. London Math. Soc. 39, 1964, 321–326.

Mendez, C. G. On sigma-ideals of sets. Proc. Amer. Math. Soc. 60, 1976, 124–128.

Mendez, C. G. On the Sierpiński-Erdös and the Oxtoby-Ulam theorems for some new sigma-ideals of sets. Proc. Amer. Math. Soc. 72, 1978, 182–188.

Miller, A. W. Special subsets of the real line. In Handbook of Set-Theoretic Topology (ed. K. Kunen and J. E. Vaughan). North-Holland, Amsterdam, 1984, Chap. 5, pp. 201–233.

Morgan, J. C., II. Infinite games and singular sets. Colloq. Math. 29, 1974, 7–17.

———— On translation invariant families of sets. Colloq. Math. 34, 1975, 63–68.

———— The absolute Baire property. Pacific J. Math. 65, 1976, 421–436.

———— Baire category from an abstract viewpoint. Fund. Math. 94, 1977a, 13–23.

———— On zero-one laws. Proc. Amer. Math. Soc. 62, 1977b, 353–358.

———— On ordinally closed sets. Proc. Amer. Math. Soc. 68, 1978a, 92–96.

———— On the absolute Baire property. Pacific J. Math. 78, 1978b, 415–431.

———— On sets every homeomorphic image of which has the Baire property. Proc. Amer. Math. Soc. 75, 1979, 351–354.

———— On product bases. Pacific J. Math. 99, 1982, 105–126.

———— On equivalent category bases. Pacific J. Math. 105, 1983, 207–215.

———— On the general theory of point sets. Real Anal. Exchange 9, 1983–1984, 345–353.

———— Measurability and the abstract Baire property. Rend. Circ. Mat. Palermo (2) 34, 1985, 234–244.

———— On the general theory of point sets, II. Real Anal. Exchange 12, 1986–1987, 377–386.

Morgan, J. C., II and Schilling, K. Invariance under operation $\mathscr{A}$. Proc. Amer. Math. Soc. 100, 1987, 651–654.

Mueller, B. J. Three results for locally compact groups connected with the Haar measure density theorem. Proc. Amer. Math. Soc. 16, 1965, 1414–1416.

Mullikin, H. C. Subgroups of the reals and periodic functions. Delta (Waukesha) 6, 1976, 16–22.

Munroe, M. E. Introduction to Measure and Integration, 2nd ed. Addison-Wesley, Reading, Mass., 1971.

Mycielski, J. Independent sets in topological algebras. Fund. Math. 55, 1964, 139–147.

———— On the axiom of determinateness, II. Fund. Math. 59, 1966, 203–212.

———— Algebraic independence and measure. Fund. Math. 61, 1967, 165–169.

———— Some new ideals of sets on the real line. Colloq. Math. 20, 1969, 71–76.

Mycielski, J. and Sierpiński, W. Sur une propriété des ensembles linéaires. Fund. Math. 58, 1966, 143–147. [Sierpiński O.C. III, 681–685]

Mycielski, J., Świerczkowski, S. and Zięba, A. On infinite positional games. Bull Acad. Polon. Sci., Cl. III. 4, 1956, 485–488.

Neubrunn, T. Poznámka k merateľným transformaciam [Note on measurable transformations]. Acta Fac. Natur. Univ. Comenian. 4, 1959, 287–290.

Neumann, J. von. Ein System algebraisch unabhängiger Zahlen. Math. Ann. 99, 1928, 134–141.

——— Continuous Geometry, Princeton Math. Series, Vol. 25. Princeton University Press, Princeton, N.J., 1960.

Nikodym, O. Sur une propriété de l'opération A. Fund. Math. 7, 1925, 149–154.

——— Sur quelques propriétés de l'opération A. Spraw. Tow. Nauk. Warszaw. Wyd. III. 19, 1926, 294–298.

——— Sur la condition de Baire. Bull. Internat. Cracovie, 1929, 591–598.

Osgood, W. F. Ueber die ungleichmässige Convergenz und die gliedweise Integration der Reihen. Göttingen Nachr., 1896, 288–291.

——— Non-uniform convergence and the integration of series term by term. Amer. J. Math. 19, 1897, 155–190.

——— Zweite Note über analytische Functionen mehrerer Veränderlichen. Math. Ann. 53, 1900, 461–464.

Ostaszewski, A. J. Martin's axiom and Hausdorff measures. Proc. Cambridge Philos. Soc. 75, 1974, 193–197.

——— Absolutely non-measurable and singular co-analytic sets. Mathematika 22, 1975, 161–163.

Ostrowski, A. Über die Funktionalgleichung der Exponentialfunktion und verwandte Funktionalgleichungen. Jber. Deutsch. Math.-Verein. 38, 1929, 54–62.

Oxtoby, J. Note on transitive transformations. Proc. Nat. Acad. Sci. USA 23, 1937, 443–446.

——— The Banach-Mazur game and Banach category theorem. In Contributions to the Theory of Games, Vol. 3 (ed. M. Dresher, A. W. Tucker, and P. Wolfe), Annals of Math. Studies 39. Princeton University Press, Princeton, N.J., 1957, pp. 159–163.

——— Cartesian products of Baire spaces. Fund. Math. 49, 1961, 157–166.

——— Measure and Category, 2nd ed. Springer-Verlag, New York, 1980.

Pawelska, J. On some example of a non-measurable set. Demonstratio Math. 12, 1979, 343–350.

Pelc, A. Ideals on the real line and Ulam's problem. Fund. Math. 112, 1981, 165–170.

Perron, O. Irrationalzahlen. Walter de Gruyter, West Berlin, 1960.

Pettis, B. J. Remarks on a theorem of E. J. McShane. Proc. Amer. Math. Soc. 2, 1951, 166–171.

Piccard, S. Sur les ensembles de distances des ensembles de points d'un espace Euclidien. Mémoires de l'Université de Neuchâtel, Vol. 13. Secrétariat de l'Université, Neuchaâtel, Switzerland, 1939, 212p.

Pol, R. Remark on the restricted Baire property in compact spaces. Bull. Acad. Polon. Sci. Sér. Sci. Math. 24, 1976, 599–603.

Popruženko, J. Sur une propriété d'une classe de mesures abstraites. Colloq. Math. 4, 1957, 189–194.

Prikry, K. Kurepa's hypothesis and σ-complete ideals. Proc. Amer. Math. Soc. 38, 1973, 617–620.

——— Kurepa's hypothesis and a problem of Ulam on families of measures. Monatsh. Math. 81, 1976, 41–57.

Pu, H. W. Concerning non-measurable subsets of a given measurable set. J. Austral. Math. Soc. 13, 1972, 267–270.

Rademacher, H. Eineindeutige Abbildungen und Messbarkeit. Monatsh. Math. 27, 1916, 183–291.

——— Über eine Eigenschaft von messbaren Mengen positiven Masses. Jber. Deutsch. Math.-Verein. 30, 1921, 130–132.

Ray, K. C. and Lahiri, B. K. An extension of a theorem of Steinhaus. Bull. Calcutta Math. Soc. 56, 1964, 29–31.

Riesz, F. Sur le théorème de M. Egoroff et sur les opérations fonctionnelles linéaires. Acta Litt. Sci. Szeged 1, 1922, 18–25.

——— Elementarer Beweis des Egoroffschen Satzes. Monatsh. Math. 35, 1928, 243–248.

Rindung, O. Et bevis for muligheden af at dele de reelle tal i mere end taellelig mange disjunkte og kongruente maengder med positivt ydre Lebesgue-mal [A proof of the possibility of dividing the real numbers into more than countably many disjoint and congruent sets with positive outer Lebesgue measure]. Mat. Tidsskr. B, 1950, 16–17.

Rogers, C. A. Hausdorff Measures. Cambridge University Press, Cambridge, 1970.

Rosenthal, J. Nonmeasurable invariant sets. Amer. Math. Monthly 82, 1975, 488–491.

Rothberger, F. Eine Äquivalenz zwischen der Kontinuumhypothese und der Existenz der Lusinschen und Sierpińskischen Mengen. Fund. Math. 30, 1938, 215–217.

Royden, H. L. Real Analysis, 2nd ed. Macmillan, New York, 1968.

Rozycki, E. P. On Egoroff's theorem. Fund. Math. 56, 1964, 289–293.

Ruziewicz, S. Une application de l'équation fonctionnelle $f(x + y) = f(x) + f(y)$ à la décomposition de la droite en ensembles superposables, non mesurables. Fund. Math. 5, 1924, 92–95.

——— Sur une propriété des fonctions arbitraires d'une variable réelle. Mathematica (Cluj) 9, 1935, 83–85.

——— Sur une propriété de la base hamelienne. Fund. Math. 26, 1936, 56–58.

Ruziewicz, S. and Sierpiński, W. Sur un ensemble parfait qui a avec toute sa translation au plus un point commun. Fund. Math. 19, 1932, 17–21. [Sierpiński O.C. III, 90–94]

——— Un théorème sur les familles de fonctions. Mathematica (Cluj) 7, 1933, 89–91.

Saks, S. Theory of the Integral, Hafner Press, New York, 1937.

Saks, S. and Sierpiński, W. Sur une propriété générale des fonctions. Fund. Math. 11, 1928, 105–112. [Sierpiński O.C. II, 678–684]

Samuels, P. A topology formed from a given topology and ideal. J. London Math. Soc. (2) 10, 1975, 409–416.

Sander, W. Ein Beitrag zur Baire-Kategorie-Theorie. Manuscripta Math. 34, 1981a, 71–83.

——— A decomposition theorem. Proc. Amer. Math. Soc. 83, 1981b, 553–554.

——— A duality principle. Proc. Amer. Math. Soc. 84, 1982, 609–610.

Sander, W. and Slipek, K. A generalization of a theorem of Banach. Boll. Un. Mat. Ital. (5) 16-A, 1979, 382–384.

Scheeffer, L. Zur Theorie der stetigen Funktionen einer reellen Veränderlichen. Acta Math. 5, 1884, 279–297.

Schilling, K. On absolutely Δ_2^1 operations. Fund. Math. 121, 1984, 239–250.

Schilling, K. and Vaught, R. Borel games and the Baire property. Trans. Amer. Math. Soc. 279, 1983, 411–428.

Scorza-Dragoni, G. Sul princìpio di approssimazióne nélla teoria dégli insièmi e sulla quasicontinuità délle funzióni misuràbili. Rend. Sem. Mat. R. Univ. Roma (4) 1, 1936, 53–58.

Sélivanowski, E. Sur les propriétés des constituantes des ensembles analytiques. Fund. Math. 21, 1933, 20–28.

Semadeni, Z. Sur les ensembles clairsemés. Rozprawy Mat. 19, 1959, 39p.

Semadeni, Z. Periods of measurable functions and the Stone-Cech compactification. Amer. Math. Monthly 71, 1964, 891–893.

Sendler, W. On zero-one laws. Transactions Eighth Prague Conference on Information Theory, Statistical Decision Functions, Random Processes, Prague, 1978, Vol. B. D. Reidel, Dordrecht, The Netherlands, 1978, pp. 181–191.

Sierpiński, W. Oeuvres choisies, Vols. I–III. Académie Polonaise des Sciences, Institut Mathématique, Państwowe Wydawnictwo Naukowe, Warsaw, 1974–1976.

———— Sur une propriété générale des ensembles de points. C. R. Acad. Sci. Paris 162, 1916a, 716–717. [O.C. II, 120–121]

———— Démonstration élémentaire du théorème de M. Lusin sur les fonctions mesurables. Tôhoku Math. J. 10, 1916b, 81–86. [O.C. II, 141–145]

———— Sur un problème de M. Lusin. Giorn. Mat. 55, 1917a, 272–277. [O.C. II, 166–170]

———— Sur un ensemble non mesurable. Tôhoku Math. J. 12, 1917b, 205–208. [O.C. II, 183–186]

———— L'axiome de M. Zermelo et son rôle dans la théorie des ensembles et l'analyse. Bull. Internat. Cracovie, 1918, 97–152. [O.C. II, 208–255]

———— Sur un ensemble ponctiforme connexe. Fund. Math. 1, 1920a, 7–10. [O.C. II, 286–288]

———— Sur la question de la mesurabilité de la base de M. Hamel. Fund. Math. 1, 1920b, 105–111. [O.C. II, 322–327] Czy "Basis" Hamela jest mierzalne? Prace Mat.-Fiz. 31, 1920, 17–22.

———— Sur un problème concernant les ensembles mesurables superficiellement. Fund. Math. 1, 1920c, 112–115. [O.C. II, 328–330]

———— Sur l'équation fontionnelle $f(x + y) = f(x) + f(y)$. Fund. Math. 1, 1920d, 116–122. [O.C. II, 331–336]

———— Sur les fonctions convexes mesurables. Fund. Math. 1, 1920e, 125–128. [O.C. II, 337–340]

———— Démonstration de quelques théorèmes fondamentaux sur les fonctions mesurables. Fund. Math. 3, 1922, 314–321. [O.C. II, 474–480]

———— Une remarque sur la condition de Baire. Fund. Math. 5, 1924a, 20–22. [O.C. II, 520–521]

———— Sur l'hypothèse du continu $(2^{\aleph_0} = \aleph_1)$. Fund. Math. 5, 1924b, 177–187. [O.C. II, 527–536]

———— Sur une propriété des fonctions de M. Hamel. Fund. Math. 5, 1924c, 334–336. [O.C. II, 541–543]

Sierpiński, W. Sur l'ensemble de distances entre les points d'un ensemble. Fund. Math. 7, 1925, 144–148. [O.C. II, 567–571]

——— Sur une propriété des ensembles (A). Fund. Math. 8, 1926a, 362–369.

——— La notion de dérivée comme base d'une théorie des ensembles abstraits. Math. Ann. 97, 1926b, 321–337. [O.C. II, 598–615]

——— La propriété de Baire de fonctions et de leur images. Fund. Math. 11, 1928a, 305–307. [O.C. II, 705–707]

——— Sur une décomposition d'ensembles. Monatsh. Math. 35, 1928b, 239–242. [O.C. II, 719–722]

——— Remarque sur le théorème de M. Egoroff. Spraw. Tow. Nauk. Warszaw. Wyd. III. 21, 1928c, 84–87. [O.C. II, 667–670]

——— Sur une fonction transformant tout ensemble non dénombrable en un ensemble de deuxième catégorie. C. R. Acad. Sci. Paris 188, 1929a, 613–614. [O.C. II, 770–771]

——— Sur une décomposition du segment. Fund. Math. 13, 1929b, 195–200.

——— Sur un théorème de MM. Banach et Kuratowski. Fund. Math. 14, 1929c, 277–280. [O.C. II, 761–764]

——— Remarque sur les suites infinies de fonctions (Solution d'un problème de M. S. Saks). Fund. Math. 18, 1932a, 110–113.

——— Sur les translations des ensembles linéaires. Fund. Math. 19, 1932b, 22–28. [O.C. III, 95–100]

——— Un théorème concernant les transformations continues des ensembles linéaires. Fund. Math. 19, 1932c, 205–210. [O.C. III, 107–112]

——— Remarque sur un théorème de M. Fréchet. Monatsh. Math. 39, 1932d, 233–238.

——— Sur une propriété de fonctions de deux variables réelles, continues par rapport à chacune de variables. Publ. Math. Univ. Belgrade 1, 1932e, 125–128.

——— Un théorème équivalent a l'hypothèse du continu. Bull. Acad. Roumaine 16, 1933a–1934, 103–107.

——— Sur un problème de M. Ruziewicz concernant l'hypothèse du continu. Bull Acad. Sci. Math. Nat. Belgrade 1, 1933b, 67–73. [O.C. III, 123–128]

——— Sur l'ensemble de valeurs d'une fonction mesurable à valeurs distinctes. Fund. Math. 20, 1933c, 126–130.

——— Sur une certaine suite infinie de fonctions d'une variable réelle. Fund. Math. 20, 1933d, 163–165. (Correction in Fund. Math. 24, 1935, 321–323.)

Sierpiński, W. Sur un théorème de recouvrement dans la théorie générale des ensembles. Fund. Math. 20, 1933e, 214–220. [O.C. III, 129–134]

——— Sur les constituantes des ensembles analytiques. Fund. Math. 21, 1933f, 29–34. [O.C. III, 135–139]

——— Sur un problème de M. Ruziewicz concernant les superpositions des fonctions mesurables. Spraw. Tow. Nauk. Warszaw. Wyd. III. 26, 1933g, 12–14.

——— Sur l'équivalence de deux conséquences de l'hypothèse du continu. Studia Math. 4, 1933h, 15–20.

——— Sur les ensembles partout de deuxième catégorie. Fund. Math. 22, 1934a, 1–3. [O.C. III, 181–183]

——— Sur la dualité entre la première catégorie et la mesure nulle. Fund. Math. 22, 1934b, 276–280. [O.C. III, 207–210]

——— Sur une propriété des ensembles linéaires quelconques. Fund. Math. 23, 1934c, 125–134. [O.C. III, 215–223]

——— Sur un problème concernant les familles indénombrables d'ensembles de mesure positive. Spraw. Tow. Nauk. Warszaw. Wyd. III. 27, 1934d, 73–75.

——— Sur deux ensembles linéaires singuliers. Ann. Scuola Norm. Sup. Pisa (2) 4, 1935a, 43–46.

——— Les superpositions des fonctions. Časopis Mat. Fys. 64, 1935b, 73–79.

——— Sur un problème de M. Ruziewicz concernant les superpositions de fonctions jouissant de la propriété de Baire. Fund. Math. 24, 1935c, 12–16. [O.C. III, 243–247]

——— Sur le produit combinatoire de deux ensembles jouissant de la propriété C (Solution d'un problème de M. Szpilrajn). Fund. Math. 24, 1935d, 48–50. [O.C. III, 252–254]

——— Sur les suites infinies de fonctions définies dans les ensembles quelconques. Fund. Math. 24, 1935e, 209–212. [O.C. III, 255–258]

——— Un théorème de la théorie générale des ensembles. Fund. Math. 25, 1935f, 546–550.

——— La base de M. Hamel et la propriété de Baire. Publ. Math. Univ. Belgrade 4, 1935g, 220–224. [O.C. III, 273–276]

——— Un théorème de la théorie générale des ensembles et ses applications. Spraw. Tow. Nauk. Warszaw. Wyd. III. 28, 1935h, 131–135.

——— Sur un ensemble linéaire non mesurable complètement homogène. Spraw. Tow. Nauk. Warszaw. Wyd. III. 28, 1935i, 154–155.

——— Sur les fonctions de classe 1. C. R. (Dokl.) Acad. Sci. URSS. N.S. 3, 1936a, 47–48.

Sierpiński, W. Sur les fonctions semi-continues. C. R. (Dokl.) Acad. Sci. URSS. N.S. 4, 1936b, 3–4.

———— Remarque sur les translations d'ensembles. Fund. Math. 26, 1936c, 59–60.

———— Un théorème concernant les translations d'ensembles. Fund. Math. 26, 1936d, 143–145. [O.C. III, 299–301]

———— Sur un problème concernant les fonctions de première classe. Fund. Math. 27, 1936e, 191–200.

———— Sur une décomposition de la droite. Publ. Math. Univ. Belgrade 5, 1936f, 44–51.

———— Sur un problème concernant les fonctions semi-continues. Fund. Math. 28, 1937a, 1–6. [O.C. III, 329–333]

———— Sur une décomposition du segment en plus que $2^{\aleph_0}$ ensembles non mesurables et presque disjoints. Fund. Math. 28, 1937b, 111–114. [O.C. III, 339–342]

———— Sur une décomposition du segment. Fund. Math. 29, 1937c, 26–30.

———— Le théorème de M. Lusin comme une proposition de la théorie generale des ensembles. Fund. Math. 29, 1937d, 182–190. [O.C. III, 356–363]

———— Sur une proposition de la théorie générale des ensembles équivalente au théorème de M. Lusin. Spraw. Tow. Nauk. Warszaw. Wyd. III. 30, 1937e, 69–74.

———— Sur le rapport d'une certaine propriété métrique a la théorie générale des ensembles. Spraw. Tow. Nauk. Warszaw. Wyd. III. 30, 1937f, 182–187.

———— Sur un problème concernant les fonctions mesurables. Ann Sci. Univ. Jassy 24, 1938a, 154–156.

———— Fonctions additives non complètement additives et fonctions non mesurables. Fund. Math. 30, 1938b, 96–99. [O.C. III, 380–382]

———— Sur une relation entre deux conséquences de l'hypothèse du continu. Fund. Math. 31, 1938c, 227–230. [O.C. III, 386–388]

———— Un théorème concernant la convergence des fonctions sur les ensembles dénombrables. Fund. Math. 31, 1938d, 279–280.

———— Remarque sur les suites doubles des fonctions continues. Fund. Math. 32, 1939a, 1–2. [O.C. III, 392–393]

———— Sur les ensembles concentrés. Fund. Math. 32, 1939b, 301–305. [O.C. III, 399–403]

———— Sur les fonctions inverses aux fonctions satisfaisant à la condition de Baire. Mathematica 15, 1939c, 198–200. [O.C. III, 409–410]

———— Sur une proposition de Mlle S. Piccard. Comment. Math. Helv. 18, 1946, 349–352.

Sierpiński, W. Sur une proposition qui entraîne l'existence des ensembles non mesurables. Fund. Math. 34, 1947, 157–162. [O.C. III, 521–525]

—— Sur les translations des ensembles linéaires. Fund. Math. 35, 1948, 159–164. [O.C. III, 585–590]

—— Sur un problème de A. Mostowski. Colloq. Math. 2, 1949a, 67–68.

—— Sur une décomposition de la droite. Comment. Math. Helv. 22, 1949b, 317–320.

—— Sur quelques propositions qui entraînent l'existence des ensembles non mesurables. Spraw. Tow. Nauk. Warszaw. Wyd. III. 42, 1949c, 36–40. [O.C. III, 617–620]

—— Sur les suites doubles de fonctions. Fund. Math. 37, 1950, 55–62. [O.C. III, 628–635]

—— On the congruence of sets and their equivalence by finite decomposition. Lucknow University Studies No. 20, Lucknow University, Lucknow, India, 1954.

—— Hypothèse du continu, 2nd ed. Chelsea, New York, 1956.

—— Cardinal and Ordinal Numbers. Polska Akademia Nauk. Monografie Matematycne, Vol. 34. Państwowe Wydawnictwo Naukowe, Warsaw, 1958.

Sierpiński, W. and Szpilrajn, E. Remarque sur le problème de la mesure. Fund. Math. 26, 1936a, 256–261. [O.C. III, 302–306]

—— Sur les transformations continues biunivoques. Fund. Math. 27, 1936b, 289–292. [O.C. III, 307–310]

Sierpiński, W. and Zygmund, A. Sur une fonction qui est discontinue sur tout ensemble de puissance du continu. Fund. Math. 4, 1923, 316–318. [O.C. II, 497–499]

Sikorski, R. On an unsolved problem from the theory of Boolean algebras. Colloq. Math. 2, 1949, 27–29.

—— Boolean Algebras, 2nd ed. Academic Press, New York, 1964.

Sindalovskiĭ, G. H. O ravnomernoĭ shodimosti semeĭstva funkciĭ, zavisjaščih ot nepreryvno menjajuščegosja parametra [On the uniform convergence of a family of functions depending on a continuously varying parameter]. Vestnik Moskov. Univ. 5, 1960, 14–18.

Slagle, J. The genesis and exodus of topological concepts. M.S. thesis, California State Polytechnic University, Pomona, Calif., 1988.

Smítal, J. On the functional equation $f(x + y) = f(x) + f(y)$. Rev. Roumaine Math. Pures Appl. 13, 1968, 555–561.

Smith, K. A. Generalizations of Steinhaus' theorem. Master's thesis. Murdock University, Perth, Australia, 1977.

Steinhaus, H. Nowa własność mnogośći G. Cantor'a [A new property of the Cantor set]. Wektor 6, 1917, 105–107.

Steinhaus, H. Sur les distances des points des ensembles de mesure positive. Fund. Math. 1, 1920, 93–104.

Stout, L. N. Independence theories and generalized zero-one laws. Proc. Amer. Math. Soc. 66, 1977, 153–158.

Suckau, J. W. T. On uniform convergence. Amer. J. Math. 57, 1935, 549–561.

Swingle, P. M. Connected sets of Van Vleck. Proc. Amer. Math. Soc. 9, 1958, 477–482.

——— Algebras and connected sets of Vitali. Portugal. Math. 18, 1959, 69–85.

Szpilrajn, E. O mierzalności i warunku Baire'a [On measurability and the Baire property]. Comptes-Rendus du I Congrès des Mathematiciens des Pays Slaves, Warsaw, 1929. Książnica Atlas T.N.S.W., Warsaw, 1930, pp. 297–303. [English translation available from J. C. Morgan II.]

——— Sur un ensemble non mesurable de M. Sierpiński. Spraw. Tow. Nauk. Warszaw. Wyd. III. 24, 1931, 78–85.

——— Sur certains invariants de l'opération (A). Fund. Math. 21, 1933, 229–235.

——— Remarques sur les fonctions complètement additives d'ensembles et sur les ensembles jouissant de la propriété de Baire. Fund. Math. 22, 1934, 303–311.

——— Sur une classe de fonctions de M. Sierpiński et la classe correspondante d'ensembles. Fund. Math. 24, 1935, 17–34.

——— O zbiorach i funkcjach bezwzględnie mierzalnych [On absolutely measurable sets and functions]. Spraw. Tow. Nauk. Warszaw. Wyd. III. 30, 1937, 39–68. [English translation available from J. C. Morgan II.]

Szpilrajn-Marczewski, E. Sur deux propriétés des classes d'ensembles. Fund. Math. 33, 1945, 303–307.

Tannery, J. and Molk, J. Eléments de la théorie des fonctions elliptiques, Vol. 1, Gauthier-Villars Paris, 1893.

Tarski, A. Sur la décomposition des ensembles en sous-ensembles presque disjoints. Fund. Math. 12, 1928, 188–205.

——— Sur les classes d'ensembles closes par rapport à certaines opérations élémentaires. Fund Math. 16, 1930, 181–304.

——— Drei Überdeckungssätze der allgemeinen Mengenlehre. Fund. Math. 30, 1938, 132–155.

Taylor, A. D. On saturated sets of ideals and Ulam's problem. Fund. Math. 109, 1980, 37–53.

Taylor, A. E. General Theory of Functions and Integration. Blaisdell, New York, 1965.

Tolstov, G. Zamečanie k teoreme D. F. Egorova [Remark on a theorem of D. F. Egorov]. Dokl. Akad. Nauk SSSR, 6, 1939, 309–311.

Trjitzinsky, W. J. Théorie métrique dans les espaces où il y a une mesure, Memorial des Sciences Mathématiques 143. Gauthier-Villars, Paris, 1960.

Trzeciakiewicz, L. Remarque sur les translations des ensembles linéaires. Spraw. Tow. Nauk. Warszaw. Wyd. III. 25, 1932, 63–65.

Tsuchikura, T. Quelques propositions équivalentes à l'hypothèse du continu. Tôhoku Math. J. (2) 1, 1949, 69–76.

Tumarkin, L. A. Ob odnom lemme i mnozhestvah pervoi i vtoroǐ kategoriǐ v obobščennom smysle [On a lemma and sets of the first and second category in the generalized sense]. Vestnik Moskov. Univ. Ser. I Mat. Meh. 26, 1971, No. 6, 48–50.

Ulam, S. Zur Masstheorie in der allgemeinen Mengenlehre. Fund. Math. 16, 1930, 140–150.

───── Einige Sätze über Mengen II-er Kategorie. Ann. Soc. Polon. Math. 10, 1931, 123–124.

───── Über gewisse Zerlegungen von Mengen. Fund. Math. 20, 1933, 221–223.

───── Stefan Banach (1892–1945). Bull Amer. Math. Soc. 52, 1946, 600–603.

───── Problems in Modern Mathematics. Wiley, New York, 1964a.

───── Combinatorial analysis in infinite sets and some physical theories. SIAM Rev. 6, 1964b, 343–355.

Vaidyanathaswamy, R. The localisation theory in set-topology. Proc. Indian Acad. Sci. Sect. A Math. Sci. 20, 1944, 51–61.

───── Set Topology, 2nd ed. Chelsea, New York, 1960.

Viola, T. Ricerche assiomatiche sulle teorie délle funzióni d'insième e déll'integrale di Lebesgue. Fund. Math. 23, 1934, 75–101.

Vitali, G. Sul problèma délla misura déi gruppi di punti di una rètta. Nota. Bologna 1905, 2p. [Also: Bologna, Tip. Gamberini & Parmeggiani, 1905a, 5p. (3 pages of text).]

───── Una proprietá délle funzióni misuràbili. Reale Istit. Lombardo (2) 38, 1905b, 599–603.

Vivanti, G. Sugli aggregati perfetti. Rend. Circ. Mat. Palermo 13, 1899, 86–88.

Vleck, E. B. Van. On non-measurable sets of points, with an example. Trans. Amer. Math. Soc. 9, 1908, 237–244.

Volkmann, B. Gewinnmengen. Arch. Math. 10, 1959, 235–240.

Vrkoč, I. Remark about the relation between measurable and continuous functions. Časopis Pěst. Mat. 96, 1971, 225–228.

Wagner, E. Convergence in category. Rend. Accad. Sci. Fis. Mat. Napoli (4) 45, 1978, 303–312.

——— Sequences of measurable functions. Fund. Math. 112, 1981, 89–102.

Wagner, E. and Wilczyński, W. Convergence of sequences of measurable functions. Acta Math. Acad. Sci. Hungar. 36, 1980, 125–128.

Walsh, J. T. Marczewski sets, measure and the Baire property. Ph.D. dissertation, Auburn University, Auburn, Ala., 1984.

Walter, W. A counterexample in connection with Egorov's theorem. Amer. Math. Monthly 84, 1977, 118–119.

Weiss, W. A. R. A solution to the Blumberg problem. Bull. Amer. Math. Soc. 81, 1975, 957–958.

——— The Blumberg problem. Trans. Amer. Math. Soc. 230, 1977, 71–85.

Weston, J. D. A counter-example concerning Egoroff's theorem. J. London Math. Soc. 34, 1959, 139–140.

——— Addenum to a note on Egoroff's theorem. J. London Math. Soc. 35, 1960, 366.

White, H. E., Jr. Topological spaces in which Blumberg's theorem holds. Proc. Amer. Math. Soc. 44, 1974a, 454–462.

——— Two unrelated results involving Baire spaces. Proc. Amer. Math. Soc. 44, 1974b, 463–466.

——— An example involving Baire spaces. Proc. Amer. Math. Soc. 48, 1975a, 228–230.

——— Some Baire spaces for which Blumberg's theorem does not hold. Proc. Amer. Math. Soc. 51, 1975b, 477–482.

——— Topological spaces in which Blumberg's theorem holds, II. Illinois J. Math. 23, 1979, 464–468.

Wilczyński, W. Remark on the theorem of Egoroff. Časopis Pěst. Mat. 102, 1977, 228–229.

Wilkosz, W. Sugli insièmi non-misuràbili (L). Fund. Math. 1, 1920, 82–92.

Wolff, J. Sur les ensembles non mesurables. C. R. Acad. Sci. Paris 177, 1923, 863–864.

——— On perfect sets of points in which all pairs of points have irrational distances. Kon Akad. van Weten. Amsterdam 27, 1924, 95–96. Over perfecte punktverzamelingen waaren alle punktenparen irrationale afstanden heben. Amst. Ak. Versl. 33, 1924, 61–62.

Yates, C. E. M. Banach-Mazur games, comeager sets and degrees of unsolvability. Math. Proc. Cambridge Philos. Soc. 79, 1976, 195–220.

Young, W. H. Overlapping intervals. Proc. London Math. Soc. 35, 1902–1903a, 384–388.

———— Zur Lehre der nicht abgeschlossenen Punktmengen. Sitzungsber. Sächs. Akad. Wiss. Leipzig Math.-Natur. Kl. 55, 1903b, 287–293.

———— The progress of mathematical analysis in the twentieth century. Proc. London Math. Soc. (2) 24, 1926, 421–434.

Young, W. H. and Young, G. C. The Theory of Sets of Points. Cambridge University Press, Cambridge, 1906.

Zaubek, O. Über nicht messbare Punktmengen und nicht messbare Funktionen. Math. Z. 49, 1943–1944, 197–218.

Zermelo, E. Über das Mass und die Diskrepanz von Punktmengen. J. Math. 158, 1927, 154–167.

Index